NUMERICAL CONTROL PART PROGRAMMING

NUMERICAL CONTROL PART PROGRAMMING

James J. Childs

President, James J. Childs Associates

Alexandria, Virginia

INDUSTRIAL PRESS INC., 200 Madison Avenue

New York, N. Y. 10016

Library of Congress Cataloging in Publication Data

Childs, James J.
Numerical control part programming.

1. Machine-tools—Numerical control. 2. APT (Computer program language) I. Title.
TJ1189.C468 621.9′02 73-9766
ISBN 0-8311-1099-6

Sixth Printing

NUMERICAL CONTROL PART PROGRAMMING

Contents

Foreword

As a professional society, one of the major objectives of the Numerical Control Society is the dissemination of education and information to those active in the field, or to those about to enter it. A new, responsible, and authoritative publication which furthers this goal is therefore most welcome. Numerical Control technology, as related to machine tools, has had a significant impact around the world. As the "state of the art" continues to advance and proliferate into many new manufacturing areas, one of the most significant requirements for a successful numerical control operation is part programming. An NC industrial facility is either successful or unsuccessful depending on the capability of its part programming staff. Because part programming is a novel requirement, most part programmers acquire their necessary education via some formal training program; either through an academic environment (classroom or correspondence course) or the equipment builder's or programming service facility.

The publication of this text is of particular importance in view of the advances being made in the computer-aided design and computer-aided manufacturing part programming fields. Coverage of the basic principles of computer operation and its relation to part programming is felt to be most appropriate, and a welcome extension to the manual part programming concepts which are also well covered in this text.

This book is a natural sequel to Mr. Child's widely accepted PRINCIPLES OF NUMERICAL CONTROL.

William H. White
Executive Director, Numerical Control Society

Foreword for British Readers

Numerical Control Part Programming is a no-nonsense guide to NC Programming principles and as such fills a gap in a field where recent publications have tended to be rather specialized. The subject matter is of direct interest to persons who plan to use and install NC and who want an introduction to the subject of part programming. The book should be acceptable as a teaching base for the Polytechnics and other places offering practical training, and with the current plans of these establishments for installing and using the NC equipment purchased through the Department of Trade and Industry scheme, its appearance is timely.

The technical emphasis is on answering the question "what" rather than "why," but even intending specialists will find the book useful as a preparation for learning about computer programming before they go on to read detailed reference manuals. Readers will find that the quality of the diagrams and the structuring of the text are both good and the questions at the end of sections are clearly asked and answered, so that students and teachers alike will be helped.

Although the EIA standards are given pride of place, with the ISO discussed only briefly, the coverage offered in the text should satisfy the reader's understanding and educational requirements. The conventions for control system commands that have been referred to are standard, but the reader should be wary if he has older or nonstandard control systems. All the computer-aided programming discussion is based on the APT language and this is admirably laid out.

I was asked to write this foreword with a view to establishing the value of the book for British readers. In my opinion it is a book which should be useful to a wide range of people in Britain, particularly those at the start of their NC Education.

DR. I. D. NUSSEY
Chairman, British Numerical Control Society.
Application Programmes Manager, IBM, United Kingdom, Ltd.

Preface

In a little over twenty years the use of numerical control has grown from a single experimental milling machine at the Massachusetts Institute of Technology to over 25,000 units operating in the United States. And while this figure is relatively small when compared to the total number of machine tools in the field, approximately 50 percent of the money spent on machine tools such as drills, mills, and lathes goes for equipment that is numerically controlled. The accelerating demand for numerical control machine tools has put a severe strain on the manpower resources required to operate and maintain such equipment. In fact, many people in this field are convinced that the number of numerical control machine tools would be much greater if more trained personnel, particularly in the part programming area, were available.

The requirement for numerical control part programming has shifted the machining skill and technology from the machine shop floor to the part programmer's office. The part programmer, in most cases, is now responsible for determining the exact cutting sequences and holding devices, in addition to selecting the proper tools, cutting speeds, and feeds. This does not obviate the role of the machine operator, however, since he must also have an understanding of part programming to carry out the programmer's machining plan.

While it is possible to accumulate the necessary knowledge required for part programming without formal guidance, the most efficient means is by attending a formal training course or by undertaking a careful and sequential review of academic information, such as that contained in this book. The text has, therefore, been designed primarily for use in the classroom and includes a goodly number of exercises and questions. The "self-learner" will also find the text beneficial since special consideration has been given to a gradual build-up of the technology, with practice exercises and answers helping the student to put his knowledge to a test.

In addition to a detailed study of manual part programming, the text covers the role of the computer in part programming, an important consideration since approximately 55 percent of all numerical control machine tools are being programmed with the aid of a computer; and this percentage is expected to rise significantly. The particular computer part programming language selected for illustration is the APT, or *A*utomatically *P*rogrammed *T*ool, language. Although other computer part programming languages have considerable merit and are worthy of investigation, the APT is the most widely used. And by developing an understanding of the computer's role in part programming as related to the APT language, the reader may readily adapt his learned talents to another language which may be more suitable for a particular numerical control installation.

Knowledge of part programming need not, and should not, be restricted to the part programmer. It is incumbent upon anyone connected with a numerical control operation to become familiar with the principles of part programming. Because of the organization of the chapter material, coupled with the Appendix information, it is not necessary for the reader to have a strong "hands-on" machining or computer background in order to understand and appreciate the text. Therefore, *Numerical Control Part Programming* not only offers the groundwork for future part programmers, but also serves as a highly useful guide to anyone connected with a numerical control operation, including shop supervisors, managers, methods engineers, design engineers, computer programmers, and machine tool operators.

The author wishes to extend sincere appreciation to the following persons for their assistance in the preparation of this text: Eric Carlson, Maryanne Colas, John Combest, Graham Garratt, Holbrook Horton, Trudy McCarthy, Karl Moltrecht, Lloyd Murphy, Paul Schubert, and Fred Stemp.

A Note on English and SI Unit of Measurements

The measurement values used in this book are in English units, but this fact should be no obstacle to readers working with the International System of Units (SI). The SI is the modernized metric system of measurement. As many readers will know, there is a worldwide move to adopt this system, and growing interest in it in the United States.

Most of the values in this book are simple, such as positional and size dimensions, and feed rates. In many instances, a straight substitution of the SI equivalent may be made if the reader needs to know the order of size. In general, the examples presented are not designed to give absolute values but rather to demonstrate principles.

The two SI units likely to be most needed are those of length and time. The basic unit of length is the meter (m). Some useful conversions are 1 inch = 0.0254 meter = 2.54 centimeters = 25.4 millimeters. All these figures are exact.

Feed rates are given in this book in inches per minute. The basic SI unit of time is the second (s), and there is no change here from the traditional time unit common to both metric and English systems. Although the minute is outside the SI, it is nevertheless recognized as having to be retained because of its practical importance.

CHAPTER 1

What Is Numerical Control?

Numerical control, also referred to as NC, is a form of automation, but not the kind of automation by which thousands of identical parts, such as crankshafts for automobiles or electric light bulbs, are produced automatically by special machines. It is rather the kind that is applied to conventional machine tools, such as lathes, mills, and drills. These are the types of machines used in the average machine shop for producing different parts in *relatively small quantities.*

The difference between the usual type of machine tool and a numerically controlled machine tool is that for a conventional machine tool the motions are controlled by the operator, generally using handwheels, whereas for a numerical control machine tool they are controlled by an electronic unit. The electronic unit receives its instructions from a 1-inch-wide punched tape. The instructions on the tape are prepared by a person called a part programmer who plans the motions and operation of the machine. Tapes may also be prepared by the operator; however, it has been found generally more efficient to have the operation of the machine and the tape preparation handled by different people.

Figure 1–1(A) shows a numerical control milling machine. Its conventional counterpart is shown in Fig. 1–1(B). Note that motors have replaced the handwheels on the numerical control machine. The motors, which move the table and saddle at right angles to each other, are connected to the electronic control unit, shown on the right, by cables. The tape reader, which reads the coded instructions on the punched tape, is shown in the window of the control unit. The machine in Fig. 1–1(A) is also equipped with an automatic device for raising and lowering the

Courtesy of Bridgeport Machines, Inc.

FIG. 1–1(A). This numerical-control milling machine is operated by motors that are controlled by the electronic system shown at the right.

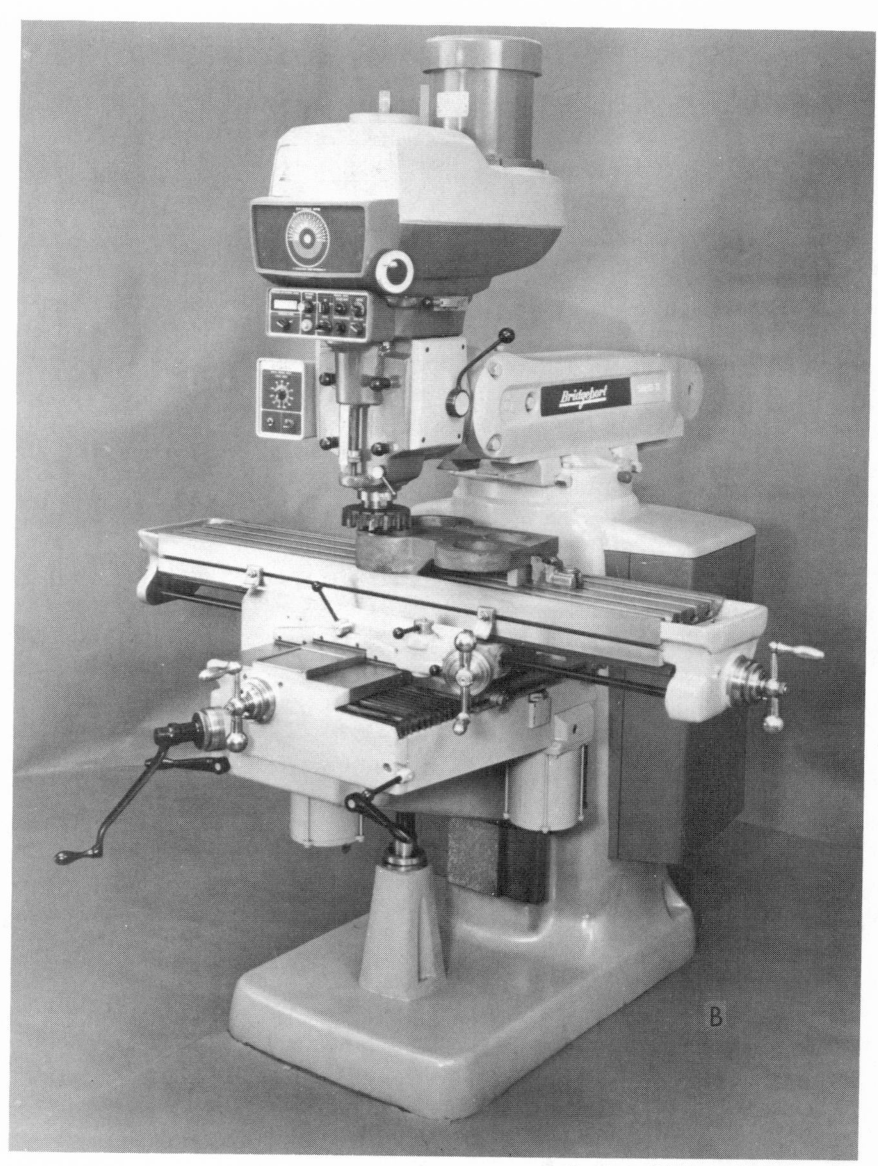

Courtesy of Bridgeport Machines, Inc.

FIG. 1–1(B). A conventional milling machine. The workpiece is moved to the required positions by using the handwheels shown.

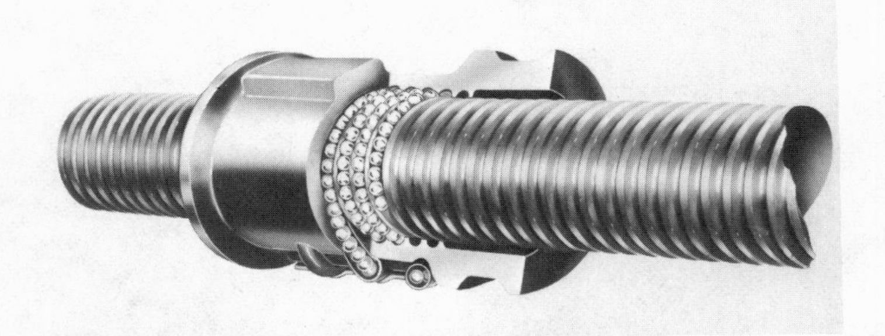

FIG. 1–2. This ball-bearing-nut lead screw is used with numerical control machines. It is more accurate and offers less friction than the kind found on conventional machines.

spindle, such as would be required for drilling. Just as handwheels are attached to threaded lead screws on a conventional machine, the motor shafts are connected to the lead screws either directly or through gearing. In the NC machine, however, the lead screw is generally more accurate and is equipped with a ball-bearing nut to reduce friction and wear. See Fig. 1–2. The ball-bearing nut is attached to the moving part of the machine, such as the table.

Numerical control has also been applied to other operations, such as drafting and electrical wiring machines. However, the great majority of applications fall in the metal-cutting category. Several different machine tool applications are shown in Fig. 1–3 through Fig. 1–7.

How Did Numerical Control Get Started?

Numerical control began a few years after World War II with a requirement for stronger and lighter jet airplane structures. Instead of riveting together a large number of parts, it was less expensive to machine a wing or fuselage part from a solid piece. The one-piece structure was also stronger and weighed less. As aircraft traveled faster, the amount of machining required to build an airplane went up in almost the same ratio as the maximum speed of the airplane. A Mach 2 (approximately 1400 mph) airplane, for example, required twice the amount of machining as did a Mach 1 airplane. Since most of the machining operations involved milling, the Air Force decided to run an experiment and convert a conventional-type tracing-milling machine to a numerically controlled milling machine. This machine, believed to be the first successful numerical control machine, was demonstrated at the Massachusetts Institute of Technology in 1952. See Fig. 1–8. Since the prime interest for the Air Force's funding of the MIT project was to develop machines that would produce parts in far less time, and of better quality, it was not long (1956) before large milling

Courtesy of Moore Special Tool Co., Inc.

FIG. 1–3. This NC jig borer is capable of obtaining positioning accuracies of ±.0001 inch. The machine shown is equipped with a vertically positioned rotary table which allows for the machining of various sides of the same part.

Courtesy of Ex-Cell-O Corp.

FIG. 1–4. This numerical control *machining center* is capable of performing milling, drilling, boring, and tapping operations. Coded instructions on the tape also specify the tool to be used, and an automatic tool selector replaces the selected tool in the spindle. This numerical control machine also has the added feature of being able to load and unload parts as well as to rotate the part for accessibility to the cutting tool—all being done automatically from instructions on the tape.

machines were being installed in aircraft companies throughout the country. One such numerical control machine designed for milling large jet-airplane wing sections is shown in Fig. 1–9.

However, it was not until approximately 1960 that the use of numerical control really began to grow, and twelve years later (1972) there were approximately 24,000 numerical control machines operating in the field. The types of numerical control machines vary considerably—from fabric-cutting machines for automobile upholstery, to hole-drilling machines for electronic circuit boards. Most of the machines are much smaller than the one shown in Fig. 1–9, and look more like the one shown in Fig. 1–1(A). Although there are many different applications of numerical control, the most popular are the metal-cutting machines, such as drills, mills, and lathes. It is estimated that approximately 85 percent of the numerical control machines in the field are of this type.

Advantages of Numerical Control

Numerical control is superior to conventional equipment mainly because:

1. Parts can be produced in less time and therefore at less cost.
2. Parts can be produced more accurately.

Other secondary reasons which are also very important and which have made numerical control so popular are:

1. The operator is less likely to make an error, and therefore there will be less scrap.
2. The jig or template sometimes required to guide a cutter is eliminated since the coded punched holes on the tape instruct the machine where to move. It is generally less costly to prepare a tape that will guide the cutting tool than to manufacture steel guiding fixtures such as those shown in Fig. 1–10. It is also much easier to store a tape than it is to store metal fixtures.
3. Inspection time is reduced since it is expected that every part made from the same tape will be practically identical.
4. Numerical control machines have greater capability than the con-

Courtesy of South Bend Lathe

FIG. 1–5. Horizontal numerical control engine lathe.

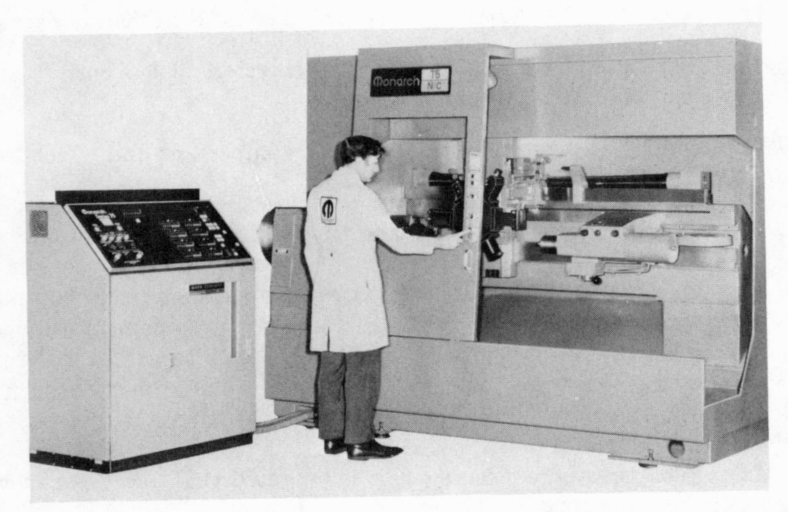

Courtesy of Monarch Machine Tool Co.

FIG. 1–6. The machine shown is a numerically controlled lathe, although, at first glance, it might be hard to identify. Many NC machines have been especially designed to make the best use of the NC concept and therefore may not resemble their conventional counterparts. This NC lathe has a vertical bed, while most conventional lathes have horizontal beds. The popularity of NC lathes is increasing.

ventional types, and this allows the engineers more freedom to design parts. The very complex part shown in Fig. 1–11 would be extremely difficult or even impossible to machine economically without numerical control, yet it is an essential part required for a rocket engine.

How Does a Numerical Control Machine Work?

The motions of a numerical control machine are very much like its conventional counterpart, only with the numerical control machine, drive motors replace the operator's handwheels. Since in most drilling and milling operations the workpiece must be moved, it is necessary to control both the worktable *and* the saddle on which the worktable sits. It is possible to move the workpiece to any desired position by coordinating the motions of the worktable and the saddle, which travel at right angles to each other. See Fig. 1–12. Movement of the machine-tool spindle in the vertical direction may also be controlled by attaching a lead screw and drive motor to it.

There are two ways of controlling the motions of a numerically controlled machine. One is by an *open-loop* system and the other by a *closed-loop* system. In the open-loop system, pulses generated by the

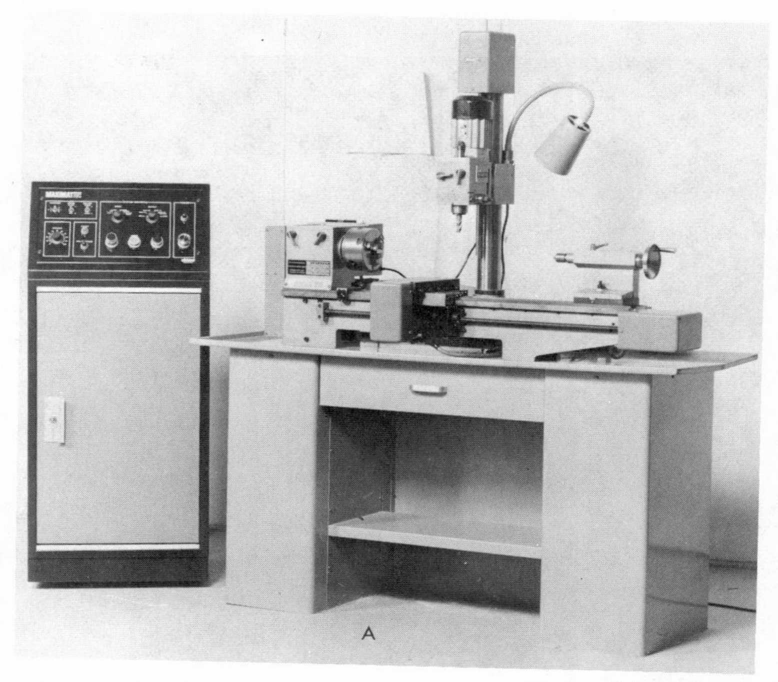

Courtesy of American Edelstaal, Inc.

Courtesy of Clausing Corp.

FIG. 1–7. (A) This low-cost NC machine combines both milling and turning capabilities. (B). The price of NC machine tools is continually dropping. This NC lathe is one of a number of new low-cost types.

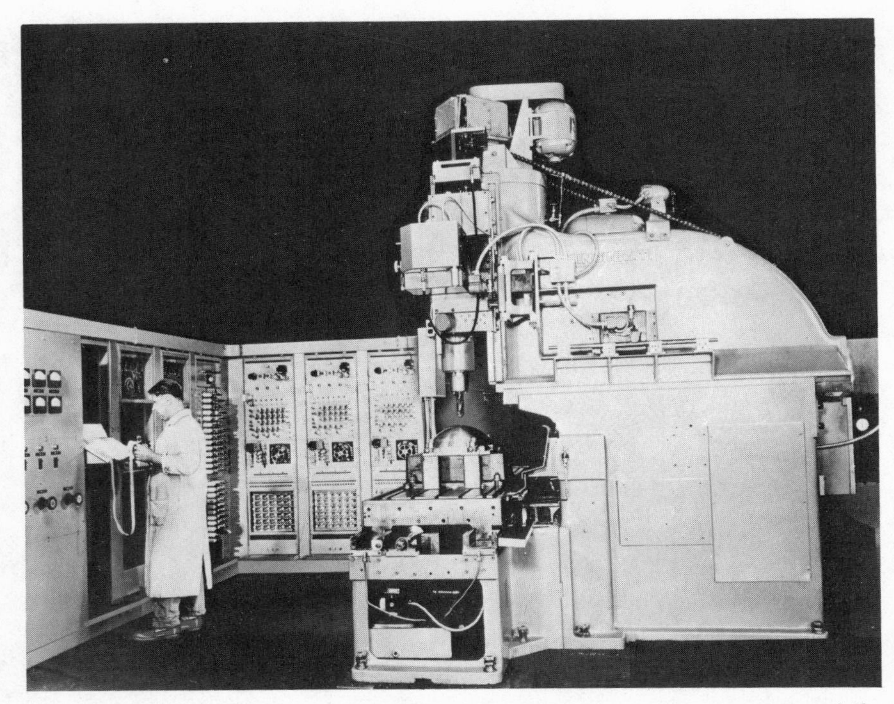

FIG. 1–8. Believed to be the first NC machine, the one shown above was successfully demonstrated at MIT in 1952.

electronic unit are fed to a special type of motor that rotates a fixed amount with each pulse. This type of motor is referred to as a *stepping motor*. Usually, each pulse rotates the motor 1.8 degrees. It would, therefore, take 200 pulses to make the motor rotate 360 degrees, or one complete revolution. The motor is generally connected directly to a lead screw with five threads per inch. Thus, five revolutions of the motor and screw are required to move the table 1 inch, or one revolution moves the table $\frac{1}{5}$ or .200 inch. Each pulse rotates the lead screw $\frac{1}{200}$ revolution which causes the table to move $\frac{1}{200} \times .200$ or .001 inch. The speed at which the motor rotates and, consequently, the feed rate of the table, is controlled by the rate at which the pulses are fed to the motor. For example, to move the table at a feed rate of 10 inches per minute (ipm) 10,000 pulses would have to be fed to the motor over a time period of 1 minute. By coordinating the rate of the two series of pulses that are fed to the table and saddle motors at the same time, it is possible to have the table move to any point under the spindle, or along any desired path. Figure 1–13 illustrates this arrangement. The straight line, or rotary path, of con-

Courtesy of Republic Aviation Div. of Fairchild Hiller

FIG. 1–9. The numerical control machine shown in the upper picture has a worktable that is 30 feet long by 12 feet wide. The lower picture shows a wing part for an Air Force supersonic fighter being removed from its holding fixture, after being machined.

Courtesy of Pratt & Whitney Co., Inc.

FIG. 1–10. The steel plates shown are used as a guide for drilling holes. The material to be drilled is clamped beneath one of these plates, and the drill is guided through each of the holes shown in the plates and then through the material to be drilled.

Courtesy of Rocketdyne Div., North American Aviation, Inc.

FIG. 1–11. The part shown being machined on a numerical control machine would be very difficult or impossible to machine on a conventional machine.

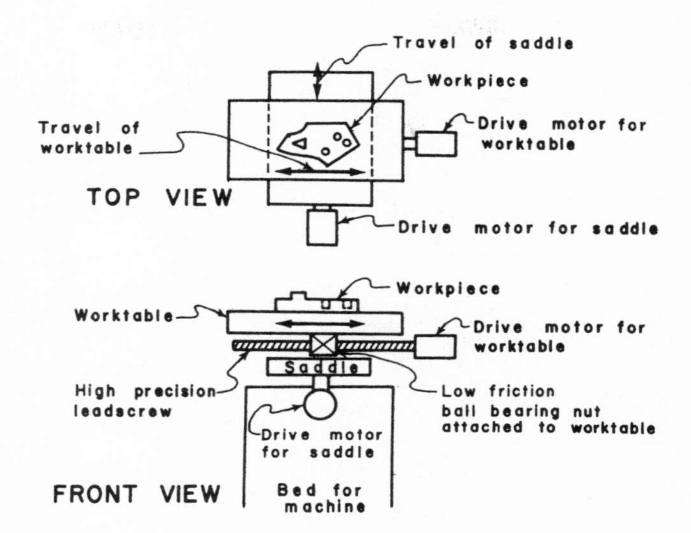

FIG. 1–12. In the sketch shown the saddle is actually attached to the bed and slides along it on *ways*. In turn, the table is attached to the saddle and also slides on *ways*, at a right angle to the saddle. The ways have not been shown in the illustration above.

trolled movement of any major part of the machine, such as the table, or saddle, is called an *axis*.

The *closed-loop* system is quite a bit more complex and requires that the position of the table, or saddle, or other controlled part of a machine be constantly checked and that information about its position be fed back

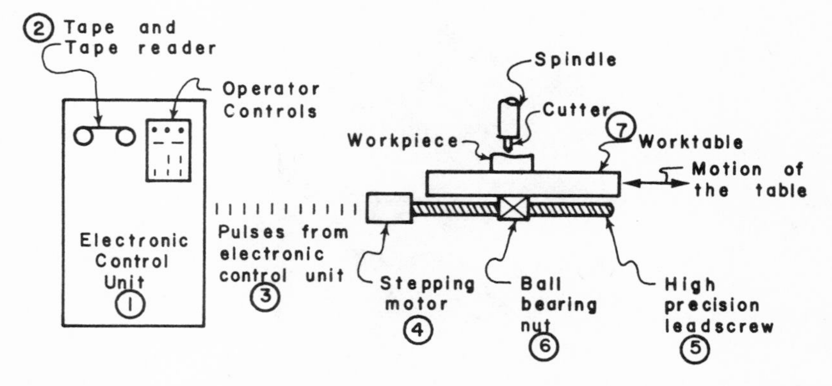

FIG. 1–13. Pulses are generated by the electronic control unit (1) in accordance with the instructions punched on the tape (2). These pulses (3) drive a stepping motor (4) which rotates a high-precision lead screw (5). The lead screw moves a ball-bearing nut (6) which is attached to the moving part of the machine tool, such as the worktable (7) or saddle.

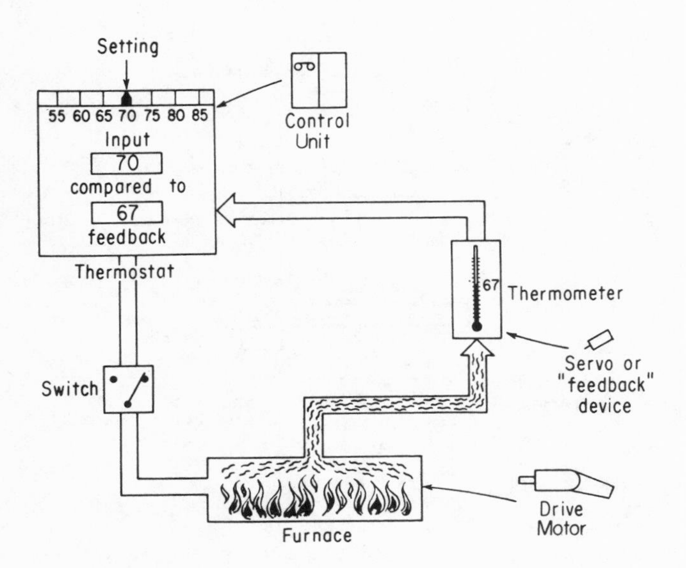

FIG. 1–14. The *closed loop* NC system works on the same *feedback* principle as the home heating system. The feedback device may be compared to the thermometers, the drive motor may be compared to the furnace, and the control system may be compared to the thermostat.

Courtesy of Kearfott Division, General Precision, Inc.

FIG. 1–15. This unit is called a resolver. It measures only about 1½ inches overall, is generally connected to the lead screw of a machine, and feeds back the position of the major machine component, such as the slide or saddle.

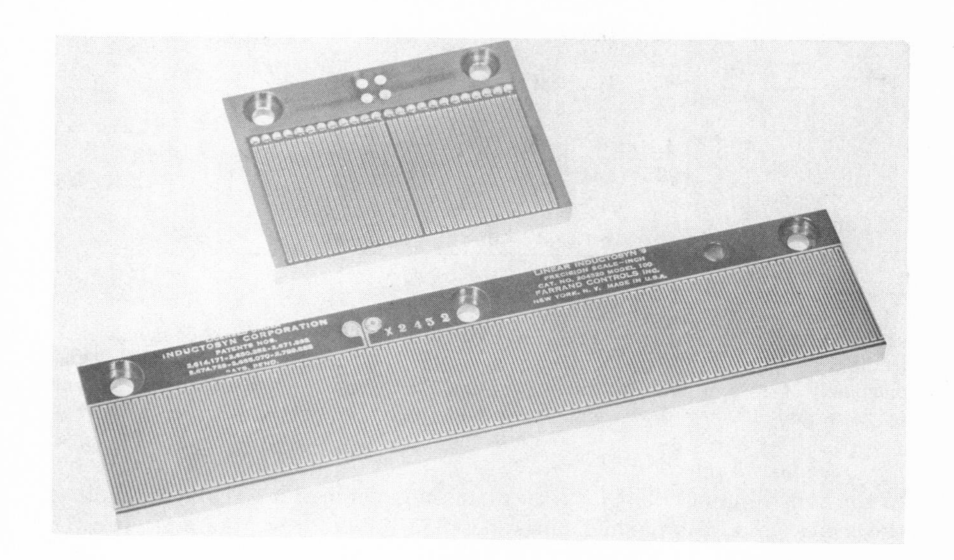

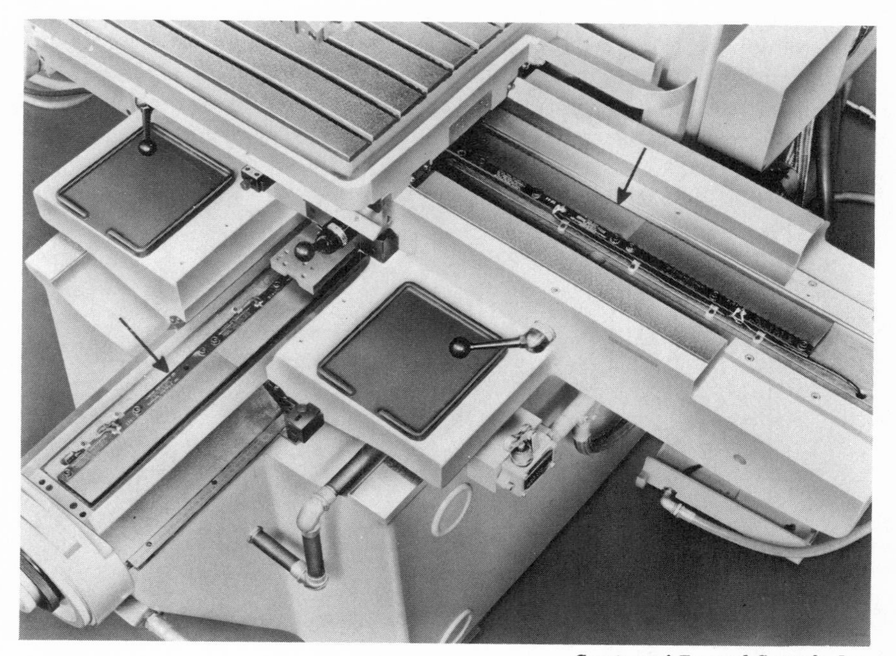

Courtesy of Farrand Controls, Inc.

FIG. 1–16. Top: Industosyn feedback devices. There are two elements, one of which passes over the other. Bottom: two pairs of the longer elements (see arrows) are shown in place on the machine tool.

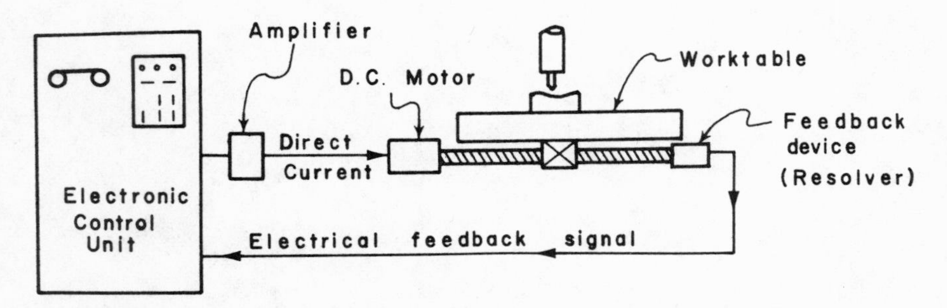

FIG. 1–17. The closed-loop arrangement consists of a feedback device that signals back to the control system the position of the worktable, or of another controlled part of the machine tool. Also, instead of a stepping motor, closed loop systems use a direct current motor in conjunction with an amplifier, as shown above, or an hydraulic arrangement as shown in Fig. 1–18.

to the control unit. Thus, the feedback position can be compared against the *command* position, which is the position called for on the tape. Power is fed to the drive motors until the feedback position agrees with the command position, and then the drive motors stop. This is also called a *servo* system and is not too unlike that used with an automatic pilot aboard an airplane. For example, the pilot sets a required course, and a measuring device senses whether the airplane is on course. When this sensing device does not agree with the course that is set, small electric-drive motors move

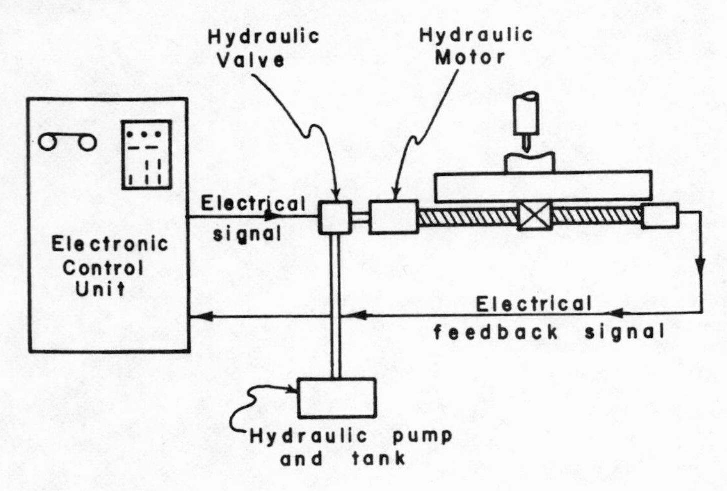

FIG. 1–18. The difference between a hydraulic and an electric system is in the drive unit or the "muscle" part of the NC system. In the above sketch a hydraulic valve and motor, together with a pump and tank, create the required hydraulic pressure and replace the DC amplifier and drive motor as shown in Fig. 1–17.

Courtesy of Boston Digital Corp.

FIG. 1–19. Tape reader and reels shown mounted in control unit.

the controls of the airplane to bring it back on course. When the airplane gets back on course, the small drive motors stop just as the drive motors stop when the table of the machine tool moves to the position called for on the tape. This servo-loop principle may also be compared with the conventional home-heating system. Referring to Fig. 1–14, the thermometer senses the temperature of the room, and in concept, this is very similar to what the servo or feedback device that is attached to the machine does. The main difference is that the thermometer measures temper-

Courtesy of Allen-Bradley Co.

FIG. 1–20. An electronic circuit board. These may be readily replaced when one of the components fails. This reduces the downtime by allowing the machine to operate while a maintenance technician checks for the faulty component on the circuit board.

ature, whereas the feedback device measures position. If the temperature agrees with the setting on the thermostat, nothing happens—just as nothing will happen if the table of a machine is at the position called for on the tape. Should the setting on the thermostat be raised, and therefore not agree with the reading on the thermometer, a switch automatically closes and starts a furnace. The furnace is the *power* in the home-heating system, just as the drive motor is the power in a numerical control system. Next, the heat generated by the furnace raises the temperature of the room and is recorded by the thermometer just as the feedback device records the position of the table of the machine tool. When the thermometer records the same temperature as that set on the thermostat, the switch automatically opens and the furnace stops—just as the drive motor stops when the feedback device records the same position as that called for on the tape.

In addition to the *feedback* concept, another significant difference between the closed-loop and open-loop systems is the type of drive motors used. Instead of a stepping motor, a feedback numerical control system uses a direct-current motor or a hydraulic motor. The speed of the electric motor is governed by the amount of current passing through it, rather than by the rate of pulses, such as in the stepping motor. If a hydraulic motor is used, its speed is governed by the amount of fluid passing through it.

There are several types of feedback devices in use with closed-loop systems. One is called a *resolver* and is usually connected to the lead screw through a gearbox. See Fig. 1–15. Since the resolver is actually measuring the rotary position of the lead screw and not the position of a

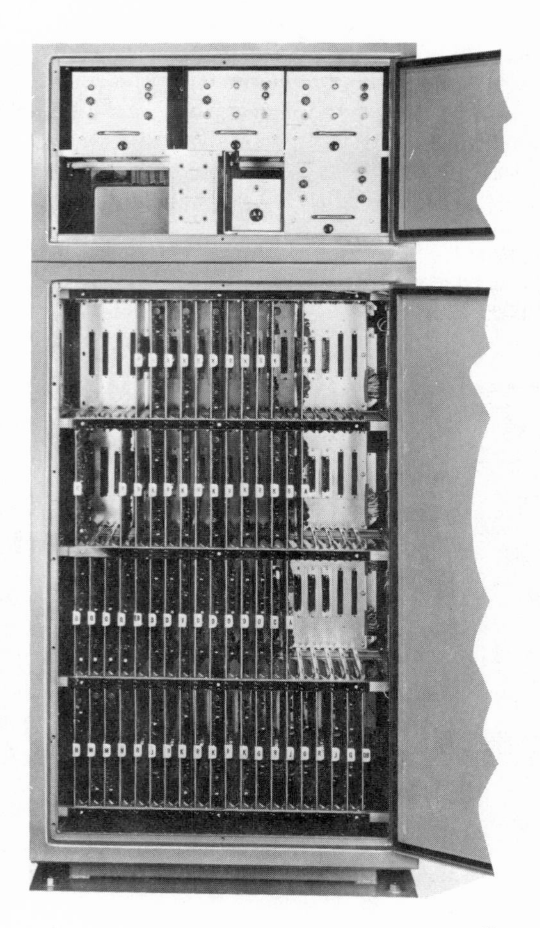

Courtesy of Allen-Bradley Co.

Fig. 1–21. A rear view of a control cabinet showing inserted circuit boards.

moving part of the machine tool, such as the table or saddle, the fit between the lead screw and the ball-bearing nut must be very accurate. This is the reason it is called a high-precision, ball-bearing lead screw. There are other feedback devices, called *Inductosyns*, which measure the position of the moving part of the machine tool directly. They are somewhat more expensive than resolvers, especially on larger machines; however, the result is generally a more accurately machined part since the position of the table is measured directly. Figure 1–16 shows an Inductosyn scale and slide (top of figure) fitted to the saddle and table of a milling machine (bottom of figure).

The functional arrangement of a closed-loop system using a variable-speed, electric D-C motor is shown in Fig. 1–17. The feedback signal is generated by the resolver, and this is fed back to the control system where it is compared to the input signal generated by the coded information on the tape. There is one such *loop* for every axis on the machine tool.

To control hydraulic motors, a valve is regulated by an electrical signal from the control unit. When it is desired to increase the rpm of the

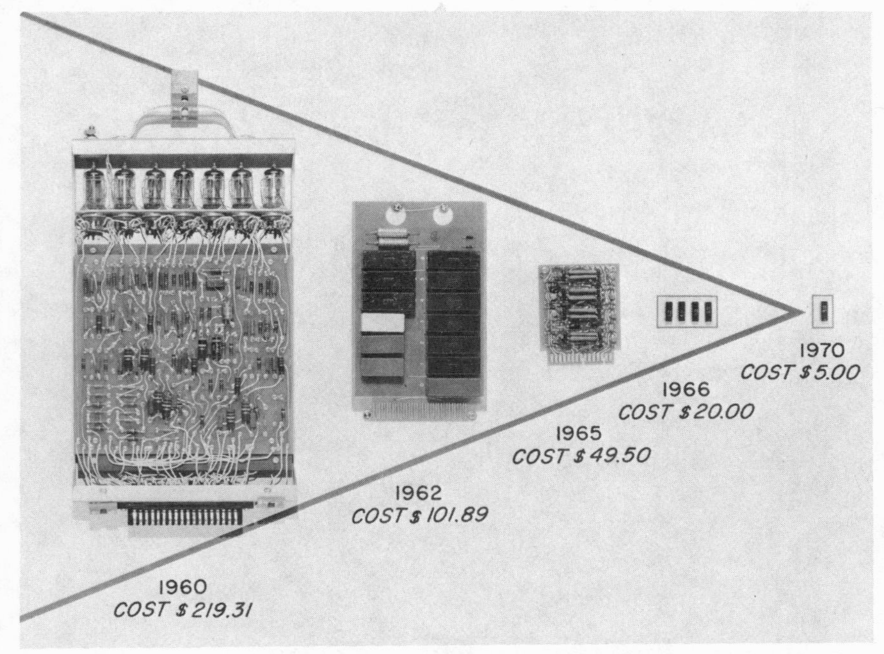

Courtesy of Industrial Controls Div., Bendix Corp.

Fig. 1–22. The integrated circuit shown at the right (1970) has replaced the entire circuitry shown on the large circuit board on the left (1960). In addition to being smaller and less expensive, the integrated circuit is also far more reliable and lasts longer.

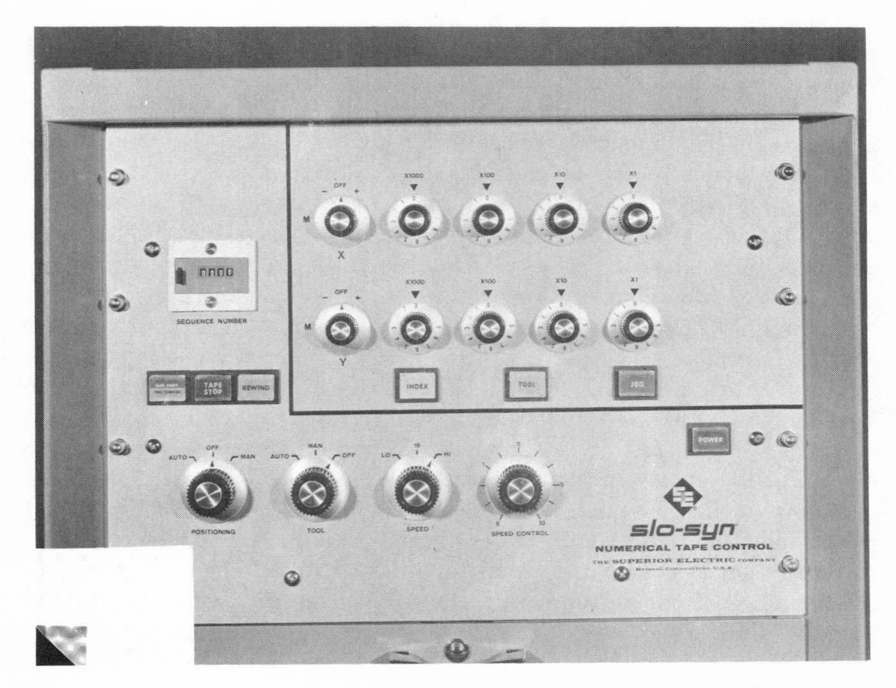

Courtesy of Superior Electric Co.

FIG. 1–23. Numerical control machines may be moved to precise locations by means of dials and push buttons as well as by the tape. The control panel above allows the operator to set a distance and then move the machine that precise distance. "Jogging" may also be accomplished by pressing a button as long as the operator desires motion. When the operator releases the jog button the machine stops.

motor, the valve is opened wider to allow more fluid to pass through the motor; when the rpm is to be decreased, the valve is closed. In the hydraulic arrangement shown in Fig. 1–18 the feedback portion and the control unit are the same as in the electric drive arrangement. The significant difference is in the part of the system that furnishes the muscle, or in other words, the drive package. Also, there are systems that use hydraulic cylinders in place of the rotary hydraulic motor and the lead screw, but these are not as common as the lead screw type.

Major Parts of a Numerical Control Unit

Tape Reader. The tape reader, the only moving part of the control unit, serves to read the coded punched holes in the tape and pass the instructions on to the internal electronic system of the control unit (see Fig. 1–19). Readers are of numerous types and vary considerably as to the speed with which they can read the tape. This ranges from 20 to 500 lines per second. The speed required depends on the type of machine tool being controlled, or the application, if it is other than a machine tool. Also, while there are still a good number of tape readers that read the holes in the tape with mechanical tabs and index fingers that are electrically wired, most readers now being manufactured are of the photoelectric type which are actuated by light showing through holes in the tape. The tapes may be handled in what is known as the *tumble box* or may be wound on reels. The reel arrangement is the more popular.

Circuitry. Most numerical control units are designed so that they may be maintained without too much difficulty, although maintenance technicians do need special training. The electronic circuitry is generally in the form of circuit boards (see Fig. 1–20) which are inserted into specific slots in the control unit and may be readily replaced when one becomes faulty (see Fig. 1–21). The faulty circuit board can then be analyzed to determine the exact component that failed—without the machine's being shut down while the search for the faulty component is being made. Most control units now employ *integrated circuits* (also known as ICs) which are extremely small and compact and thus have reduced the cost and space requirements of control units. A comparison of circuitry used over the past ten years is shown in Fig. 1–22.

Control Panel. In addition to the tape, most numerical control ma-

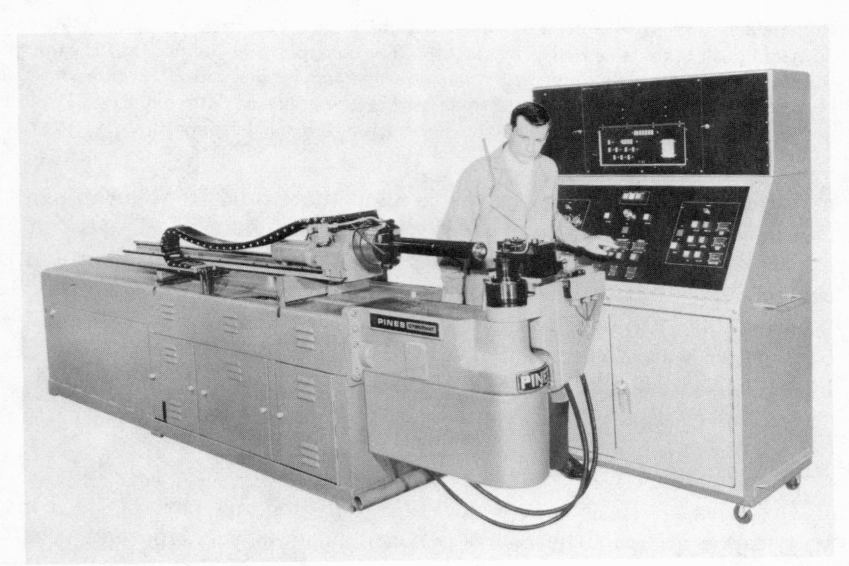

Courtesy of Teledyne Pines.

Fig. 1–24. A tube may be bent automatically with this NC machine. The length and angle of the bend are controlled from instructions on the tape.

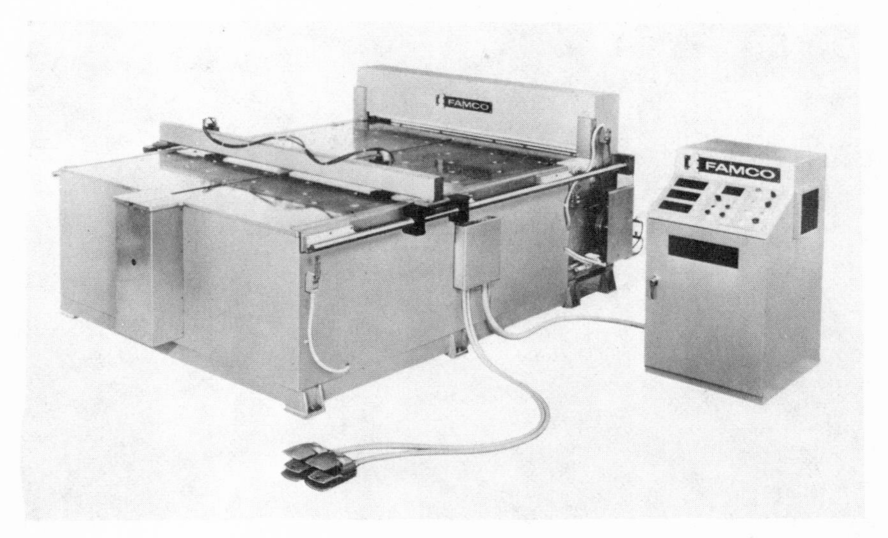

Courtesy of Famco Machine Co.

Fig. 1–25. This numerically controlled shear positions sheet stock so that it may be cut to a precise length. In the picture above the cutter is at the rear and the sheet holder at the front part of the machine.

chines can be operated manually—not *manually* in the sense of using handwheels, but rather by pushing buttons and turning knobs. As an example, the operator's panel on the control unit shown in Fig. 1–23 allows the operator two means of manually moving the machine. One method allows the operator to push a *jog* button. As long as the operator holds the *jog* button down, the machine moves at the feed rate which has been set by a dial. In the example shown, another way of moving the machine manually is to dial-in the distance to be moved; the machine will then move the precise distance each time the *index* button is pressed. Manual controls are normally used only for set-up, and the actual machining operation is generally controlled by the tape.

Numerical Control Applications

Theoretically, any machine that requires controlled motion is a candidate for numerical control. As noted earlier in this chapter, NC has been applied all the way from simple drilling machines to automatic drafting machines and even cloth-cutting machines. However, its greatest use is in the metal cutting area. Of these applications, approximately 85 percent fall under the category of drilling machines, milling machines, lathes, and

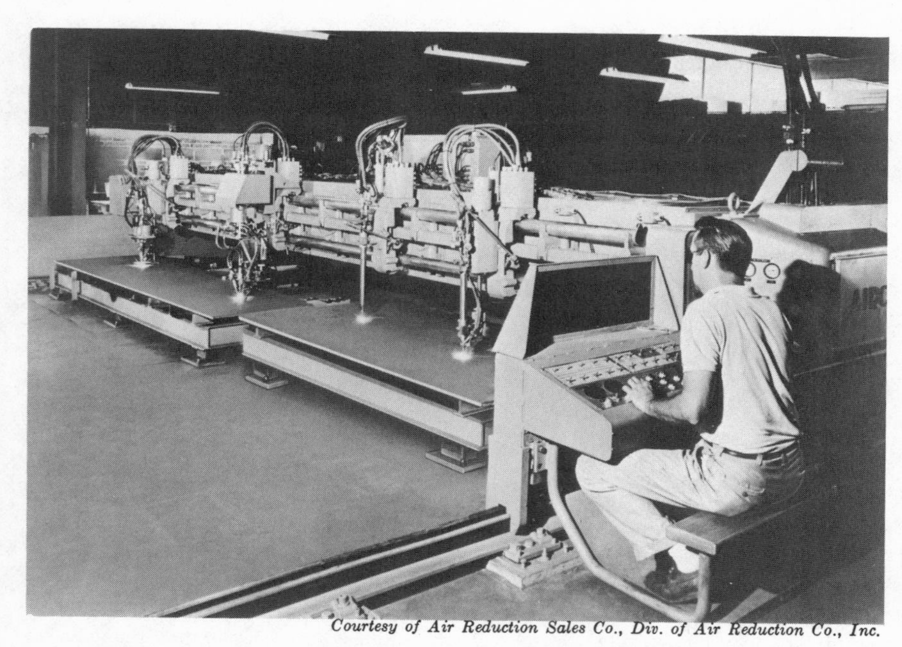

Courtesy of Air Reduction Sales Co., Div. of Air Reduction Co., Inc.

FIG. 1–26. Torch cutting machines such as this one are being used in shipyards to cut steel plate.

boring machines. Whether a particular numerical control application will become popular depends on whether it can be justified from a dollars-and-cents standpoint and not on whether it can be built or not. The reason that numerical control machine tools, especially the cutting type, have become so popular is that they can outproduce their conventional counterparts generally by at least two to four times. When the anticipated savings are weighed against the cost of the numerical control machine, it then becomes evident whether or not the machine would be a good buy. A number of different numerical control applications are shown in Figs. 1–24 through 1–30.

Numerical Control and Jobs

There are some people who feel that numerical control, like any other type of automation, will put people out of work. If this were true, there would be hardly any work for anybody in today's age of automation and advanced technology. With the population of the world increasing at an extremely fast rate it is essential that *more* jobs be made available as time progresses. In fact, because of automation and new technologies, more and better jobs *are* being created. Automation also improves our standard

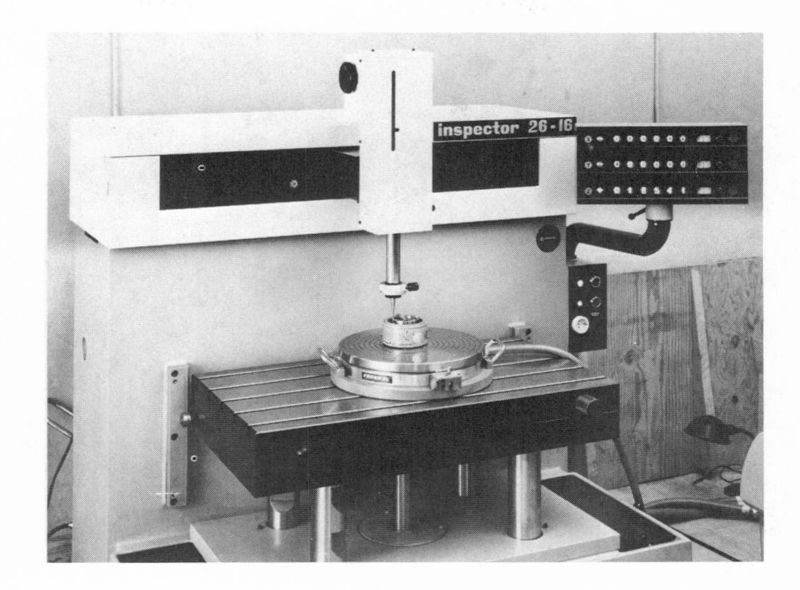

Courtesy of Ferrand Controls Inc.

FIG. 1–27. NC machines are also used to inspect parts. Shown here is a NC inspection machine being used to check out the coordinates of a part.

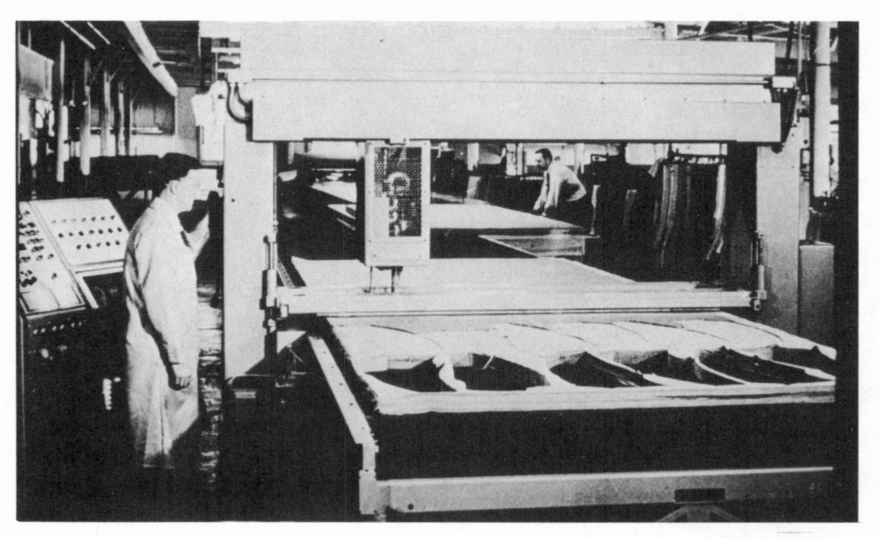

Courtesy of Cincinnati Milacron

FIG. 1–28. Numerical control is also being applied to fabric cutting. The cutting knife, capable of cutting through a stack of material, is directed by a numerical control unit.

Courtesy of Paul Dosier Associates

FIG. 1–29. A multihead high-speed router for contouring softer materials, such as aluminum, plastics, and wood.

Courtesy of Volkart Bros.

FIG. 1–30. The control system at the left guides a part around a fine wire that cuts through metal via the electro-discharge method (EDM).

of living. Imagine how many automobiles there would be if there were no high-production machinery to produce them. If each automobile had to be handmade, it would probably cost nearly $100,000, instead of around $3200, and most of us would be doing a good deal of walking.

Automation, as well as progress, however, does have a price tag. In order to use it properly one must be trained to meet its technical demands. Numerical control, as an example, while outproducing conventional equipment, requires trained operators, maintenance personnel, and technicians (part programmers) to prepare the punched tapes.

QUESTIONS CHAPTER 1

1. How does numerical control differ from the mass production type of automation?
2. What is the function of the part programmer?
3. How does the operation of an NC machine differ from that of a conventional machine tool?
4. Name at least four applications of NC that lie outside the metal cutting field?
5. What government organization was the most responsible for starting NC in the United States?
6. When, and where, was the first NC machine demonstrated?
7. The majority of NC applications fall into what category?
8. Approximately how many NC machines were in the field in 1972?
9. What is the rule of thumb regarding the amount of machining time versus the speed of aircraft that prompted the government's support of NC?
10. What are the two key advantages of NC?
11. What are four secondary advantages of NC?
12. What are the two types of systems used for controlling the motions of NC machines?
13. How many pulses are required to move the table of an open-loop NC machine a distance of 3.500 inches.
14. How many pulses per minute would be required to move an open-loop NC machine table at the rate of 18 ipm?
15. What is the feedback device that fits on the end of a lead screw called?
16. What is the main difference between a closed-loop and an open-loop NC system?
17. What happens when the input signal and the feedback signal on a closed-loop system are in agreement?
18. What is an Inductosyn?
19. What is the most popular type of tape reader?

20. What is the common abbreviation for the term *integrated circuit*?
21. What major component of an NC system may the furnace of a home heating system be compared to?
22. If it is agreed that numerical control and automation will create more jobs and improve our standard of living, what is the price tag?

CHAPTER 2

Numerical Control Machine Movements

Identifying the Motions

For the control unit to feed electrical instructions to the proper drive motor, the particular drive motor, and the motion that it controls, must be identified. For example, to move the worktable shown in Fig. 2–1 electrical signals must be directed to the motor that drives this worktable and not to the motor that drives the saddle or to the motor that drives the head of the machine. Each movement has to be individually identified for the electrical instructions to be properly directed.

In line with standard practice the worktable movement would be described as an X motion; the saddle movement would be described as a Y motion and would be at right angles to the X motion. Since the worktable rides on the saddle, it is possible by a combination of X and Y motions to direct the workpiece to any desired position under the spindle in order to perform a machining operation, such as drilling. A vertical movement of the head, or spindle, on this type of machine would be a Z motion. These letter notations (X, Y, and Z), which apply to the motions of a numerically controlled machine, have been standardized and are followed by all numerical control machine tool builders.

For example, to move the worktable 5.000 inches to the left, as shown in Fig. 2–2, the coded instruction on the tape would be x + 5000. The (The decimal point is normally omitted on the tape.)[1] The plus (+)

[1] This applies to an *incremental* system where the distance to be moved, from one point to the next, is described on the tape. This differs from an *absolute* system in which the coordinates of the point to be moved *to* are described, and in which the sign would depend on the location of the point. Absolute and incremental systems are described later in the text.

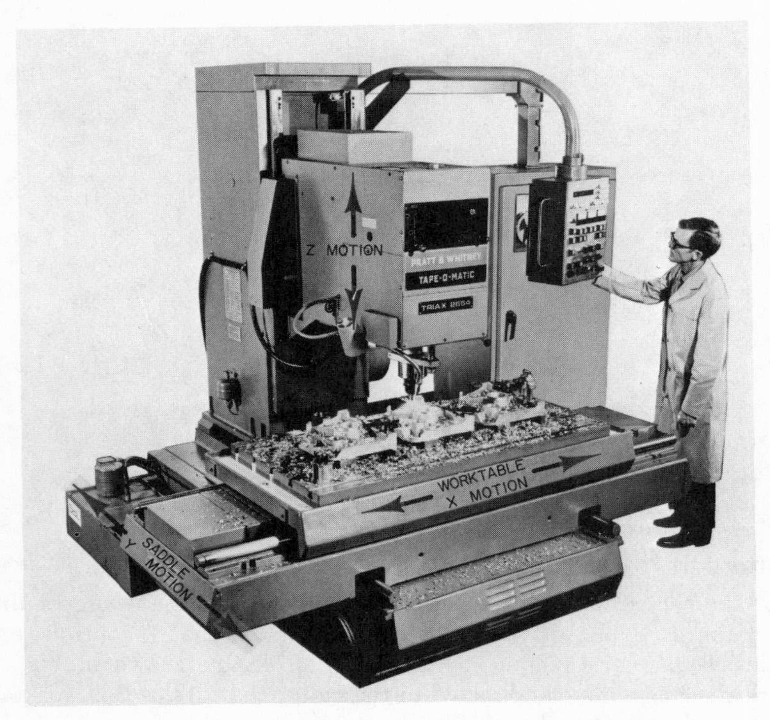

Courtesy of Pratt & Whitney Machine Tool, Div. of Colt Industries

Fig. 2–1. Movements of the worktable, saddle, and head are identified as X, Y, and Z motions, or *axes*.

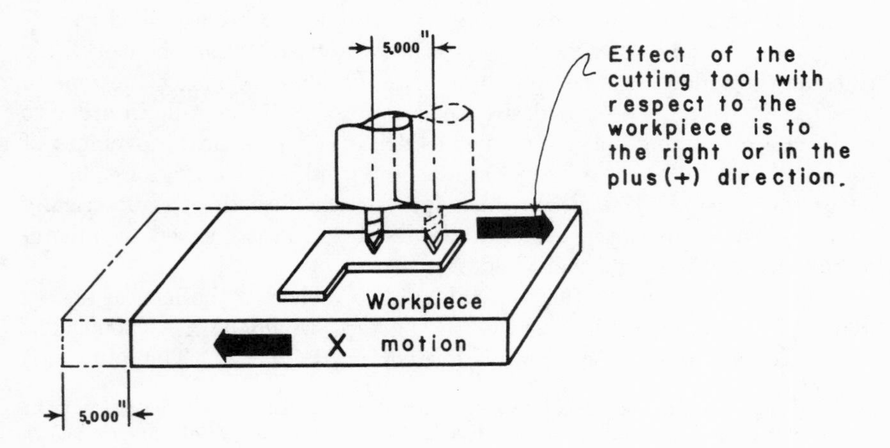

Fig. 2–2. Moving the table to the *left* 5.000 inches has the same *effect* as though the cutting tool were moved to the *right* 5.000 inches with respect to the workpiece. This motion would be considered a plus (+) motion.

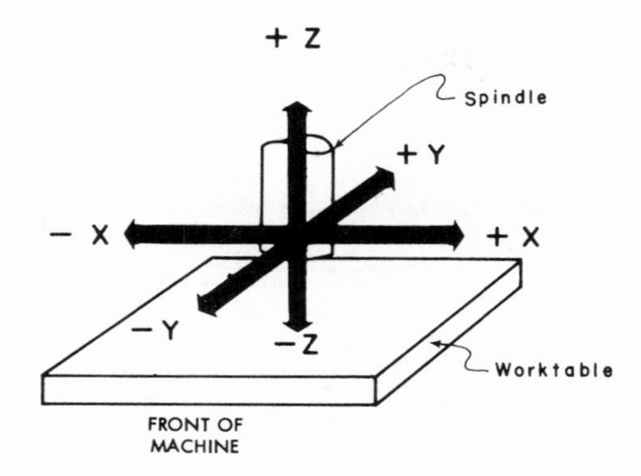

FIG. 2–3. The three axis motions, X, Y, and Z, with their plus (+) and minus (−) signs are shown above. The (+) and (−) signs denote whether the movement of the spindle relative to the worktable is to the right or left (X axis), forward or backward (Y axis), or up or down (Z axis).

sign means that the worktable is to move to the *left*, which would be the same as if the table were fixed and the spindle were moved to the *right*. A minus (−) sign in front of 5000 (−5000) would mean that the worktable is to move to the *right*, which would be the same as if the table were fixed and the spindle were moved to the left. In other words if the spindle is stationary, as is the case with the type of machine shown in Fig. 2–1, then the effect of the cutting tool with respect to the work is opposite to the direction of table movement (see Fig. 2–2). It is the *relative motion of the cutting tool with respect to the workpiece* that determines the plus (+) or minus (−) sign. Both the machine tool operator and the man who plans the job always consider the tool to be moving about the work, even if the workpiece actually moves and the cutting tool does not. A move of the cutting tool with respect to the workpiece to the *right* is described

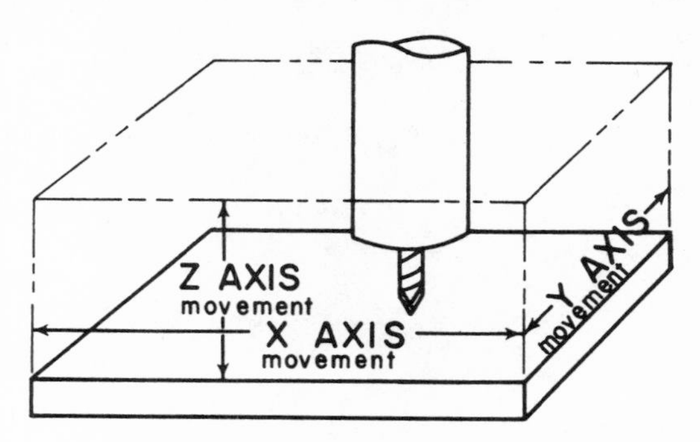

FIG. 2–4. By coordinating the motions of the X, Y, and Z axes, it is possible to move the cutting tool to any point within an imaginary cube sitting on the worktable.

by a plus (+) sign and a move of the cutting tool to the *left* is described by a minus (−) sign, as shown below.

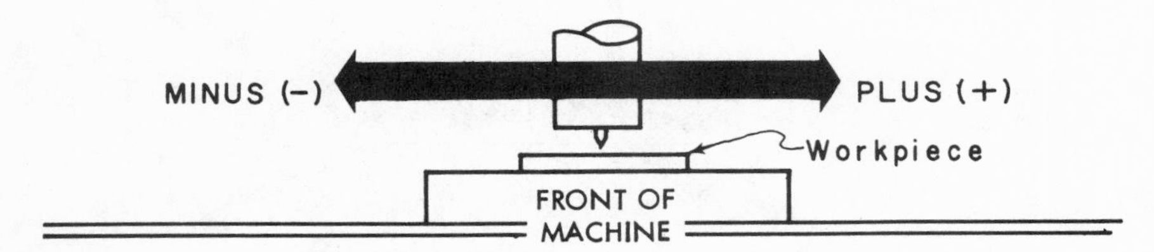

In the case of the *Y* motion a plus (+) movement would mean that the saddle shown in Fig. 2–1 would move forward—toward the operator, and would have the same *effect* on the workpiece as though the spindle were moving *back*. A plus (+) *Z* motion would mean that the *spindle* is to move *up* and away from the workpiece and a minus (−) *Z* motion would

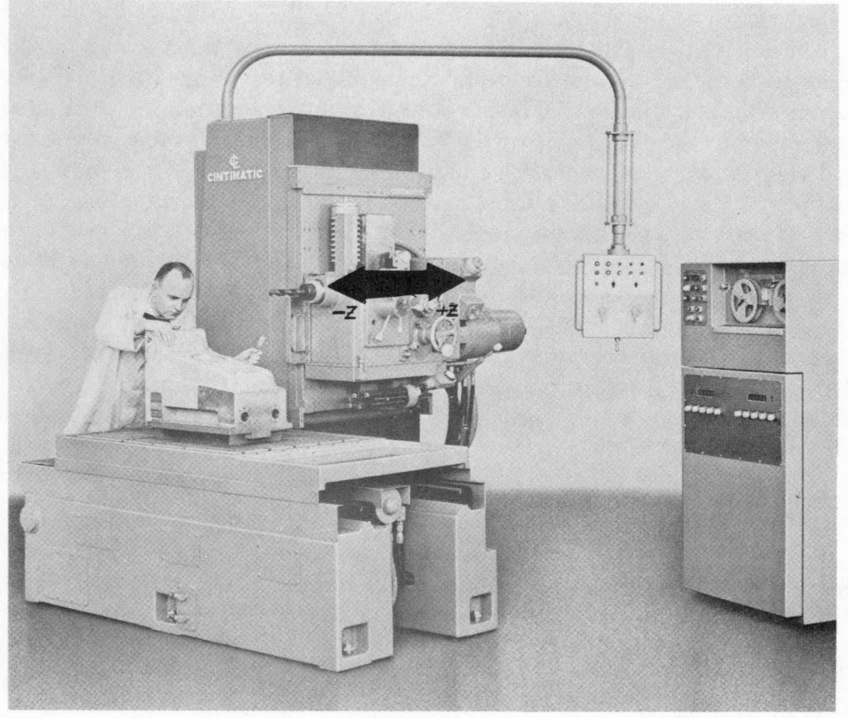

Courtesy of Cincinnati Milacron

FIG. 2–5. The axes of a horizontal spindle machine differ from those of a vertical spindle machine. In the case of the horizontal spindle machine shown, the *Z* axis is horizontal and parallel to the motion of the spindle, or head.

Courtesy of De Vlieg Machine Co.

FIG. 2–6. To observe the working surface of a horizontal spindle machine the viewer stands at the rear or alongside of the machine, and looks forward. The X, Y, and Z axes have the same relationship to each other, only turned up at 90 degrees.

mean that the spindle is to move down and toward the workpiece. The relative X, Y, and Z motions of the cutting tool with respect to the workpiece are shown in Fig. 2–3.

Axis Designations

The imaginary line that the worktable, saddle, spindle, or other machine part moves along is called an *axis.* Referring to Fig. 2–3, the X motion would be along the X *axis*; the Y motion along the Y *axis*; and the Z motion along the Z *axis.* These axes are at right angles to each other, with the Z axis being perpendicular to the plane formed by the X and Y axes.

The axes designations that have been assigned to various machine tool movements are all standardized and have been agreed to by the numerical control equipment builders. Otherwise it would be extremely confusing if two people tried to discuss the motions of a machine with each other without knowing which axis of motion was being referred to.

As with the X axis movement just described, the distance to be moved along any axis may be noted on the tape as a letter, such as x, y, or z, together with a number.

For example, to move the cutter, with respect to the workpiece, 7.000 inches in the plus Y direction the instruction on the tape would read y +7000. To move the *spindle* down toward the workpiece 3.000 inches, the instruction on the tape would be z −3000. Again the decimal points are usually not shown. By combining X, Y, and Z motions one can move the cutting tool to any position within an imaginary cube sitting on the worktable. See Fig. 2–4.

Not every motion of a machine qualifies as an *axis*. For example, if the amounts of the X and Y motions of the machine shown in Fig. 2–1 are controlled by tape instructions but the amount of Z motion is set and controlled by the operator, then the Z movement would *not* be a Z axis. This machine would then be described as a *two-axis* machine (X and Y axis movements only). The spindle could be automatically controlled to move up and down, as would be required in a drilling operation. However, the amount of the vertical movement would be set by the operator prior to the machining operations. If, on the other hand, there were a drive motor attached to the spindle, whether an open-loop stepping motor or a closed-loop system with tape instructions controlling the distance that the spindle moves, then this would be designated as a *Z axis* and the machine would be noted as a *three-axis* machine. Most of the lower cost numerical

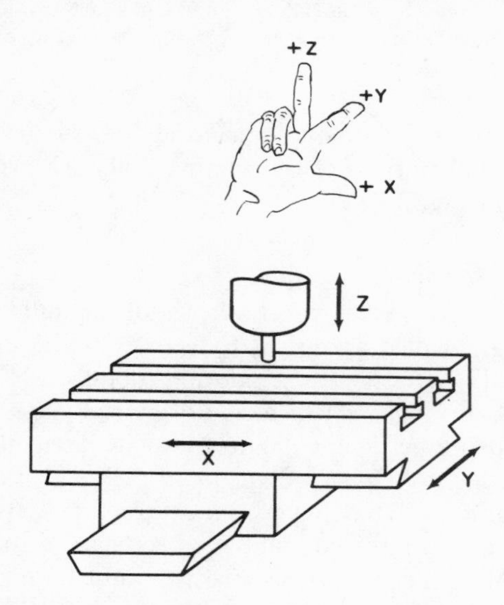

FIG. 2–7. The right-hand rule establishes the plus (+) X, Y, and Z directions. Minus (−), or negative, directions are the opposite to those shown.

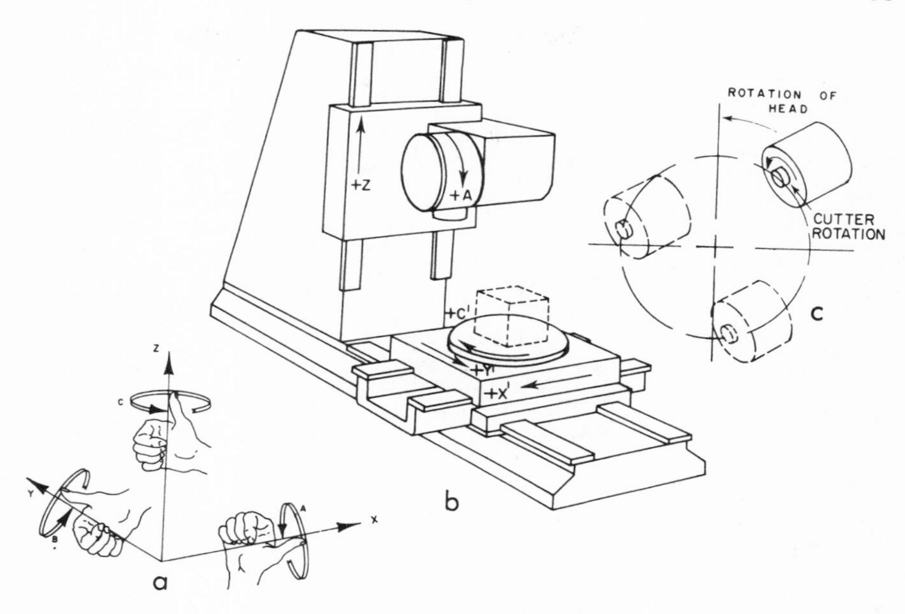

FIG. 2–8. The right-hand rule (a) also applies to controlled rotary motion. The two rotary motions of the machine (b) are shown as A and C'. The A motion applies to the spindle head, and the C' motion applies to the rotary table on which the workpiece sits. The reason for designating the motion as C' rather than C is that the workpiece is rotating in a clockwise direction, producing the same effect as if the head of the machine were rotating in the counterclockwise direction, as shown in (c). Actually, the head does not rotate: it is the effect of the cutter with respect to the workpiece that matters. The same applies to the X' and Y' directions. In order, for example, for the cutter to have a $+X$ motion with respect to the workpiece, the table must move in the opposite direction, which is shown as X'.

control machines are of the two-axis type, although the three-axis type is becoming more popular.

The machine shown in Fig. 2–1 is of the most common type in that the spindle is in the vertical position and the table lies in a horizontal plane. There are, however, many machines that have the spindle in the horizontal position. A machine of this type is shown in Fig. 2–5. The workpiece may sit on the horizontal table, as shown in Fig. 2–5, or be held in a vertical position as shown in Fig. 2–6. In both cases, the X-Y plane, or working surface, is in a vertical plane instead of a horizontal plane as in the case of the machine in Fig. 2–1. Also, the Z motion of the spindle is now horizontal instead of vertical. It will be noted that the relationship of the X, Y, and Z axes are the same. It will also be observed that the axes are described when looking at the machine from the front in the case of a vertical machine and from the back in the case of a horizontal machine.

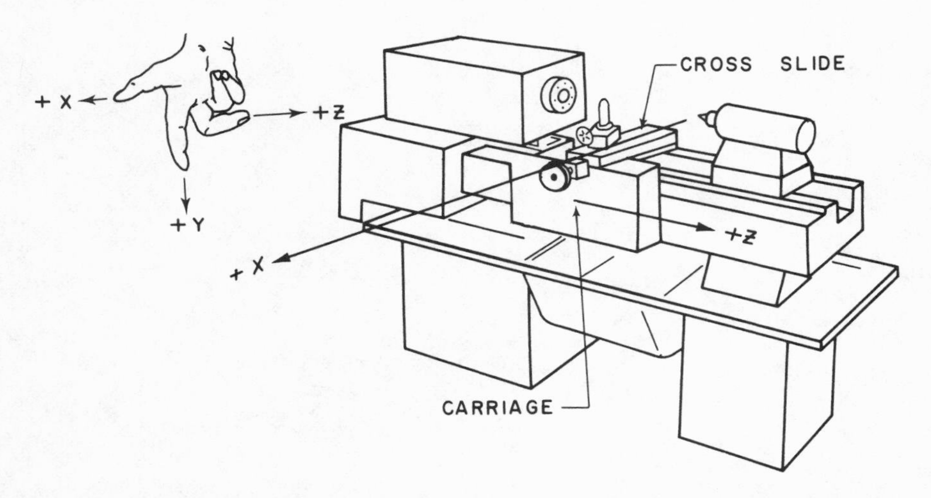

FIG. 2–9. The axes on a lathe also conform to the standard right-hand rule. The horizontal travel of the carriage is the Z axis, and the horizontal travel of the cross slide is the X axis.

There are many different configurations of numerical control machines, and some even have *rotary* axis motions. In order to establish a standard way of describing these motions and assign axis designations to the various motions, two *right-hand rules* have been developed. One rule makes use of the thumb, forefinger, and second finger of the right hand (see Fig. 2–7).

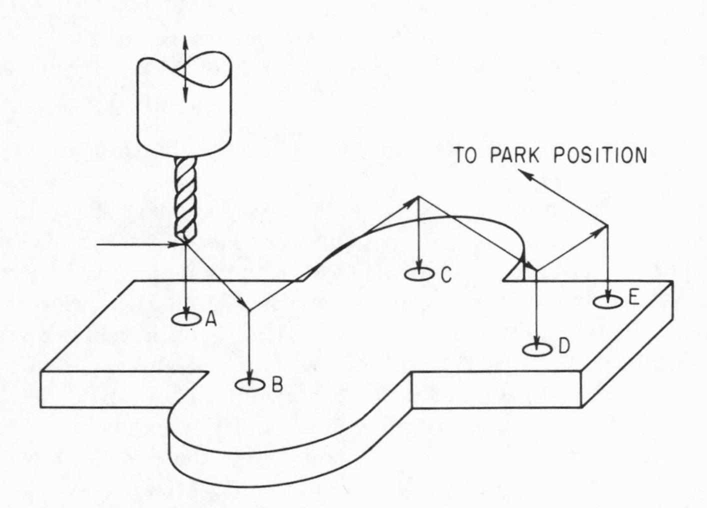

FIG. 2–10. Point-to-point numerical control involves moving a cutter, or other tool, to a specific point and then performing an operation.

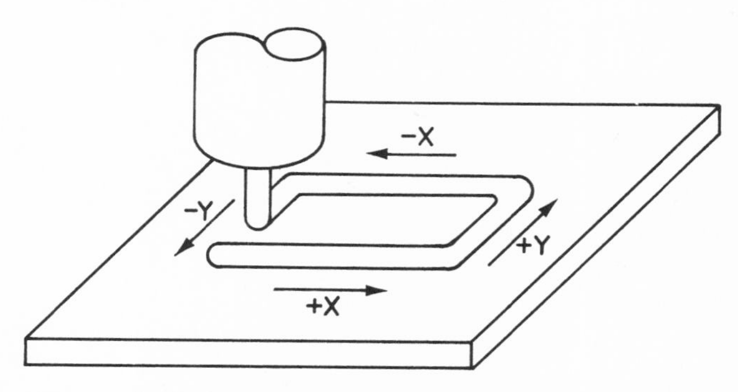

FIG. 2–11. An example of *picture-frame milling* in which only one axis is moved at a time.

If the thumb is pointed in the plus (+) X direction, the plus (+) Y direction will be in the direction of the forefinger and the plus (+) Z direction will be perpendicular to the plane formed by the X and Y axes. The minus (−) directions are opposite to those shown. This axis rule would apply to any numerical control machine. It should also be noted that the line running through the center of the spindle will normally be in the direction of the Z axis.

A second right-hand rule is shown in Fig. 2–8. In this case the rule applies to *rotary motions*. As with the requirements for X, Y, and Z designations, each rotary motion must be under drive-motor control to be called an axis. The rotation of the cutting tool on the spindle, for example, would *not* be a rotary *axis* motion. On the other hand, a rotary table,

Courtesy of the U.S. Naval Research Laboratory

FIG. 2–12. The part shown was machined on a point-to-point numerical control machine having picture-frame milling capability. All milling cuts were made along either the X or Y axis.

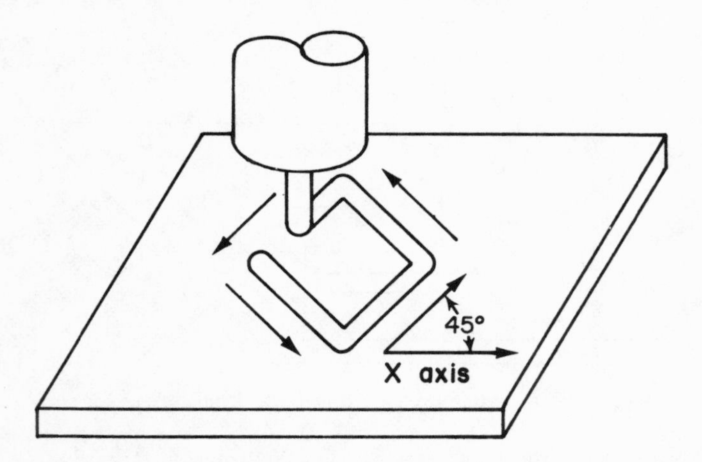

FIG. 2–13. Some point-to-point machines have the capability of milling at a 45-degree angle to the X and Y axes.

controlled from tape instructions and having a stepping motor or closed loop drive, would be considered an axis. (If it is an indexing table, which moves a specific amount on a command signal from the control unit, it would *not* be considered an axis.) Referring again to Fig. 2–8, if the thumb is pointed in the direction of the positive axis, the curled fingers point in the plus (+) rotary direction. For example, if the thumb is pointed in the direction of the plus (+) X axis, a rotary motion in the

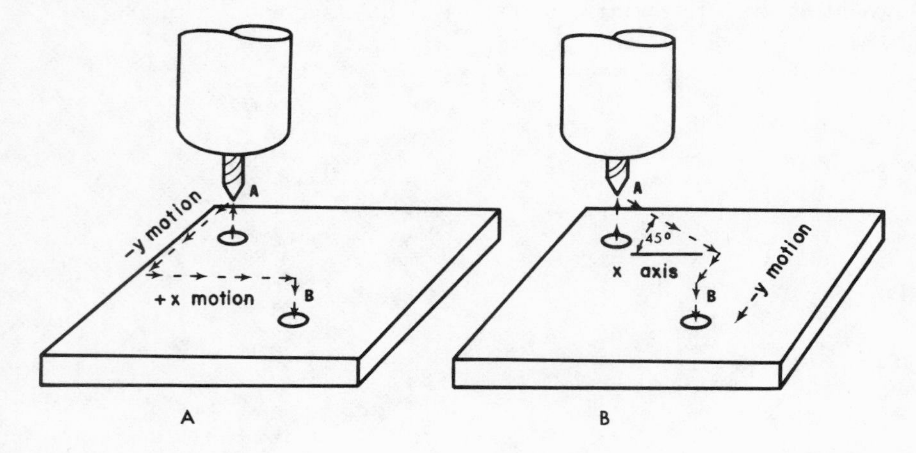

FIG. 2–14. Two different means of moving from one point to another are shown in the illustration. At A the cutting tool moves first along the Y axis, then along the X axis when going from point A to point B. At B the cutter first moves at a 45-degree angle until it is lined up with the Y axis, and then it moves down to point B.

counterclockwise direction—indicated by the curled fingers when looking toward the tip of the thumb—would be a plus (+) *a* axis. The same condition would hold if the thumb were pointed in the direction of the plus (+) *Y* axis; only in this case the rotary motion would be a plus (+) *b* axis. The *c* axis would be a rotary motion about the *Z* axis.

The axis standard also applies to lathes. Movement of the saddle of the lathe shown in Fig. 2–9 is in the *Z* axis. A cross-slide motion in the direction of the operator's position would be a plus (+) *X* axis motion. The position of the fingers on the right hand is also shown in Fig. 2–9. The *Y* axis is generally not used with lathes.

Point-to-Point Numerical Control

There are two types of numerical control, each used for a different kind of machine operation. One kind is called *point-to-point*, or *positioning* numerical control, and the other is called *contouring*, or *continuous path* numerical control. With point-to-point numerical control the cutting tool, or worktable, is moved to a specified position, as called for on the tape, and an operation is performed, such as the drilling of a hole. After the operation is performed, the tape reader then accepts the next instruction

Courtesy of Strippit, Div. of Houdaille Industries

Fig. 2–15. Holes are punched in sheet metal parts with the machine shown above. The NC control unit automatically moves the piece of sheet metal to the precise spot to be punched. The instruction on the tape also calls out the particular punch that is to be used. The different punches are positioned in a rotary drum.

from the tape and the machine moves to the next position. This process continues until all the holes have been drilled, or until the operations have been performed at the specified points.

In Fig. 2–10 the cutting tool moves from a start position to a point noted as A. The machine stops at this point and the spindle moves down and drills a hole at this position. After the hole is drilled, the spindle retracts and the cutting tool is next moved to position B where the cycle is repeated. The machine continues on its travel to points C, D, and E, and the drilling cycle is repeated at each point. After all the holes have been drilled, the machine is directed to a park position and automatically stopped.

Undoubtedly the most common form of the point-to-point numerical control machine is a drill press. Similar machines are often referred to as *machining centers* since they have the capability of milling, tapping, and boring, as well as drilling. As a matter of record, most point-to-point machines have this combined capability. Examples of machines of this type are shown in Figs. 1–1(A) and 2–1. The milling capability is somewhat restricted, however, because machining cuts can generally be made

Courtesy of Dixon Automatic Tool

FIG. 2–16. Screws are positioned and then driven to a specified depth, or all the way, with this point-to-point NC screwdriver. An automatic feed mechanism keeps the screws flowing to the screwdriver head.

Courtesy of Edlund, Div. of Monarch Machine Tool Co., Universal Instruments Corp.

Courtesy of Universal Instruments Corp.

FIG. 2–17. Electronic circuit boards are manufactured with the aid of point-to-point NC equipment. The machine shown in (A) is capable of drilling up to 30 holes a minute in 16 boards at the same time. Components are automatically inserted into the boards by the NC point-to-point machine shown in (B).

Courtesy of Spacerays Co.

FIG. 2–18. This point-to-point system is capable of positioning a powerful light ray (laser) which can pierce metal.

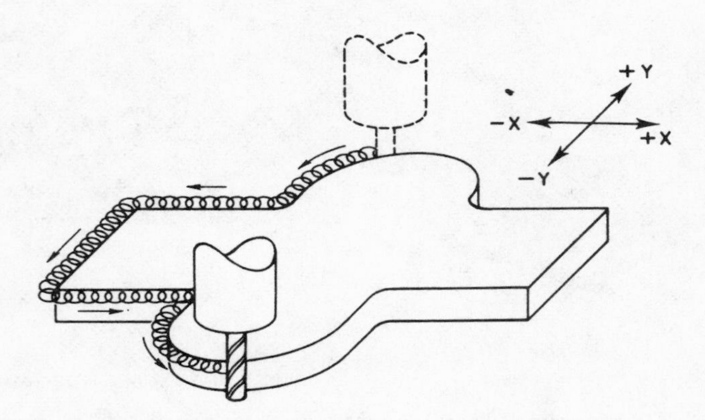

FIG. 2–19. The milling cutter in the illustration must follow a contour, or continuous path, in order to machine the configuration shown. This is accomplished by coordinating the distance movements along the X and Y axes.

only along one axis at a time. That is, only one drive motor will be turning at any particular time. This type of machining is called *picture frame* milling, since a combination of four right angle cuts form a rectangular pattern that looks like a picture frame. See Figs. 2–11 and 2–12. Some machines also have the capability of moving in an accurate 45-degree motion. In this case both lead screws are rotated at precisely the same rpm. See Fig. 2–13. With this type of control the 45-degree cut is the only angular motion that can be made. Even when the cutter is going through air in traveling from one point to the next, it moves along one axis at a time or in a combination of an axis and a 45-degree movement. See Fig. 2–14.

Point-to-Point Applications

In addition to drilling machines and machining centers, there are quite a few other applications of point-to-point numerical control. Many of these applications lie outside the metal-working industry. Figures 2–15 through 2–18 describe a number of different NC point-to-point applications.

Courtesy of R. K. Le Blond Machine Tool Co.

FIG. 2–20. The contour on the shaft, shown being machined on a NC lathe, is accomplished by coordinated motions along the X and Z axes.

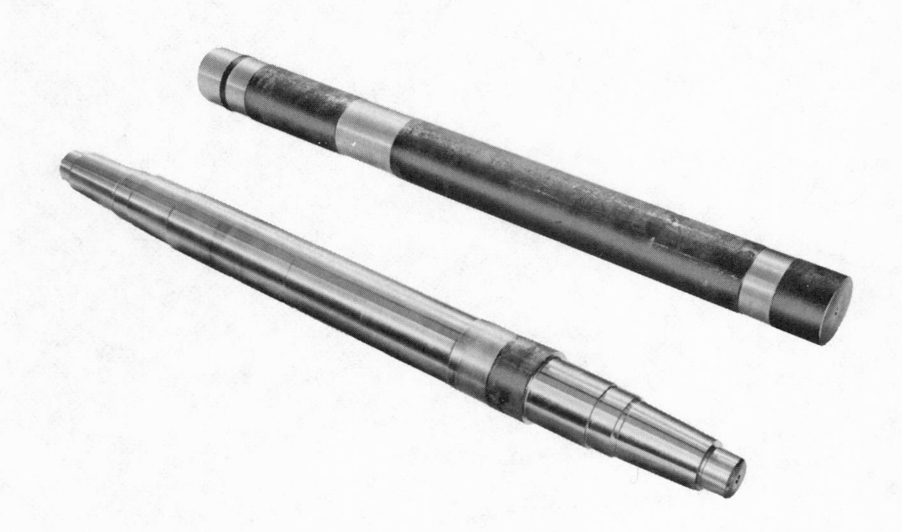

Courtesy of R. K. Le Blond Machine Tool Co.

FIG. 2–21. Contoured lathe parts machined on a two-axis numerical control lathe.

Contouring Numerical Control

Contouring, or *continuous path*, numerical control differs from point-to-point numerical control in that the *path* of the tool must be continuously controlled. With point-to-point numerical control it is not too important to control the path of the tool when moving from one point to the next since it is usually traveling in air. Control *is* important with contouring, however, since the cutter, or tool, is normally performing a cutting operation while it is moving along its path. This is particularly true with contour milling machines that are required to cut arcs and move at angles other than 45 degrees. See Fig. 2–19. The path of the cutting tool must be precise and very accurate in order to meet the tolerance requirements for the part.

In addition to contour milling, another popular application of continuous path numerical control is in operating the lathe. In this case the single-point tool must be guided along a path that would form the contour of the part. Figure 2–20 illustrates the turning of a shaft on a numerically controlled lathe. In this case coordinated motion along both the X and Z axes is required. Two contoured parts, machined on a numerical control lathe, are shown in Fig. 2–21.

Figure 2–13 shows the milling cutter moving in two axes. This is called

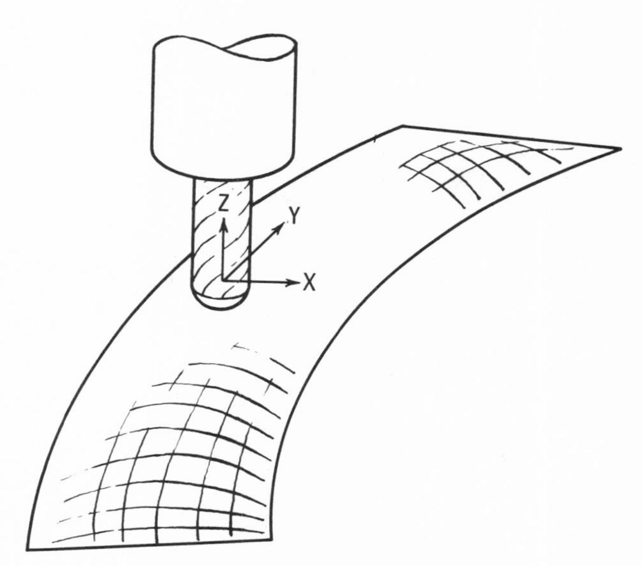

FIG. 2–22. A three-dimensional cut is achieved by the coordinated motion of the three axes (X, Y, and Z). As with the two-dimensional movements, this coordination is automatically achieved by the numerical control unit.

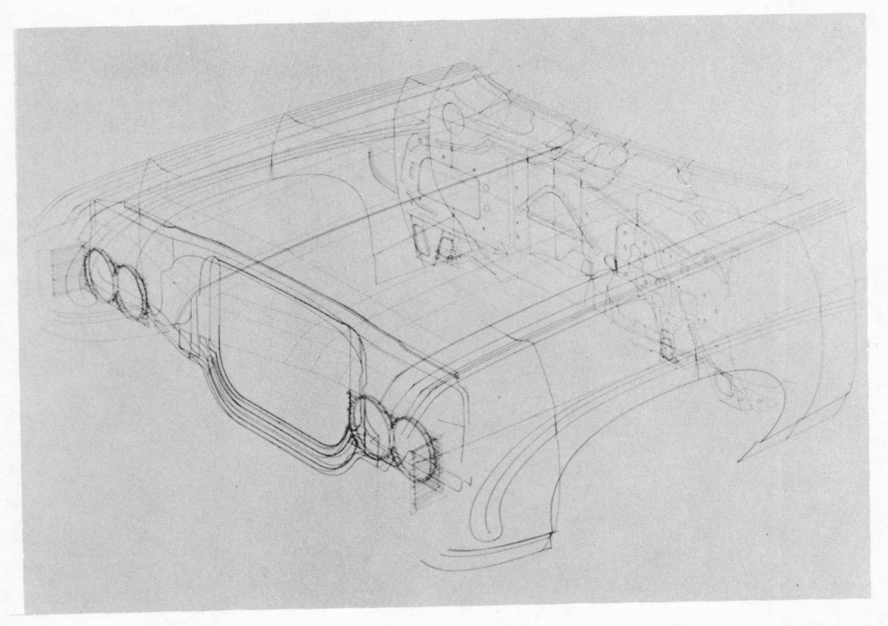

Courtesy of Universal Drafting Machine Co.

Fig. 2–23. The NC contour drafting machine shown in Fig. 2–24 handles the drafting requirements for both architectural and automotive engineers, as shown in the sketches above.

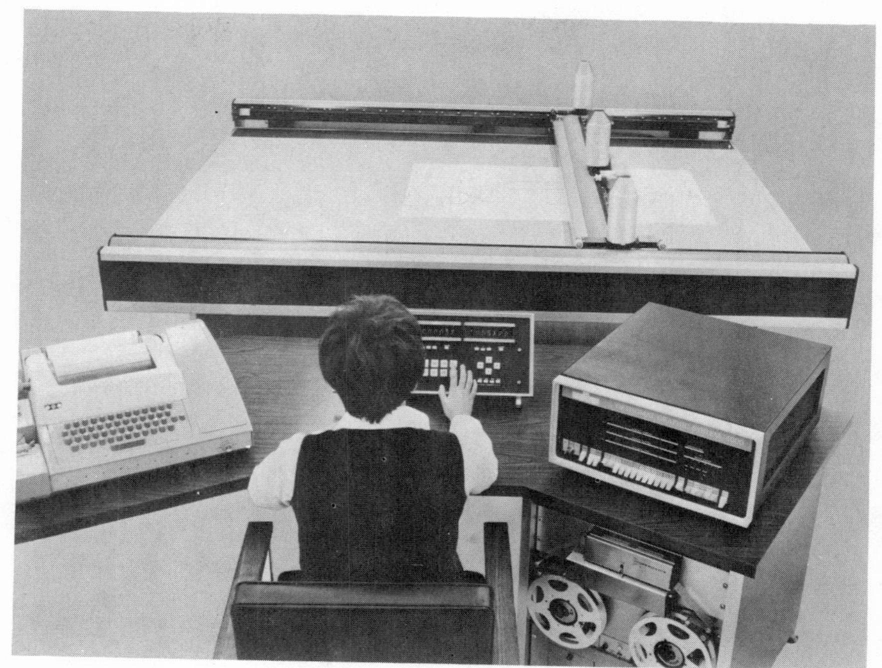

Courtesy of Universal Drafting Machine Co.

FIG. 2–24. The two-axis NC drafting machine shown above is capable of plotting complex mathematical curves as well as the draftings shown in Fig. 2–23.

a *two-dimensional* cut, or move. It is also possible to move in *three dimensions* by cutting along the *X*, *Y*, and *Z* axes at the same time, providing, of course, that the machine tool can be dimensionally controlled in the *Z* axis and is therefore a three-axis machine. A three-dimensional move is shown in Fig. 2–22. Three-dimensional movements are very helpful when machining the surfaces of forging dies or molds.

It should be pointed out that contour machines are generally capable of performing point-to-point operations, but it is usually not practical to perform complex contour movements on a point-to-point machine. Of course, contouring machines cost more, and the calculations required for preparing the tape are generally more involved and time-consuming because of the more complex moves.

As with point-to-point equipment, contouring numerical control also offers a large variety of applications in addition to metal cutting. A number of these are shown in Figs. 2–23 through 2–25.

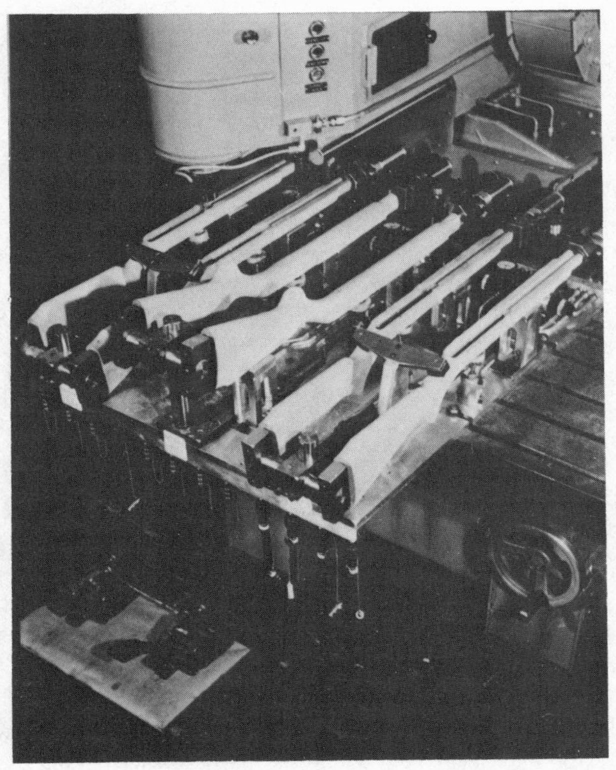

Courtesy of Ekstrom Carlson & Co.

FIG. 2–25. Wooden gun stocks being machined on a high-speed continuous-path three-axis NC wood-cutting machine.

QUESTIONS CHAPTER 2

1. What are the designations for the three basic motions of an NC machine?
2. What is the axis designation for the vertical motion on a vertical spindle *two-axis* NC machine?
3. According to standard procedure, if the forefinger of the right hand points in the plus (+) Y direction, in which direction does the thumb point?
4. What is the axis designation for the carriage motion of a lathe?
5. What would the axis motion be if the spindle of a horizontal NC machine were to move away from the workpiece?
6. There are five axes of motion for the machine shown below. The

spindle slide moves up and down; the entire column rotates about the Y axis and the rotational direction shown is a $+B$ motion. The entire column also moves along the bed of the machine and this is the X axis motion. What are the two axis designations for the spindle assembly as described by the arrows?

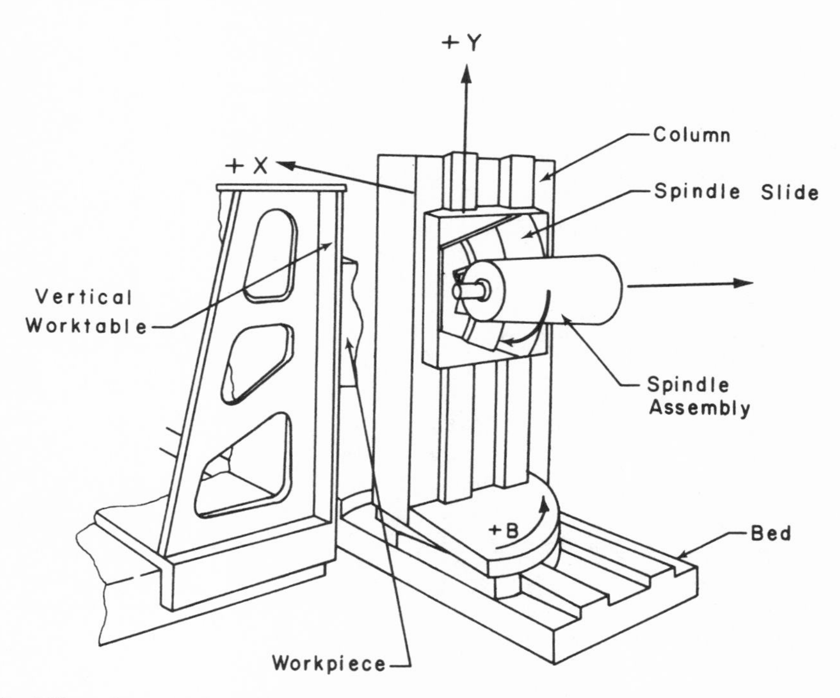

7. What is the difference between a point-to-point and a contouring control system for a machine tool?
8. What are the two possible routes that a point-to-point system may take in moving from one point to another?
9. What is meant by the term *picture frame milling*?
10. Note whether the following NC applications are of the point-to-point or contouring type.
    a. Drill
    b. Spot welder
    c. Flame cutter
    d. Shear
    e. Cloth cutter
    f. Continuous path milling machine

CHAPTER 3

Coordinate System

In order to instruct the NC machine as to precisely where or how far it should move, it is necessary that there be some kind of a locating system. The *locating system* used with numerical control machines is similar to that used when identifying locations by noting *streets* and *avenues*. Consider Fig. 3–1. If someone were standing at the intersection of *0 Street* and *0 Avenue* and were instructed to go the intersection of First Street and Second Avenue, he would go up one block and over two blocks, as shown in Fig. 3–1. If this person were next instructed to go to the intersection of Second Street and Third Avenue he would go up two blocks. Again, if he were instructed to go to Fifth Street and First Avenue, he would go to the right three blocks and down two. Instead of being directed to a particular location he could be given an instruction to move a distance of two blocks toward, say, the lower numbered streets, from the corner of First Avenue and Fifth Street, as an example. This would then leave our urban traveler at the corner of First Avenue and Third Street.

The numerical control machine works in very much the same way. However, instead of streets and avenues the intersections are referred to as x and y coordinates. This follows the same concept as datum line dimensioning. The x coordinates replacing the numbers of the streets would lie along the X axis and the y coordinates replacing the numbers of the avenues would lie along the Y axis. Also, instead of blocks, the measurement would be in inches. See Fig. 3–2. In this case the spindle would start at a point where the coordinates would be $x = 0$ inches and $y = 0$ inches. Instructions on the tape would then instruct the spindle, or cutting tool, to move to a point where the coordinates are $x = +2$

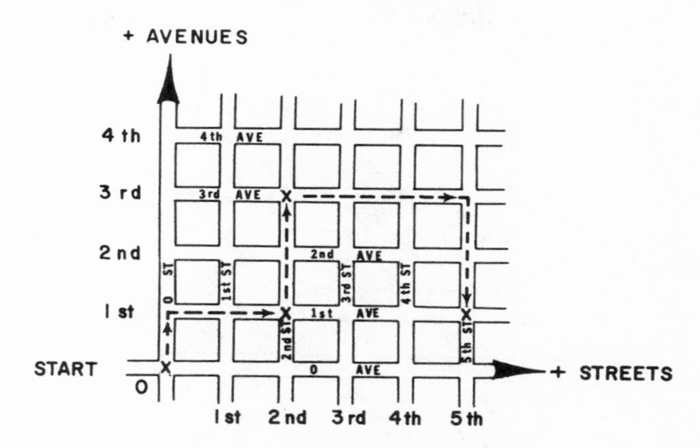

FIG. 3–1. Directing a numerical control machine is very much like instructing a person to go to the intersection of a street and an avenue. The numerical control system is based on the same concept.

inches and $y = +1$ inch. Next the cutting tool would be instructed to move to a point whose coordinates are $x = +2$ inches and $y = +3$ inches. As in the example of streets and avenues, the next to last move would be to coordinates $x = +5$ and $y = +1$. The last motion could be described, not by the x coordinate of $+3$ inches and the y coordinate of $+1$ inch, but rather by a *move* of 2 inches in the minus (–) X direction. The first three moves, in which the coordinate locations are noted, are described as *absolute* moves; and the last move, which describes the *distance* to be moved *from* a point, is described as an incremental move. Most NC machines operate on *either* an absolute system or an incremental system. This is described in greater detail later in the chapter.

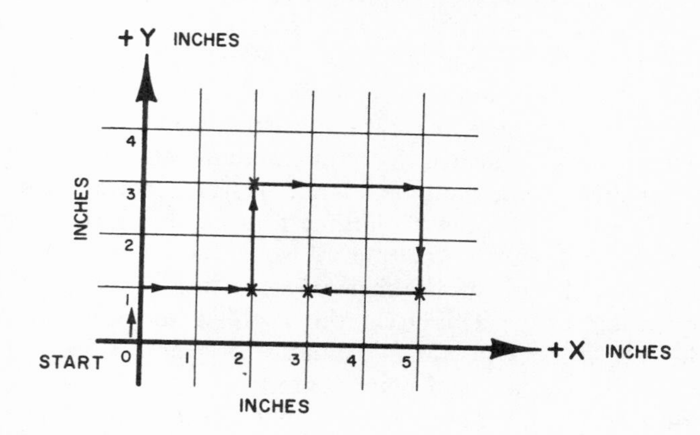

FIG. 3–2. Instead of using streets and avenues, NC instructions are noted in inches, and the distances lie along the X and Y axes.

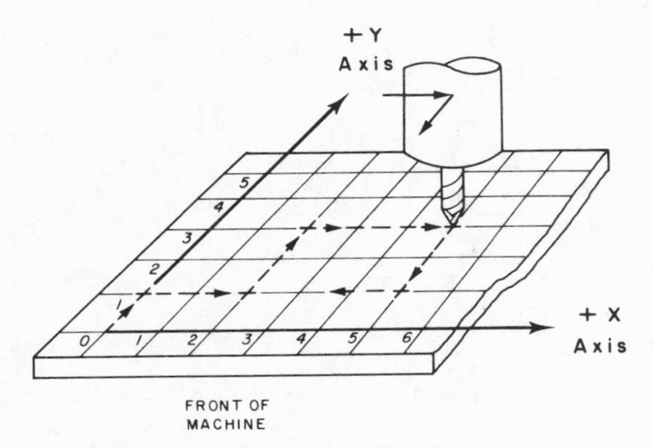

FIG. 3–3. The cutting tool is moved from one coordinate point to the next in accordance with the instructions noted on the tape.

As will be noted, the motions are at right angles to one another, and this is the way many numerical control machines operate. A good number of numerical control machines also operate by moving on a 45-degree angle and then on a line parallel to the X or Y axis. This is shown in Fig. 2–14. A three-dimensional sketch of the movement (see Fig. 3–2), as it would look on the machine, is shown in Fig. 3–3. The axes, which represent the motion of the spindle with respect to the worktable, and inch lines are used here for illustrative purposes and are generally not shown on the worktable of the machine.

In the illustrations shown thus far the coordinates have all been to the right of the Y axis and up from the X axis. This is the reason why all these coordinates had plus (+) signs before them. Another point to note is that coordinates are also referred to as *dimensions*, and all the dimensions shown in Figs. 3–2 and 3–3 would be plus (+) numbers. As shown in Fig. 2–3, the X and Y axes may also be described in a minus (−) direction. The two crossed axes form four sections, each of which is called a *quadrant*. The quadrants are indicated by Roman numerals in Fig. 3–4. Any coordinate in quadrant I would be $+x$ and $+y$; any coordinate in quadrant II would be $-x$ and $+y$; any coordinate in quadrant III would be $-x$ and $-y$; any coordinate in quadrant IV would be $+x$ and $-y$. A sample of coordinates with their proper signs is shown in Fig. 3–5. Since point A lies in quadrant I, the x dimension is $x = +4$ and the y dimension is $y = +4$. The dimensions, or coordinates, of point B are $x = -3$ and $y = +2$. The coordinates of point C are $x = -2$ and $y = -3$. Lastly, the coordinates of point D are $x = +2$ and $y = -4$. All of the dimensions are in *inches*. It is also possible to specify decimal parts of an inch, to the degree of accuracy of the machine and the control unit. For example,

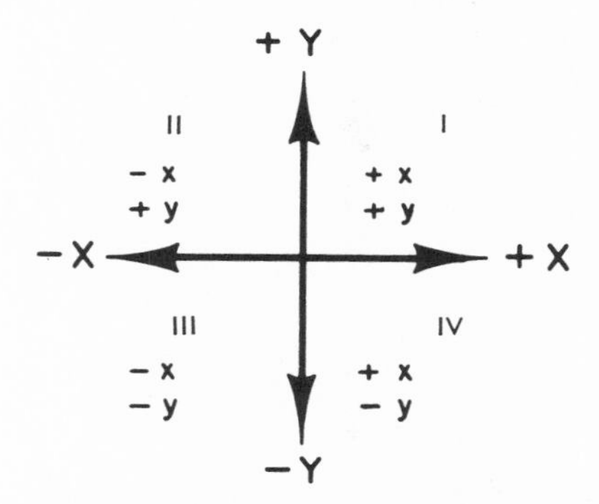

FIG. 3–4. The coordinate system is divided into four quadrants which are numbered in a counterclockwise direction as Roman numerals I, II, III, IV.

point *E*, shown in Fig. 3–5, would have coordinates $x = +1.500$ and $y = +5.000$. Most NC machines, at least the point-to-point kind, can handle up to three places to the right of the decimal point; that is, to the one-thousandth of an inch. Point *F*, shown in quadrant IV, would have coordinates $x = +3.251$ and $y = -2.254$. It is, therefore, possible to position the cutting tool to any point, over the entire grid pattern, to within 0.001 inch.

A point may also be described that does not lie on the plane formed by the *X* and *Y* axes by noting its *z* dimension, which is measured along the

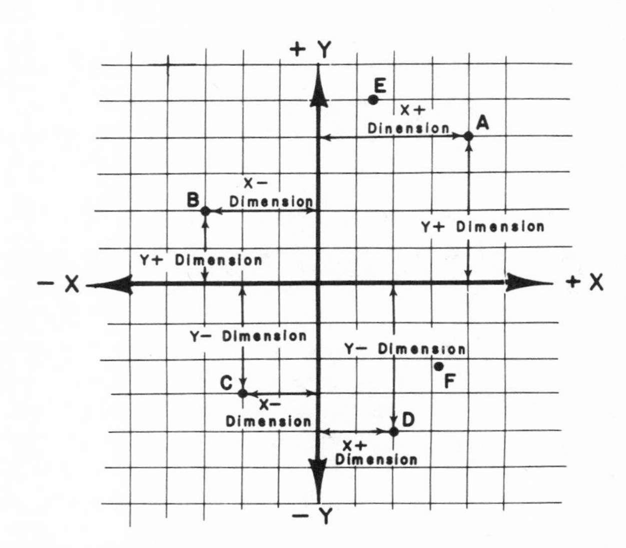

FIG. 3–5. Coordinate positions may be described by denoting the *x* and *y* dimensions, which are the distances from the *Y* and *X* axes, respectively.

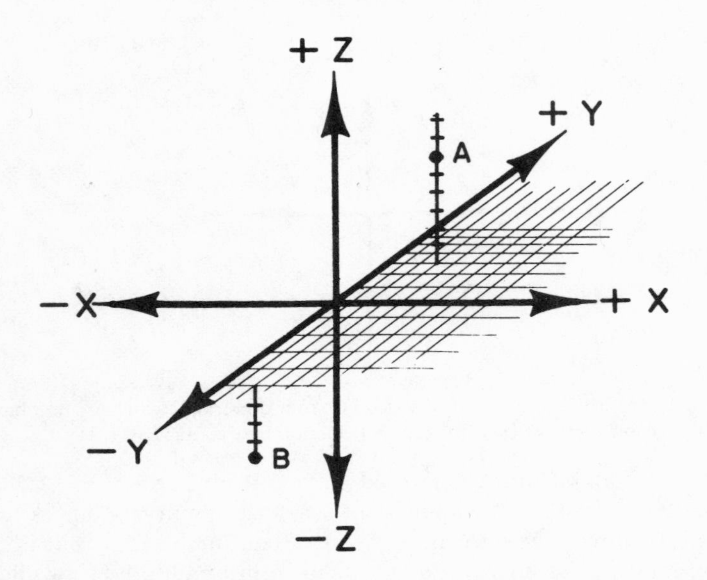

FIG. 3–6. Point A would be described by coordinates $x + 3.000$, $y + 3.000$, $z + 6.000$. Point B would be described by coordinates $x + 2.000$, $y - 5.000$, and $z - 4.000$. If the z coordinate is not noted, it is assumed that a point lies on the $X - Y$ plane and that the z dimension is 0.

Z axis. Figure 3–6 shows two points that do not lie on the X-Y plane. Point A lies 6.000 inches above a point whose coordinates are $x = +3.000$, $y = +3.000$. The x, y, and z coordinates for this point would be $x = +3.000$, and $y = +3.000$, and $z = +6.000$. Also, considering the signs of the axes, the coordinates of point B would be $x = +2.000$, $y = -5.000$, and $z = -4.000$. If the z coordinate is not noted, it is assumed that the point lies on the X-Y plane and the z coordinate would therefore be zero.

Fixed Zero Versus Full Floating Zero

The point where the X, Y, and Z axes cross is called the *zero* point. Also, at this point the x, y, and z coordinates or dimensions are all equal to zero. Numerical control machines may be of the type where the zero point is either fixed, or where it may be moved about. If the machine is of the fixed-zero type, the zero point is usually in the lower left-hand corner. This means that all of the coordinates will be $+x$ and $+y$. It is generally possible to *shift* this zero point over the full travel range of the machine by manual switches or dials. However, it should be noted that, in most systems of the fixed-zero type, machining is still restricted to quadrant I. See Fig. 3–7. With some systems the zero shift feature is mostly used for adjustment after the workpiece is positioned, and the amount of adjustment may be limited to a relatively small movement, such as 0.050 inch.

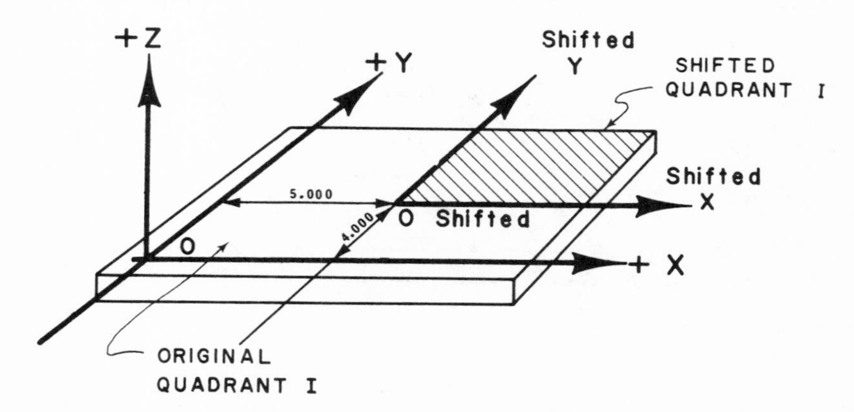

FIG. 3–7. With a *fixed-zero* system it is generally possible to *shift* the zero point a fixed amount in the X and Y axes. The work area in quadrant I, however, would generally be reduced.

Full floating zero allows the operator to set the zero point at any place over the full travel of the spindle, without regard to a fixed or initial zero point. Generally the operator may move the spindle to the desired location via the jog buttons (see Chapter 1), and then depress another button to establish the zero point. In this instance all four quadrants may be used. The person who prepares the tape (part programmer) is usually the one who determines where the zero point is to be located, and these instructions are passed on to the operator. Quite often the zero point is located *on the part* that is to be machined: either at the center of a hole that has been predrilled or at a corner that has been squared. See Fig. 3–8. In these two cases the *target point*—that is, the point where the tip of the tool is set—and the *zero point* are the same. In some systems the target

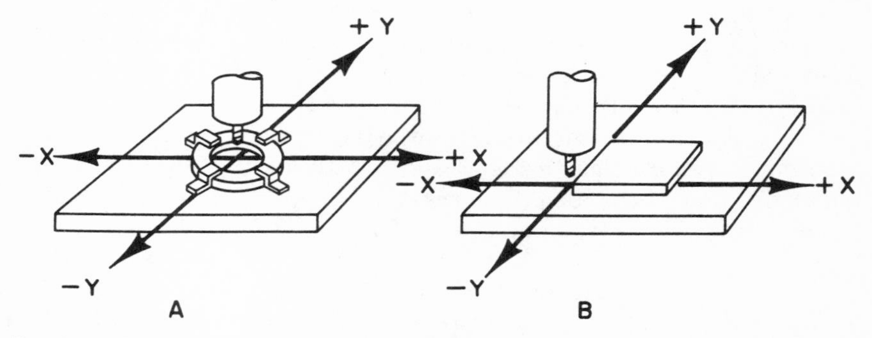

FIG. 3–8. The cutter of a *free-floating-zero* system set at the center of a prebored hole (A), and at the squared edge of a rectangular part (B).

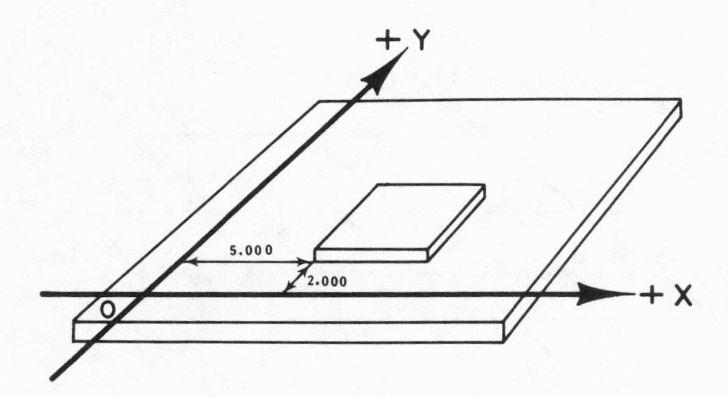

FIG. 3–9. In a fixed-zero system the workpiece must be positioned at the specified distance from the zero point. In the above illustration the part programmer has called for the workpiece to be located 5.000 inches in the $+x$ dimension and 2.000 inches in the $+y$ dimension. Whether in a fixed-zero system or free-floating system, the workpiece must still be properly aligned with the X and Y axes.

point and the zero point may be different. In these instances it may be more convenient for the operator and the part programmer to have these points at two different locations.

The significant difference between the fixed zero system and the floating zero system is that with a fixed zero system the workpiece is positioned at a specified place on the worktable and it is not necessary to *target* the cutting tool or spindle. With the full floating zero the workpiece may be positioned anywhere on the table, but the cutting tool or spindle must then be *targeted*. See Fig. 3–9.

Absolute Versus Incremental Programming

Just as there are two ways of setting up a numerically controlled machine, namely fixed-zero or full-floating-zero, there are also two ways of describing the movements. Some systems instruct the machine to move to a specific point which is described by the coordinates, and this is called an *absolute*-type system. The other type of system, which instructs the machine to move a specified distance from one point to the next, is called an *incremental*-type system. Referring to Fig. 3–10, if, with an absolute system, it is required to move from point A to point B, the *coordinates* of point B would be described. These would be $x = +8.000$, $y = +5.000$. If it were then required to move to point C, the coordinates of that point would be described. The coordinates of point C would be $x = +10.000$, $y = +2.000$. It will be noted that all the numbers are plus (+), and this is because they all lie in quadrant I.

Considering an *incremental* system, if it were required to move from

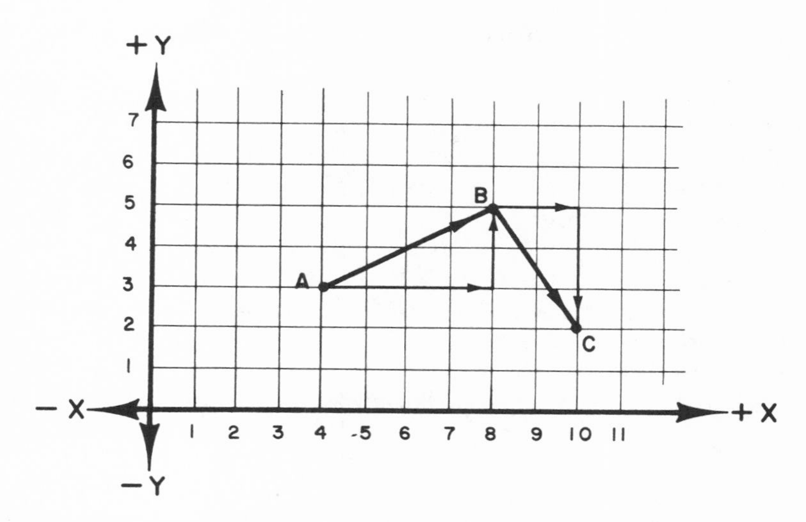

FIG. 3–10. In an absolute system the coordinates of points *A*, *B*, and *C* would be noted. In an incremental system the length of the movement, from one point to the next, would be noted.

point A to point B, the instruction on the tape would describe the x and y *distances between* point A and point B. The instruction for the x movement would therefore be $x = +4.000$ and for the y movement it would be $y = +2.000$. The sign is plus (+) for both the x and y movements because the incremental movement is to the right and in the X direction, which is a plus (+) move, and *up* and in the Y direction, which is also a plus (+) move. This holds true regardless of the course that the cutting tool may take in moving from point A to point B. If this were a contouring system, the movement would be along a straight line running between point A and point B. If it were a point-to-point system, the movement would first be parallel to the X axis and then parallel to the Y axis, or at a 45-degree angle and then parallel to an axis. Next, the movement from point B to point C would be noted as $x = +2.000$ and $y = -3.000$. In this case the y movement is minus (−) because it is *down* and in the minus Y direction. With an incremental system it does not matter which quadrant the movement is in; it is the *direction* of the movement that matters.

PRACTICE EXERCISE CHAPTER 3

1. Referring to Fig. 3–2, what would be the coordinates of a point that is four spaces to the right of the Y axis and two spaces up from the X axis?

2. A point having a minus sign for its x coordinate could be in either of what two quadrants?
3. Do quadrant designations move in a clockwise or counterclockwise direction?
4. What would be the coordinates of a point in quadrant IV that is 3.125 inches from the X axis and 2.500 inches from the Y axis?
5. A point whose coordinates are $x = -3.950$; $y = -1.666$ lies in what quadrant?
6. Consider a machine having a vertical spindle in which the horizontal plane formed by the X and Y axes lies on the worktable. Would the coordinate $x = +4.625$; $y = -4.000$; $z = +6.325$ be *above* or *below* the worktable?
7. Imagine the X-Y plane lying 5.000 inches above the worktable. How many inches above the worktable would a point be that has coordinates $x = -13.000$; $y = +6.400$; $z = -3.625$?
8. The coordinates of a point are $x = +2.000$; $y = +3.500$. A second point lies 18.000 inches directly to the right of this point and up, in the plus (+) Y direction, a distance of 6.750 inches. What are the coordinates of the second point?
9. The coordinates of point A are $x = +3.000$; $y = -2.000$; $z = +4.000$. Point B is located an incremental distance of -2.000 inches in the X axis direction, and -1.000 inch in the Y axis direction. The location in the Z axis remains the same. What are the coordinates of point B?
10. Considering the X-Y plane only, the coordinates of point A are $x = -4.900$; $y = +5.000$. Point B is located an incremental distance of 5.600 inches in the plus (+) X direction and 7.300 inches in the minus (−) Y direction. What are the coordinates of point B?

QUESTIONS CHAPTER 3

1. What would be the x and y signs for a coordinate in quadrant III?
2. What would be the coordinates of a point in quadrant II that is 3.025 inches from the Y axis and 1.675 inches from the X axis?
3. What quadrant does a point lie in that has coordinates $x = +10.685$; $y = -6.125$?
4. Considering that the plane formed by the X and Y axes lies on the worktable, what would be the Z coordinate of a point that was located 8.250 inches above the worktable?
5. The coordinates of point A are $x = +3.100$; $y = +8.000$. The incremental distance from point A to another point, B, is -2.000 in the X direction and -4.000 in the Y direction. What are the coordinates of point B?

6. Describe the essential difference between setting up a part on an NC machine having a full floating zero system and on an NC machine having a fixed zero system.
7. What is the restriction, generally, regarding the quadrant operation of a fixed zero system?
8. Point A is located at coordinates $x = +5.000$; $y = +4.000$. Point B is located at coordinates $x = -3.000$; $y = -6.000$. What are the incremental x and y distances when moving from point A to point B?

CHAPTER 4

Numerical Control Tape

History of NC Tape

After numerical control began, a little over 20 years ago, there was a good deal of argument among the electronic systems builders as to what should be used to store and transmit instructions to the control unit. Some builders proposed computer cards. Others felt that a 35-millimeter motion picture film strip would be best. Another proposal was to use a magnetic tape that was 1 inch wide, and one that was ½ inch wide. Still another proposal was for a 5-inch-wide celluloid tape with holes punched in it. The builders of numerical control equipment realized that some sort of standard would have to be established for the convenience of their customers and for the satisfactory growth of NC. Engineers from the different machine tool and control system builders met regularly to review the different proposals and choose a common standard.

The first order of business was to agree upon the form that the NC instructions were to be put on. It was decided that this should be a 1-inch-wide punched tape. Next, it was decided that the tape should have eight columns[1] of holes which were to be one-tenth of an inch apart. The tape was also to have a maximum thickness in order to be fed properly through the tape readers. The purpose of this first round of standardization was to arrange it so that the numerical control user need buy only one form of instruction material and would require only one type of device to prepare this material for the electronic control unit. Imagine the confusion and expense if a shop, for example, had five NC machines and one used computer cards; another, motion picture film; a third machine ran

[1] Also referred to as tracks, or levels.

on ½-inch magnetic tape; the fourth machine operated only on 5-inch-wide celluloid tape; and the fifth machine required five-eighths-inch-wide punched tape. Yet these were all proposed and in limited use in the early days of NC. Clearly, some form of standard tape configuration was required, and the engineers representing the control system and machine tool builders working through the Electronics Industries Association,[2] recommended the tape configuration in Fig. 4–1. Further cooperation among NC equipment manufacturers resulted in standards for the hole patterns to be used for NC machine instructions. These will also be described in this chapter.

Tape Material

While the maximum thickness of the tape has been standardized, the material has intentionally not been; for new and better material is continually being developed. The four types in general use are:

Paper tape. This is the least expensive and is popular for short runs or trial runs. Also, this tape is often used as a master and kept in the tool crib so that production tapes may be duplicated from it.

Laminated paper tape with reinforced Mylar plastic. This looks very

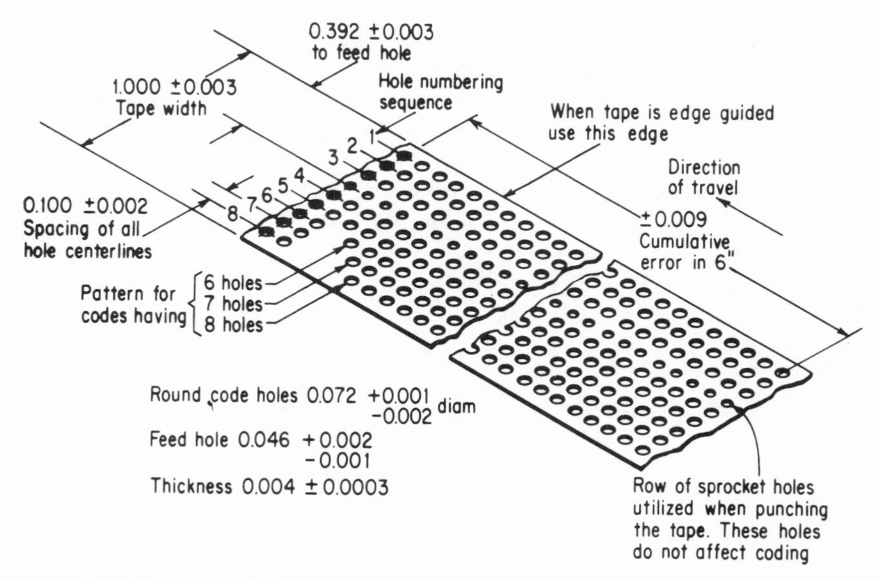

Fig. 4–1. The tape description shown above is the one recommended by NC equipment manufacturers and is universally accepted by suppliers and users of numerical control equipment.

[2] An organization responsible for many of the significant standards in the NC field. The headquarters is located at 2001 Eye Street, Washington, D.C. 20006

Courtesy of Republic Aviation Div. Fairchild Heller Corp.

FIG. 4–2. The aluminum tape with Mylar plastic coating shown is one of the strongest and most durable materials that can be used.

much like the paper tape. However, because it has a thin sheet of mylar plastic between two paper strips, it is much stronger than the paper tape. It is also more expensive. This type of tape material is used when higher production runs are required than those for which paper tape is used.

Aluminum tape with a clear mylar plastic coating. This is one of the most durable and strongest tapes and is also one of the most expensive. See Fig. 4–2.

Plastic tape. This is also a very strong tape and compares with the aluminum-mylar type for toughness, durability, and cost.

Tape Coding

As described in Fig. 4–1, there are eight columns of holes running along the length of the tape. There is also a ninth column, shown as the fourth row from the right edge of the tape in Fig. 4–1. The holes in this row, which are smaller, are used as sprocket holes and not as a part of the hole coding. Only the *eight* columns are considered for coding.

What the Columns Mean

Each column of holes on the tape either represents a numerical value or has some special meaning. The sketch shown below describes what each of these eight columns means. The columns are numbered from right to left, which follows the standards and also the *binary coded* system on which the standard is based.

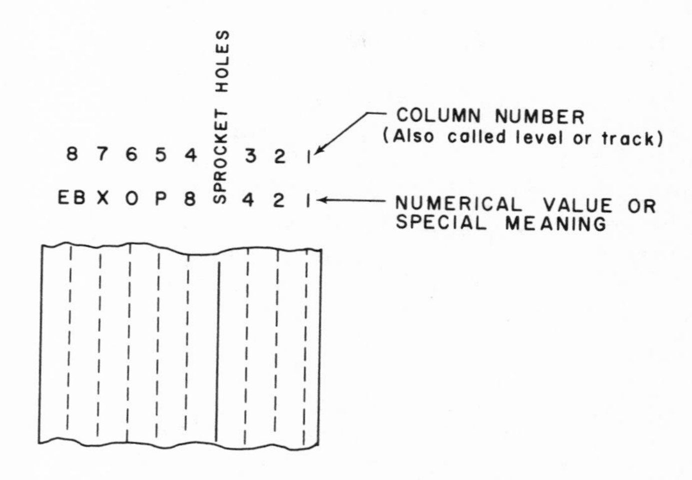

Considering columns number 1, 2, 3, and 4, above, it will be noted that column number 1 has a value of *1*; column number 2 has a value of *2*;

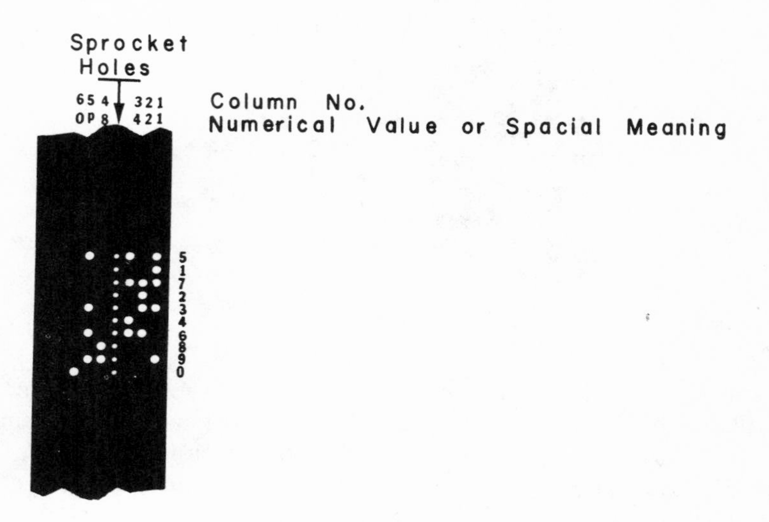

FIG. 4–3. Numerical values from 1 through 9 and 0 are expressed by a combination of holes punched in selected columns. The four right-hand columns have a numerical value of 1, 2, 4, and 8. These are combined to form any numerical value from 1 through 9. Whenever a combination results in an even number of holes, another hole is added in column number 5, or the P (parity) column, so that the number of holes across the tape would be odd. This is called *odd* parity.

column number 3 has a value of *4*; and column number 4 has a value of *8*. These four numerical values represent part of a coding system known as the *binary code*. This means that the number value of each column is the result of the number *2* raised to a power. The first number is 2^0 which is equal to the numerical value of 1. The next number is 2^1 which is equal to *2*. The third number is 2^2 which is equal to *4*. The last of the four numbers is 2^3 which is equal to *8*. By combining the numbers 1, 2, 4, and 8 it is possible to obtain any value from 1 through 9. For example, the value *3* is a combination of 1 and 2; the value *5* is a combination of 1 and 4; the value *6* is a combination of 2 and 4; the value *7* is obtained by combining 1, 2, and 4. These numbers may be obtained on the tape by combining *holes* in the appropriate first four columns. (See columns 1, 2, 3, 4 as described below.)

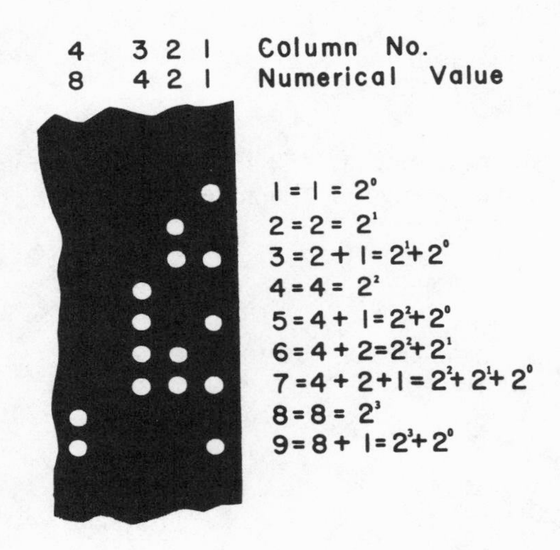

Parity

The number 5, for example, would be expressed as shown in Fig. 4–3. It will be noted that, in addition to holes being punched in columns 1 and 4, a third hole is punched in column 5., which is entitled *P*. The letter *P* stands for *parity*, and a hole is punched in this column whenever it is necessary that the total number of holes across the tape be an odd number. To express the numerical value 1, only one hole would be needed across the tape since the odd parity requirement is satisfied. There would also be no need to put a hole in the parity (*P*) column to express the numerical value 7 since the combination of 1, 2, and 4 result in an odd number of holes. The hole patterns, as they would appear on the

FIG. 4–4. In this example a numerical value of 7 is called for, but a faulty punch caused a hole to be missed in column number 2. Since only an odd number of holes across the tape will pass through the reader, the NC machine will automatically stop when it reads an even number of holes as shown above. The control unit must be equipped so that it can monitor the character hole pattern.

tape, for the numerical values 2, 3, 4, 6, 8, and 9, are also shown in Fig. 4–3. The numerical value *0* is expressed by a hole in column *6*.

The reason for the parity hole is to reduce the possibility of error or scrapping of a part. There are occasions when the chad, which is the stuff the hole is made of, does not fall out of the hole when it is punched. Consider the example shown in Fig. 4–4. If it is required to express the numerical value *7* and the chad is not punched through in column 2, the tape reader would read the numerical value *5*. This could result in a ruined part if a hole had to be drilled at $x = +7.000$ inches and the hole were drilled at $x = +5.000$ inches. A safeguard against this is to design the tape reader so that it recognizes only an odd number of holes across the tape. The tape reader would therefore automatically stop if it read the even number of holes as shown in Fig. 4–4. The NC machine would also automatically stop and a light would signal the operator that there was a parity error.

The Character

A line of holes running across the tape, such as the numerical values that have been described, is called a *character*. The letters of the alphabet can also be described by characters. Letters would be required to define an axis motion, such as x, y, or z. Letters are also used for other instructions to the NC machine, such as starting the spindle rotation or turning on the coolant. The hole arrangements across the tape for letter characters are shown in Fig. 4–5. Also shown are the hole arrangements for the / character and the + and − characters, as well as the *tab* character, which will be explained later in this chapter. It will be noted that column number 7 is used in order to describe letters a through r, and the plus (+) and minus (−) signs. The letters s through y are also described, however, without a hole in column number 7.

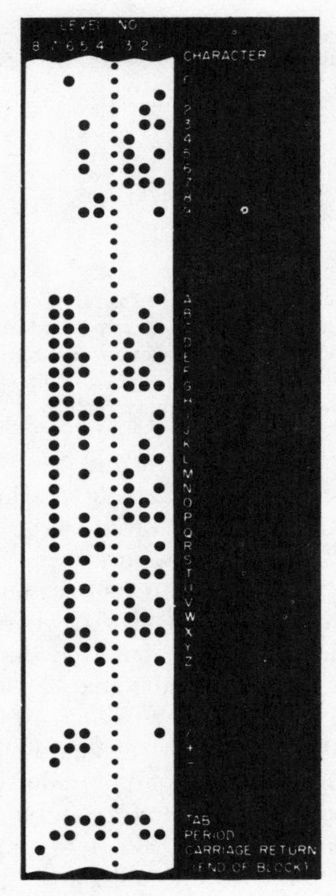

FIG. 4–5. Letters, which can be expressed as characters, are used for *dimension* words such as the x, y, and z coordinates. They are also used for other instructions to the machine, such as the code instructions for start or stop.

The Word

Just as letters in the alphabet are used to make up words, characters in numerical control tape language are used to make up NC *words*. Also, as in the English language, the NC *word* can stand on its own and has a definite meaning. A word, in numerical control, for example, would be $x+5.000$ which would mean that the x coordinate of a point (distance from the Y axis in the $+X$ direction) is 5.000 inches. This is shown as it would appear on the tape in Fig. 4–6(a). The y coordinate could also be described, and these two words would then define the position of a point. The y coordinate is shown to be $y+3.000$. This is what the instruction would look like on the tape in order to move the machine to the point

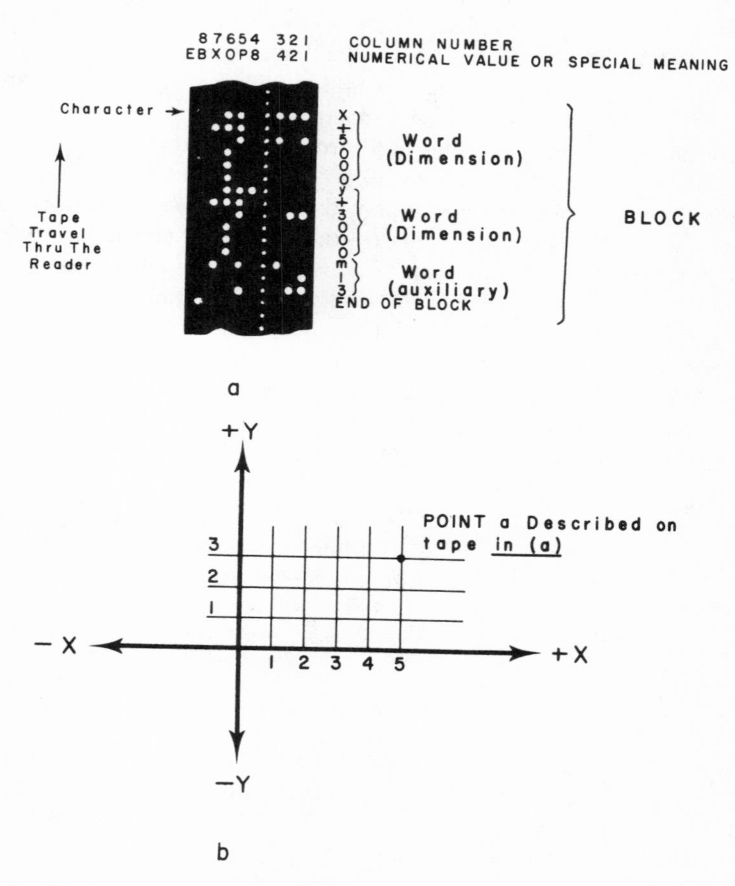

FIG. 4–6. The tape coding shown in (a) consists of two *dimension* words and one *auxiliary* word. The dimension words direct the cutting tool, or machine component, to the point shown in (b). The auxiliary word (m13) automatically turns the coolant on and starts the spindle rotation.

described by these coordinates. The point is also shown, with respect to the X and Y axes, in Fig. 4–6(b). It will be noted that the decimal point is not required on the tape since the control unit is designed to interpret 5000 as 5.000. If the control system reads 50000 it would interpret this to be 50.000 inches; 500 would be interpreted as .500 and 5 would be interpreted as .005 inch. Most control systems read two places to the left of the decimal point and three (or four) places to the right. Normally, it is three places to the right for point-to-point systems and four places to the right for contouring systems. It is because of this decimal consideration in forming words and the binary code used to form the numerical character that the code is called *Binary Coded Decimal*, more commonly known as BCD. It should be noted that for decimal numbers it is not required to show any zeros before the first significant figure in the above examples, 5. This is called *leading zero suppression.*

An alternative common practice is to drop the zeros *after* the significant figures. This is called *trailing zero suppression.* Thus, using the significant figure 5, a control unit that employed trailing zero suppression would accept the word $x-05$ as indicating $x-05.000$. The normal word for a control unit of this type is made up of two digits to the left of the decimal point and three digits to the right of the decimal point. As another example, suppose it were required to express the word $y+.050$. The word, considering trailing zero suppression, would be shown on the tape as y + 0005. Trailing zero suppression is somewhat more popular with point-to-point systems, and leading zero suppression is more popular with contouring systems. A number of examples for both are shown below:

No Zero Suppression	Control System with Leading Zero Suppression	Control System with Trailing Zero Suppression
$x+25000$	$x+25000$	$x+25$
$y-06750$	$y-6750$	$y-0675$
$z+000025$	$z+25$	$z+000025$

It will be noted that, where the dimension word is to represent a small number, the leading suppression is preferable, and where it is to represent a large number the trailing zero is preferable. A control system can have *either* leading zero suppression *or* trailing zero suppression, but not both.

Auxiliary or Miscellaneous Function

Words may be used to describe instructions other than dimensions. Assume, for example, that the point described in Fig. 4–6 is the first move of a machine and it is desired to turn the coolant on and start the spindle rotation; then an additional word would be used. This is called an *auxiliary*, or *miscellaneous, function* and the word would be a combination or a letter and a number. The letter used for an auxiliary function or word is *m*, in accordance with the standards.[3] To initiate the coolant and spindle function the word is m13, as shown in Fig. 4–6. It is important to note that the NC machine must have the capability to carry out the particular auxiliary function. The electrical wiring and mechanical apparatus must be installed on the machine in order for the auxiliary function

[3] A complete list of the standard *m* words is given in Appendix E.

to operate. Except for automatically stopping the machine and rewinding the tape, few lower cost NC machines incorporate many additional auxiliary functions. Such features as automatic tool changers and other special devices are found mostly on the higher cost machines. Other words will be described later in this chapter.

The Block

It will be noted that the three words shown in Fig. 4–6 describe a complete command for the machine to take some action. In the English language this group of words could be compared to a sentence; only in this case the words are numerical control words and, instead of being called a sentence, they are known as a *block*. *The end of the block is noted by an end of block character, which is a hole in the eighth column.* When the tape is moving through the tape-reading head, in the control unit, it stops at this character until the operation called for by the command is completed and then moves on to read the next block. While it would appear that the tape is being read a block at a time, it is actually being read character by character, and at a very fast rate. (Refer to Chapter 1.) A few tape readers do read a block at a time and are called *block readers*.

Other Numerical Control Words

The *g word:* This word, which is also called the *preparatory* word, is used to *prepare* the control unit for instructions that are to follow. As an example, according to the standards, the word *g33* is a command calling for thread-cutting on a lathe. Other words would follow that would describe the lead as well as the length of the thread to be cut. The difference between an *m* word as noted earlier and a *g* word is that an *m* word signals the control system to perform an operation immediately whereas a *g* word sets up the control unit so that it will accept and act on other words that will follow.[4]

The *t and s words:* These two words are used with machines having automatic tool changers or with machines having turrets. The *t* word calls out the particular tool that is to be brought into action and the *s* word denotes what the spindle rpm is to be. Generally the two words accompany each other in a block since a change of speed is normally required with each tool change. An example of a machining center having an automatic tool changer is shown in Fig. 4–7. When a *t* word is noted on the tape, calling for a tool change, the spindle stops; a swinging arm removes the old tool from the spindle and automatically replaces it with a new one that has been selected from the rotating drum. The underside of the rotary drum is also shown in Fig. 4–7. The spindle then auto-

[4] A complete list of *g* words is given in Appendix E.

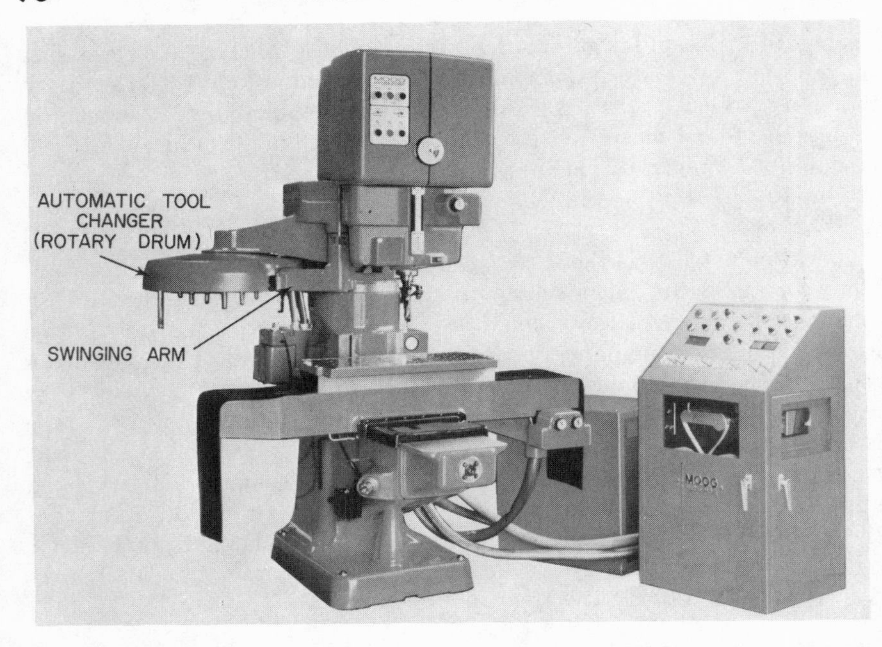

Courtesy of Moog, Inc.

FIG. 4–7. This machining center is capable of changing cutters automatically according to the *t* word on the tape. The *s* word generally accompanies the *t* word in the same block since the rpm usually has to be changed for the new cutting tool. The underside of the rotary drum is shown below the picture of the machine tool.

Courtesy of Burgmaster, Div. of Houdaille Industries

FIG. 4–8. The rotating turret on this machining center makes it possible to change tools automatically by noting a *t* word on the tape.

matically restarts in accordance with the rpm speed called out in the *s* word. The *s* word may refer to a specific rpm if the speed is handled in steps or to the "Magic Three" if the speed range is infinitely variable. The "Magic Three" rule is explained under the explanation for the *f* word, which follows in this chapter.

Figure 4–8 shows an eight-spindle turret machine. The turret is rotated until the proper cutting tool comes into position in accordance with the *t* word on the tape. As with the automatic tool changer, an *s* word is usually included in the same block on the tape as the *t* word.

The *n* word: This is also referred to as the *sequence number* and identifies the block. Although not mandatory, it is common practice to identify every block on the tape with an *n* word that is usually made up of three digits.

The first block on a tape would have a sequence number, or word, of *n*001. The sequence number word for the second block would be *n*002; for the third block *n*003, etc.

The purpose of the sequence number word is to let the operator know the number of the block that the machine is working on so that he may follow the program as prepared by the part programmer. This is ac-

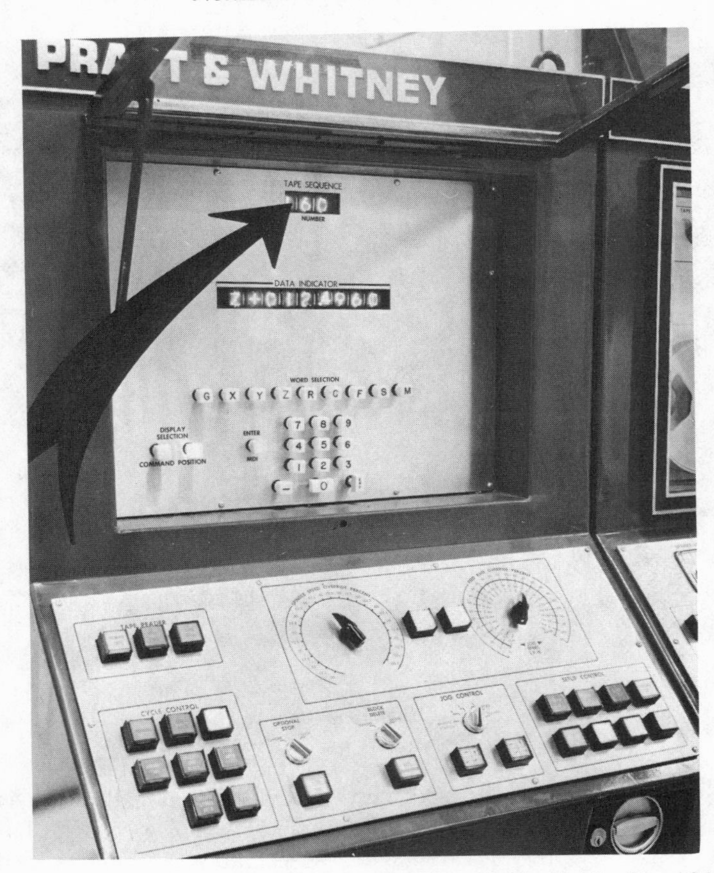

Courtesy of Pratt & Whitney Machine Tool Co., Div. of Colt Industries

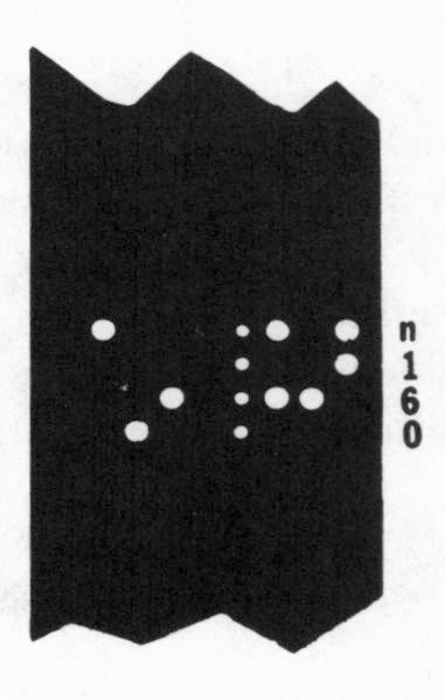

FIG. 4–9. The sequence readout is shown on the control panel. In this case it is 160, or the *n* word being read from the tape as it is shown on the panel. The characters for the *n* word, as they would appear on the tape, are also shown.

complished by a lighted readout panel that shows the sequence number of the block. Figure 4–9 shows the lighted sequence number as it is displayed for the operator on the panel of an NC control unit. The sequence number word in this case is *n*160. The sequence number, as it would appear on the tape, is also shown in Fig. 4–9. As the next block is read, the sequence number would change to *n*161. If the readout is limited to three digits—that is, *n*999 which is normal—and the number of blocks exceeded 999, then the next block could begin again with *n*001. The only condition in this case would be that the operator must remember which *set* of blocks the machine was working in. While not all control systems being manufactured have sequence number readouts, the majority do.

The *f word:* The feed rate, or the rate at which the cutter travels through the material, is specified by the *f* word. When used in conjunction with the *f* word, the feed must be given in terms of inches per minute (ipm). Frequently, on lathes and drilling machines and in the tables that provide recommended feed rates for these machines, the feed is specified in terms of inches per revolution (ipr) of the spindle. In this case, the feed rate must be converted to inches per minute (ipm), which can be done as follows:

$$f_{(\text{ipm})} = f_{(\text{ipr})} \times N$$

Where:

$$f_{(\text{ipm})} = \text{Feed, in. per min.}$$

$$f_{(\text{ipr})} = \text{Feed, in. per rev.}$$

$$N = \text{Spindle speed, rpm}$$

For example, if the drill feed is to be .010 ipr while the spindle is rotating at 120 rpm, the feed rate in ipm is calculated as follows:

$$f_{(\text{ipm})} = f_{(\text{ipr})} \times N = .010 \times 120$$

$$= 1.2 \text{ ipn.}$$

The *f* word may be expressed in any of three ways, depending on the control system. These three ways are described below. The third way, which is the simplest and most direct, is becoming the most popular.

1. The *f* word may be expressed as the ratio of the feed rate of the cutting tool in ipm divided by the distance that the cutter must travel as directed by the tape for the particular block, or:

$$\frac{\text{Feed rate of cutting tool (ipm)}}{\begin{array}{c}\text{Distance traveled (in.)}\\ \text{(as described in the block)}\end{array}}$$

In the example in Fig. 4–10(a) the cutting tool must travel 5 inches in moving from point A to point B. If the feed rate is 10 ipm, the *f* word

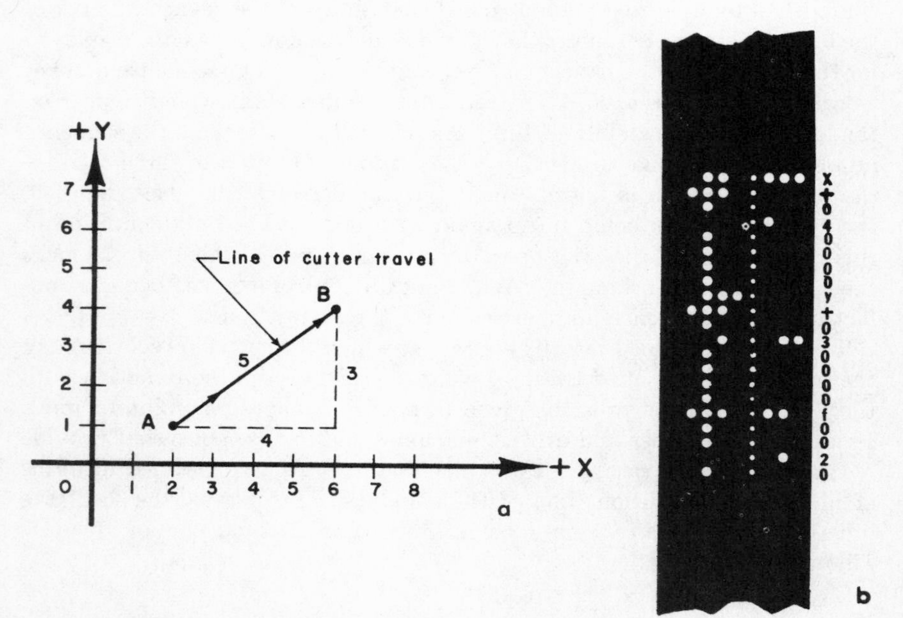

FIG. 4–10. In calculating the *f* word, the feed rate in ipm is divided by the distance of the movement in the block. In the above example the feed rate is 10 ipm and the distance is 5 inches (a). Therefore, the *f* word would be 2 and expressed as *f*0020 on the tape. The *x*, *y*, and *f* words are shown as they would appear on the tape in (b). Neither leading or trailing zero suppression has been applied in this case, and the dimension word would be two digits to the left of the decimal point and four digits to the right, as is common with some contouring systems.

will be:

$$\frac{10}{5} = 2$$

In some control systems using this form of feed-rate word the expression is noted in four digits as:

*f*0002

Also, other control systems use a multiplication factor of 10 and, in this case, the feed-rate word would be:

$$0002 \times 10 = f0020$$

It will then be written in the manuscript as 0020 in the *f* column, and on the tape it will appear as *f*0020. In situations where the feed rate is not too critical, calculation time can be saved by using the longest distance in the *X* or *Y* direction as the distance traveled, instead of the hypotenuse of the triangle. In the same example, Fig. 4–10(a), the distance would then be 4 inches instead of 5 inches and the *f* word would be 0025 in the manuscript and *f*0025 on the tape $\frac{10}{4} \times 10 = 25$). In point-to-point systems the feed rate is usually specified by the programmer or by the production planner, and the machine operator sets up the machine accordingly.

On some point-to-point machine tools, the feed rate is infinitely variable within a range that may be up to 50 ipm. However, on most point-to-point machine tools the feed rate can be varied only in steps. The number of available feeds depends upon the design of the machine; it may vary from eight to thirty, or more.

Some point-to-point systems are designed so that the feed rate is set by a tape code number, where each code number refers to a specific feed rate. For example, the system may be designed so that *f*2 would automatically set the machine to operate at a feed rate of 1 ipm, *f*3 may be 5 ipm, etc. For such systems there is no specific rule or calculated relationship between the *f* word and the feed rate.

2. The *f* word may be expressed as a coded three-digit number called the "Magic Three." There is nothing terribly magic about this; it is just a good deal easier than having to perform the division calculation, as described in the first method noted above, and has therefore been tagged with the name "magic." In this case the second and third digits represent the numerical value of the feed, and the first digit has a value that can be calculated by applying the following rules:

Rule 1. Count the number of digits to the left of the decimal point in the feed-rate number and add three (3) to this value to obtain the numerical value of the first digit.

Rule 2. If there are no numbers to the left of the decimal point in the feed rate number, count the number of zeroes to the right of the decimal point and subtract this number from three (3) to obtain the value of the first digit.

Applying Rule 1, assume that the feed rate is to be 10.0 ipm. There are two digits to the left of the decimal point, and the first digit in the *f* word is obtained by adding three plus two ($3 + 2 = 5$) to obtain five (5). In this case the *f* word would read *f*510.

Applying Rule 2, assume that the feed rate is 0.50 ipm. There are no zeroes to the right of the decimal point and the first digit in the *f* word is 3 ($3 - 0 = 3$). The *f* word would read *f*350. If the feed is to be .04 ipr, the *f* word would be *f*240, where the first digit is found by subtracting one (one zero) from three.

Remember that in all of these cases the feed rate must be specified in inches per minute (ipm), not in inches per revolution (ipr). Examples of the "Magic Three" coding are shown in Table 4–1. Note that only the first two significant numbers are given in the *f* word.

Table 4–1. "Magic Three" Feed Rate Coding

Feed Rate (ipm)	3 Digit *f* word
200	*f*620
100	*f*610
15.2	*f*515
7.82	*f*478
.64	*f*364
.00875	*f*188

The Magic Three coding formula may also be used to designate the numerical value of the *s* word when the spindle has an infinitely variable speed. For example, if the spindle speed is to be 250 rpm, the *s* word would be *s*625; if it is to be 1000 rpm, the *s* word would be *s*700. A few typical spindle speeds and the appropriate *s* words are shown in Table 4–2.

Table 4–2

Spindle Speed (rpm)	*s* Word
50	*s*550
125	*s*612
850	*s*685
1450	*s*714
3000	*s*730
550	*s*655

3. The *f* word may also be expressed directly in inches per minute (ipm). This is one of the latest NC developments and allows the programmer to describe the *f* word with the same numbers as the feed rate. A feed rate of 10 inches per minute (ipm) would be expressed as *f*010, for example. A feed rate of 2.5 ipm would be expressed as *f*0025.

TAPE FORMATS

Thus far the coding for *characters* has been reviewed, as has been the arrangement of characters to form *words*. It has also been shown how the words can be grouped in order to form a block which is a direct command to the control system and instructs the machine to move a specified distance and/or to carry out an operation. The way that the words are expressed in the block is called the *tape format*. There are three tape formats that are in general use. One is called *word address* and is popular with contouring systems. Another, the *tab sequential* format, is popular with point-to-point systems. The third is called *fixed block* and is used with some point-to-point systems.

Word Address Format

The term *word address* means that each word in the block is headed by a letter. The letter acts as an *address* and directs the particular word to

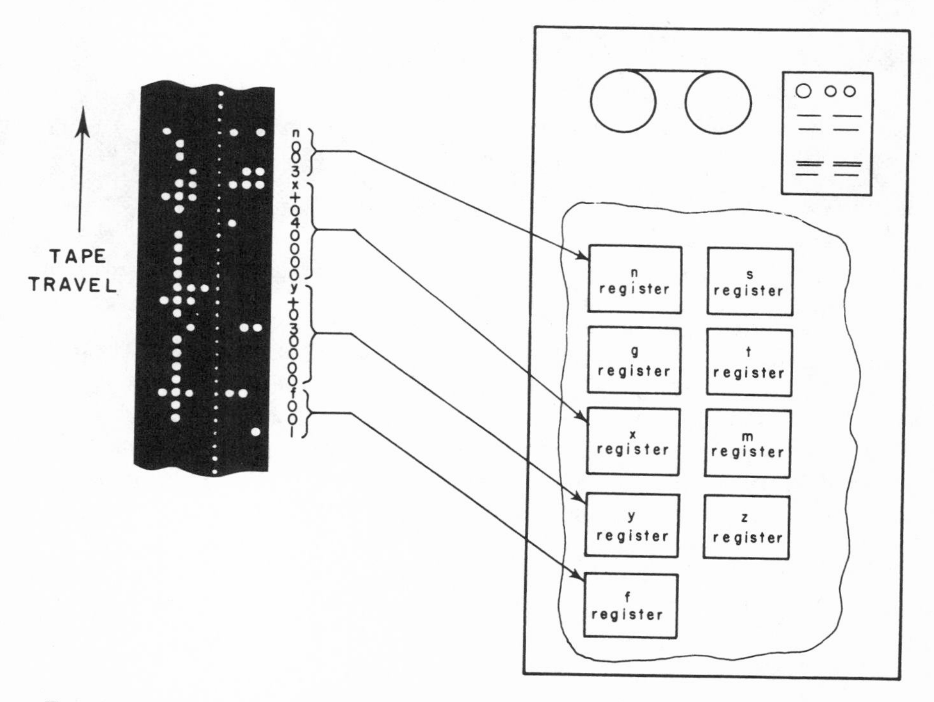

Fig. 4–11. The words are addressed to their respective storage registers. The *n* word would be directed to the *n* word register, the *x* word to the *x* word register, and so on. If there is no requirement for a word in a particular block, the word need not be shown on the tape. The control unit must be designed to house the various registers, and this would depend on the complexity of the machine that it is controlling.

Courtesy of Allen-Bradley Co.

FIG. 4–12. The registers in this control unit consist of printed circuit boards.

a specific location in the control unit. See Figs. 4–11 and 4–12. Also, since the words are addressed, it is not required to show the word if it is not to be used. If there were no z motion required, for example, there would be no need to show a z word on the tape.

Technically the words may be in any order. However, a standard has been developed which specifies that they be arranged with the sequence number word being first, and the other words in the following order.

n Sequence number word
g Preparatory word
x, y, z } Dimension words (straight line)
a, b, c } Dimension words (rotary)

f Feed rate word
s Spindle speed word
t Tool word
m Auxiliary function word

Tab Sequential Format

This format is very helpful when programming for point-to-point equipment since the majority of tapes for point-to-point machines are prepared on a device that resembles a typewriter and the particular words can be listed in even columns with the use of a *tab* key, just as on a typewriter. This will be discussed, in further detail, in Chapter 5. As far as the tape format is concerned, it may be noted that the *tab character* is used to separate the words in a block. Also, if the words are kept in a fixed order, such as the sequence number word coming first, the *g* word next, etc., then it is not necessary to note the letter addresses, since the words will be directed automatically to their proper registers in the control unit. Another feature of the tab sequential format is that if a word is not necessary in a particular block, it may be passed over by substituting a *tab* character for it. The example below shows the tab sequential format for a two-axis point-to-point machine that has the capability of calling out a feed rate *code* (not an *f* word), and auxiliary functions in words. Also, in some point-to-point machines the trailing zeros have been dropped, as well as the plus (+) sign before the *x* and *y* words. While there is no harm in showing the plus (+) sign, it is not *necessary* to do so with most NC systems since the system interprets the absence of a sign as a plus (+) sign. *The minus (−) sign, however, must be shown.*

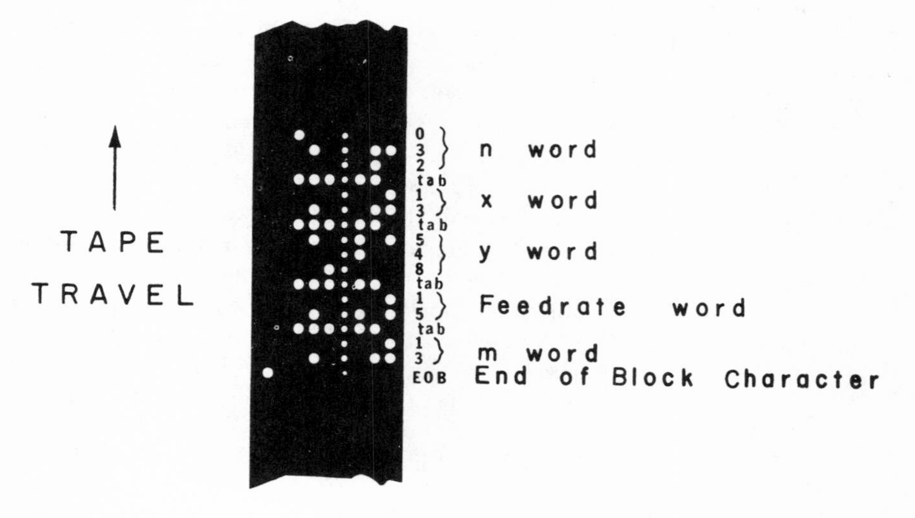

If, in the example above, both the *y* word and the *m* word were to be unchanged in the succeeding blocks, then a tab character could be substituted for these words, and the block would appear as follows:

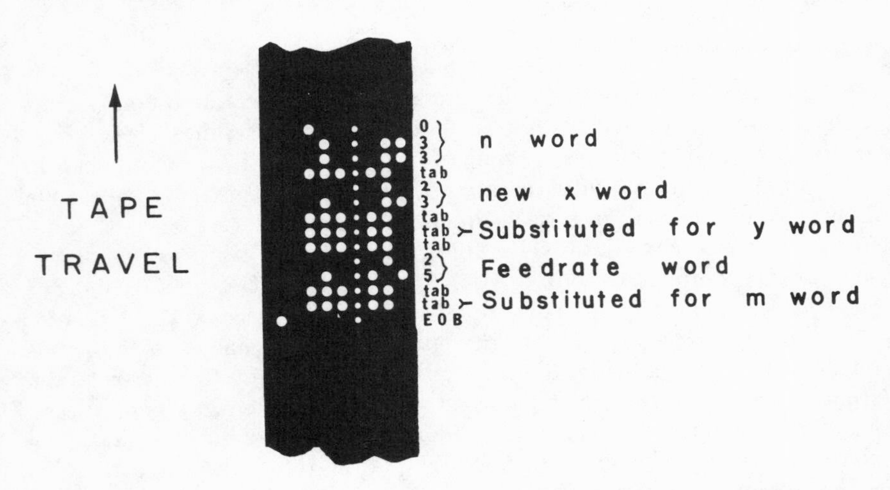

This would save tape, reduce tape reading time, and, most importantly, save tape preparation time since the overall length of the block would be reduced.

Fixed Block Format

Both the word address and tab sequential formats are classified as *variable block* formats. This means that the *length* of the block may vary and will depend on the number of characters in the block. The *fixed block* format means that the number of characters in every block is the same and the *length* of the blocks remains fixed. With this format, if a coordinate of an absolute dimensioned control system does not change, it still must be repeated in subsequent blocks. As an example, if the *x* coordinate were +3.000 in the previous block and is to remain at +3.000 in the subsequent block, the +3.000 coordinate must be described in the second block. Where there is no motion along an axis with an incremental system, it is necessary that zeros be shown so that the total number of characters will always be the same for every block. It is also necessary to show *all* plus (+) and minus (−) signs with a fixed block system, except in cases of a *fixed zero system*[5] where all coordinates are in the first quadrant. In this case the *x* and *y* dimension words will always be plus (+) and therefore need not be shown.

An example of a fixed block format having 20 characters is shown below.

[5] Refer to Chapter 3 for the description of a fixed zero system.

Each succeeding block would also have 20 characters. As with the tab sequential format the words are directed to the correct register in the control unit because of their position on the tape.

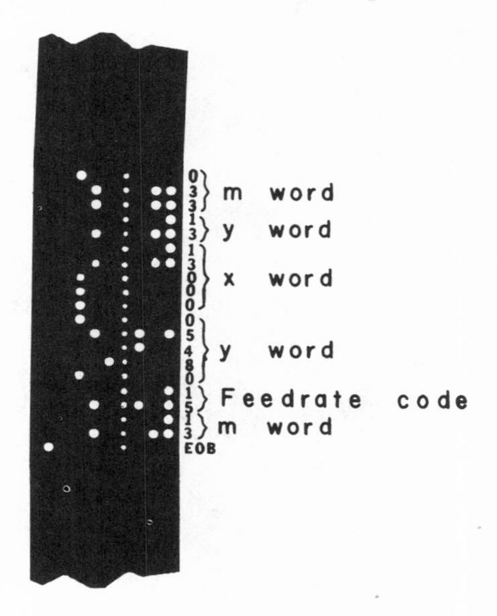

Tape Preparation Equipment

Numerical control tapes may be prepared in either one of two ways: manually, by a tape preparation unit that looks very much like a typewriter; or automatically, with the aid of a computer. Most tapes for point-to-point machines are prepared manually whereas most tapes for contouring machines are prepared by a computer. In both cases tapes *may* be prepared by either method. There is a definite trend, however, toward having the computer assist in both the part programming and tape preparation for point-to-point *and* contour programming. Whether a shop adopts computer assisted programming will depend on the *complexity* of the parts to be programmed and the number of *different* parts that have to be programmed.

Figures 4–13 and 4–14 show a different type of tape preparation unit. The keyboard is very similar to that of a standard typewriter. Each time the operator hits a key, a character is punched on the tape. At the same time the character is printed on a sheet of paper, called a printout, in the same way as on the typewriter. The characters on the paper are in Arabic letters and numerals, while on the tape they are in the form of perforated holes. The printout is used to check that the characters punched on the tape agree with the part programming instructions.

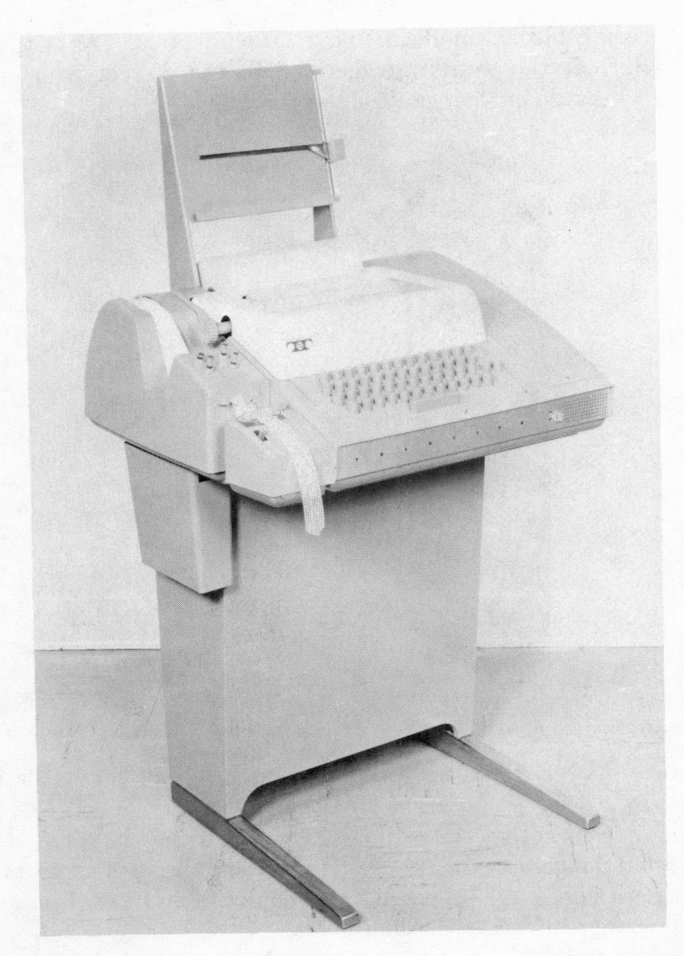

Courtesy of Teletype Corp.

Fig. 4–13. One of several types of manual tape preparation units.

In addition to preparing a tape and a printout through the action of an operator, it is possible to prepare a printout by running the tape through the machine. It can be used in this manner to check a tape. It is also generally possible to produce a second tape from an original.

EIA RS-358, Subset of USASCII

While the character code described thus far in this chapter is the most popular and the one originally recommended by the Electronics Industries

Association as RS-244 (now RS-244-A), a subset of the USA Standard Code for Information Interchange (USASCII)—referred to as RS-358—has also been recommended by the Electronics Industries Association. This subset is commonly referred to as the ASCII code and is more com-

Courtesy of Singer, Friden Div.

Courtesy of Itel Corp.

Fig. 4–14. Tape preparation units from two different manufacturers.

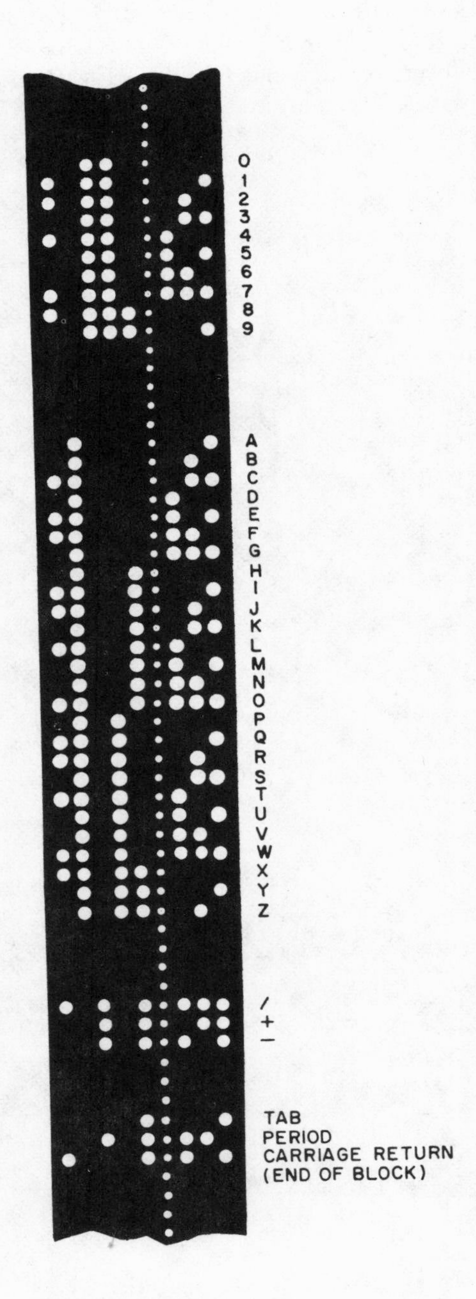

FIG. 4–15. RS-358 subset.

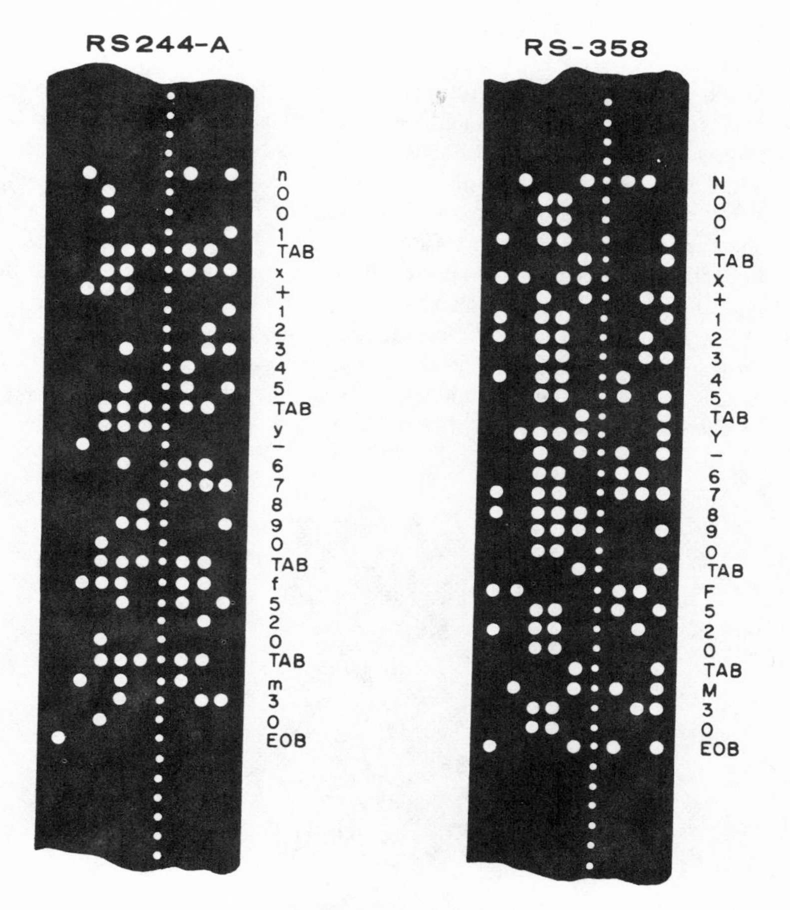

FIG. 4–16. The RS-244-A and RS-358 subset standards maintain the same format. It is the character coding that differs.

patible with the code being used in the communication and computing fields than is RS-244-A.

The tape format of the ASCII code is the same as the original EIA format (i.e., RS-244) and the revised RS-244-A format: that is, the *arrangement* of the characters to form the word address, tab sequential, and fixed block remains unchanged. It is the hole pattern of the characters running across the tape that differs.

The character coding for the RS-358 subset is shown in Fig. 4–15. A comparison of a tab sequential block, in EIA code RS-244-A, and the EIA subset RS-358 code is shown in Fig. 4–16.

PRACTICE EXERCISES CHAPTER 4

1. So that tape reader manufacturers may have a standard guide, the Electronic Industries Association established a standard for the thickness and width of the NC tape. What is the thickness specified and what are its tolerance limits?
2. What are the two other expressions for the line of holes running along the tape besides the term *column*?
3. Considering only the holes used for coding, how many columns of holes, running along the tape, are there?
4. What is the numerical value of a hole in column 3?
5. In the binary code, what is the value of 2 raised to the power of 3?
6. In the sketch of the tape shown below, note which character is correct in accordance with the number shown and which is not.

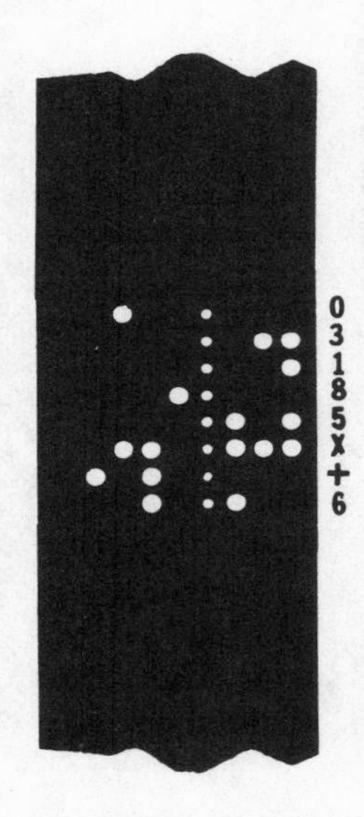

7. How does the parity check differ between the ASCII code (RS-358) and the RS-244-A code?
8. What is the word that identifies a block?
9. How is the feed rate of an NC machine usually noted?
10. The following x or y coordinates describe points to be drilled on an NC point-to-point machine. Describe the word, as it would appear on the tape, if the system had trailing zero suppression.
    a. $x = +19.000$
    b. $y = -\ \ .002$
    c. $y = -\ 6.250$

   d. $x = -\ \ .065$
   e. $y = +65.666$
11. The following words describe incremental movements, to four decimal places, as they would appear on the tape for a contouring system. Write the *full* coordinate expression considering that the system has leading zero suppression and can accommodate two digits to the left of the decimal point.
   a. $x1650$
   b. $y-1$
   c. $x-25$
   d. $z625$
   e. $y51650$
12. If there were no y word in the tab sequential EIA block shown in Fig. 4–16, how many tab characters would there be between the last character in the x word and the f character of the f word.
13. In what term is the s word expressed?
14. What does the word $n005$ denote?
15. A control system conforms to the *velocity divided by distance* rule for calculating the f word. If the x incremental move is 2.5000 inches and the y incremental move is .0001 inch and the velocity is 5 ipm, what would be the three-digit f word (using the 10 multiplier)?
16. Note the "Magic Three" words for the following:
   a. 100 ipm
   b. 25 ipm
   c. 16.9 ipm
   d. .025 ipm
   e. 3.2 ipm
   f. 3,600 rpm
17. What is meant by a *fixed block format?*
18. What piece of equipment is the most generally used for preparing tapes for point-to-point machines?
19. What are three common types of tape formats?

QUESTIONS CHAPTER 4

1. What is the spacing of the centers of the holes on an EIA tape?
2. What are the three kinds of material commonly used for NC tape?
3. What are the holes running along the length of the tape which are not used for coding called?
4. The binary code notes that a particular number may be raised to a power. What is this number?

5. Note which character code is correct and which is incorrect in the sketch shown below.

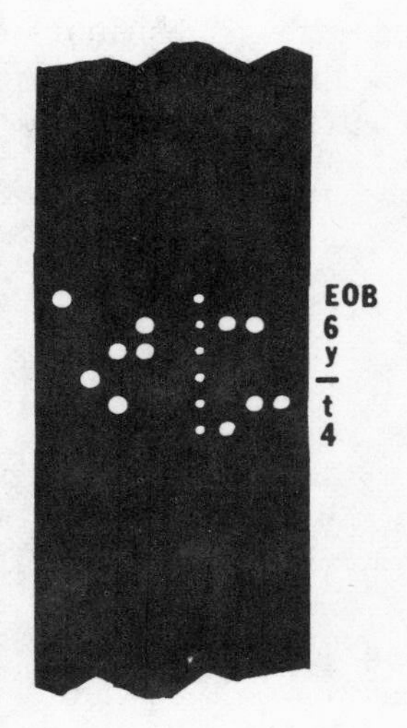

6. When is a parity hole punched in a character? (Consider RS-244-A coding.)
7. With which word is *ipm* used?
8. What does BCD mean?
9. What is the most popular tape format for contouring systems?
10. What is another term for the *g* word?
11. Convert the following "Magic Three" words into ipm or rpm, whichever seems more appropriate.
    a. *f*635
    b. *f*525
    c. *f*130
    d. *f*475
    e. *s*840
12. Considering the velocity divided by the distance ratio ×10 calculation for the *f* word on a particular control system, what is the four-digit *f* word if the cutter is to travel at 15 ipm and the distance of the straight line move is 2.250 inches?
13. What is meant by variable block format?

CHAPTER 5

Point-to-Point Programming—Manual Method

What is Part Programming?

Part programming involves the detailed, step-by-step listing of the operations that the NC machine is to perform. Usually the step-by-step instructions are listed on a prepared form called a *manuscript,* and the person who writes the instructions is known as a *part programmer.*

In *manual* part programming the tape for the NC machine is prepared directly from the manuscript on a machine that is similar to a typewriter (see Figs. 4–13 and 4–14). Tapes for *computer-assisted* part programming are prepared automatically by the computer. The manuscript used for manual part programming—the subject of the chapter—differs considerably from that used for computer-assisted part programming.

The Part Programmer

The part programmer plays a key role in numerical control. It is his responsibility to decide what the cutting sequences will be and to perform the calculations for determining the movements. Very often he must also prescribe the proper feeds and speeds for the particular material being machined. These requirements are not too unlike those of a machinist working with conventional equipment. The significant difference with numerical control is that the part programmer must *imagine* all of the operations as they would occur at the machine tool and be able to describe these operations in their proper sequence. The part programmer therefore must be familiar with the operations of the NC machine and *know ma-*

chining practices. The usual shop math requirements of algebra, geometry, and particularly trigonometry are as important to the part programmer as to the machinist working with the conventional machine tool. Sometimes the math requirements may be even more demanding for the part programmer since he generally does not have the benefit of substituting the machine motions for some of the calculations.

Two Types of Part Programming

As already mentioned, there are *two* very distinct *types* of part programming: *manual* part programming and *computer-assisted* part programming. In manual part programming the part programmer must perform every calculation and list every operation that the NC machine is to perform. Although this is not too complicated with point-to-point equipment, it can be time consuming and difficult when preparing calculations for contouring machines such as profile mills and lathes.

Computer-assisted part programming, while not particularly difficult, requires special training since the instructions to the computer, which describe the part and machining operations, must be in a very precise form to be accepted by the computer. Actually, in many cases, the math requirements for programming a part with computer assistance are far easier than for manual part programming because the computer performs the bulk of the calculations required.

The steps for preparing a tape, by the manual part programming method, are shown in Fig. 5–1. While there are more people involved in NC and more steps required than in the conventional machining method, the total time required, from blueprint to finished part, is generally shorter.

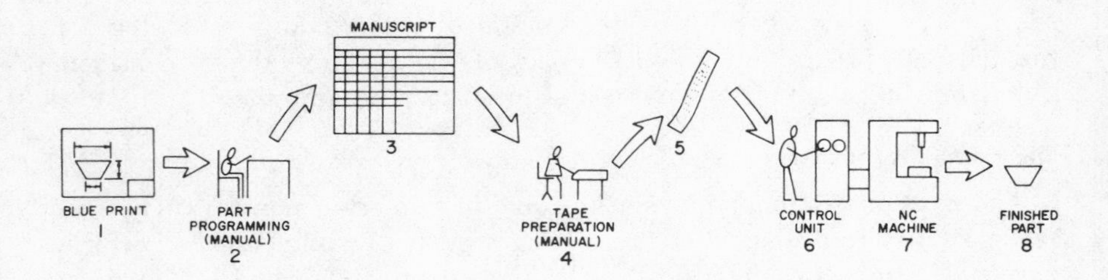

Fig. 5–1. Like conventional machining methods, the NC procedure begins with the engineering drawing or blueprint of the part (1). Next, the part programmer (2) performs the necessary calculations and lists the operations that the machine is to perform on a manuscript (3). The tape is then prepared on a machine similar to a typewriter (4) (see also Fig. 4–14). The punched tape (5) is then inserted into the tape reader on the control unit (6), where it directs the motions of the NC machine (7) that cuts the part (8).

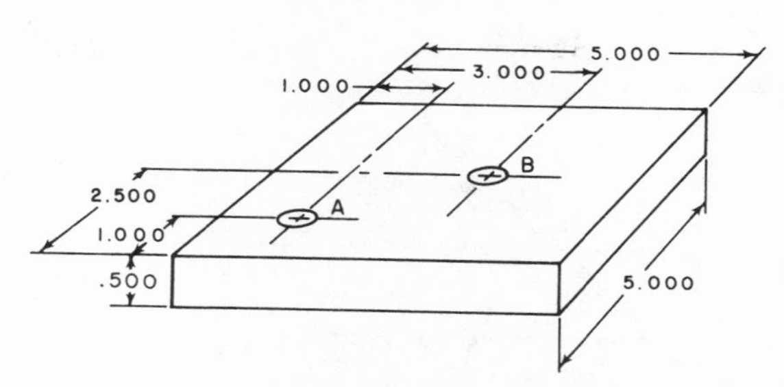

FIG. 5–2. Two holes *A* and *B* are to be drilled in the above part at the locations shown. The sides of the part are milled square prior to being positioned on the NC machine. This is necessary in order to align the part properly.

PROGRAMMING A SIMPLE POINT-TO-POINT PART—System Having Full-Floating Zero, Tab Sequential Format, Absolute Measurement

Consider the part shown in Fig. 5–2. It is a flat metal plate, and two holes (*A* and *B*) are to be drilled at the points shown. The *front, top,* and *side* views are described on the blueprint shown in Fig. 5–3.

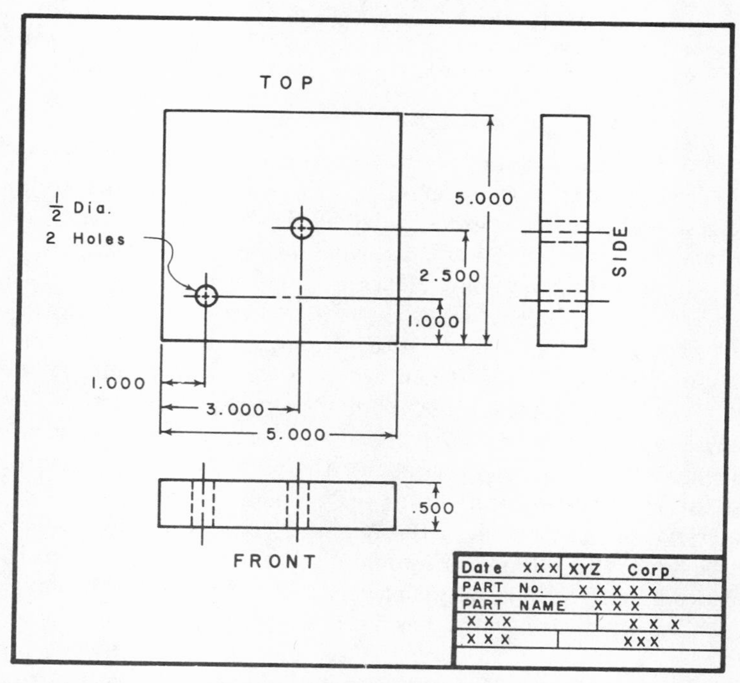

FIG. 5–3. Blueprint showing the front, top, and side views of the part shown in the three-dimensional sketch in Fig. 5–2. The two holes are to be drilled through the part. The part is shown positioned on the worktable of an NC machine in Fig. 5–4.

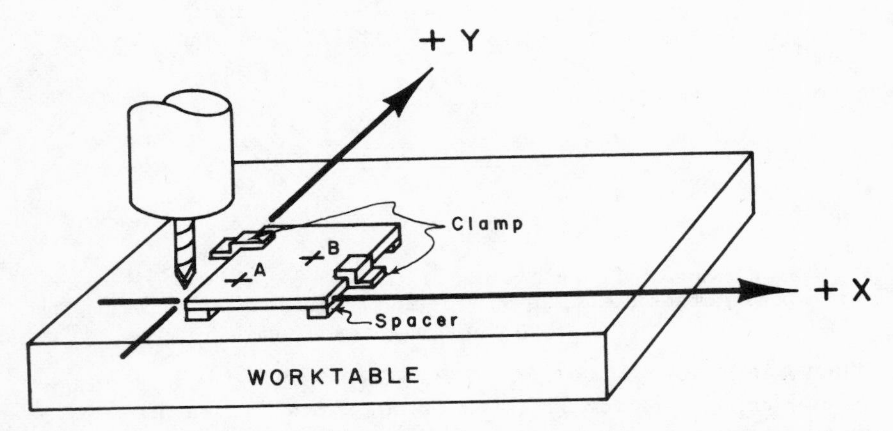

FIG. 5–4. In a full-floating zero system the operator targets the spindle to a point such as a corner of the part. Targeting is accomplished by inserting an edge finder or an optical centerscope, or centering microscope, into the spindle and moving the table until the axis of the spindle coincides with the corner. Sometimes the center of a previously machined hole is the reference point, and targeting is accomplished by indicating the hole with a dial indicator. In other instances a pin, hole, or a corner of the work-holding fixture serves as a reference surface and is targeted; the part is then located on the machine by the fixture with respect to the target or set-up point.

Figure 5–4 shows the part positioned on the worktable of the NC machine. The operator must be sure that the sides of the part are aligned with the X and Y axes in order to be consistent with the part program. The initial positioning of the spindle, with respect to the part, will depend on the type of control system. (Please refer to Chapter 3.) In a *full-floating zero* system the part may be positioned anyplace on the worktable and the cutting tool may be *targeted* at a convenient point on the part. This point is referred to as the target point or set-up point. (The procedure for establishing the target point is covered later in this chapter.) Also, with a full floating zero system, all four quadrants may be used. A convenient target or set-up point for the part shown is on the top side and at the bottom left-hand corner. This point may also be established as the *origin*, or intersection of the X and Y axes, where the x dimension would be *0* and the y dimension would also be *0*. With some full floating zero control systems this origin is established by merely pushing a *set* button after the cutting tool has been targeted by use of the manual controls.

Once the machine tool is set, the operator pushes the *start* button and the cutter is moved to point A, where it drills a hole; then to point B, where it drills the second hole; and then back to the target, or set-up point, where it is programmed to stop.[1] The operator may then remove the finished part and position a new blank piece of material in the same location as the previous piece.

It is not necessary to retarget the cutting tool for each new piece once this has been done for the first piece. It is essential, however, for the operator to position the blank piece accurately each time. This is usually done by means of a fixture, stops, or a vise. Also, the cutting tool does not need to be programmed to move back to the target point. Any convenient point is satisfactory. After a fresh piece of material has been positioned, the operator again pushes the *start* button and the machine proceeds on its cycle by drilling holes *A* and *B* and then moving back to the target or other conveniently programmed point, where it again stops. The loading and unloading cycle continues until all of the parts in the lot are completed.

What the Programmer Does

The first task of the part programmer is to consider the part as being oriented with respect to the *X* and *Y* axes. He must imagine the part as it would appear on the table of the NC machine since the machine tool operator is required to position the part as called for by the part programmer. The part programmer may indicate the axes on the original blueprint or prepare a separate sketch. Either way, the layout of the part with respect to the axes would appear as shown in Fig. 5–5.

Another way of expressing the locations of points *A* and *B*, as shown in Fig. 5–5, is to *list* the coordinates of the points. The coordinates of points *A* and *B*—that is, the distances from the *X* and *Y* axes—may be expressed as follows[2]:

	x	y
Point *A*	+1.000	+1.000
Point *B*	+3.000	+2.500

Point *A* is +1.000 inch from the *Y* axis in the plus x direction and +1.000 inch from the *X* axis in the plus y direction. Point *B* is +3.000 inches in the plus x direction and +2.500 inches in the plus y direction. Since both

[1] Unless a drill with a special starting point is used, it is generally necessary that each hole be started with a center drill which is then replaced by a conventional drill. This would mean repeating the machining cycle. However, the only dimension that would change would be the z dimension which refers to the depth movement. And this can either be programmed with a three-axis machine, or set by the operator with a two-axis machine.

[2] Note that x is the distance from the *Y* axis measured parallel to the *X* axis, and similarly, y is the distance from the *X* axis measured parallel to the *Y* axis.

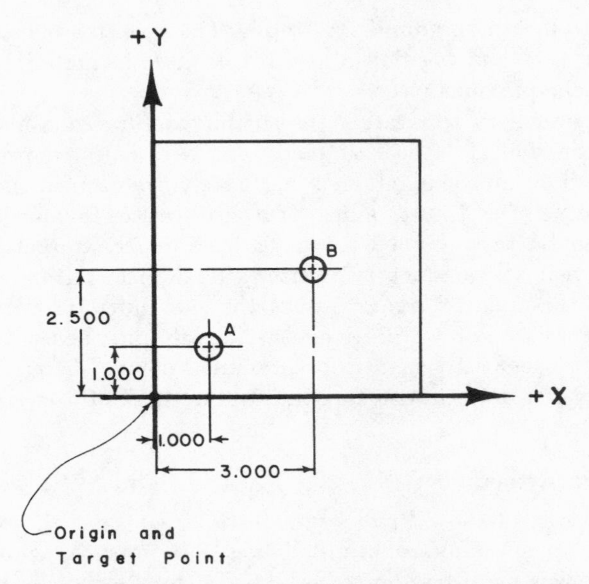

FIG. 5–5. Prior to preparing the manuscript the part is positioned (on paper) with respect to the X and Y axes.

points lie in the first quadrant, all of the signs are plus (+). (Please refer to Chapter 3.) If the coordinates of more points had to be described, they could all be listed in the x and y columns, as those for points A and B have been.

When a part program is written in *manuscript* form, the coordinates for each point are listed together with the necessary tape words for instructing the machine, such as start and stop.

'The coordinates of the various points are noted as x and y words for a two-axis system and x, y, and z words for a three-axis system. Other words that might be used include the sequence number (n word), the preparatory word (g word), the feed rate word (f word), and the auxiliary word (m word). (Please refer to Chapter 4.)

Typical Manuscript Form

As described in Chapters 3 and 4, there are various combinations of formats and systems available. To illustrate *all* of these, in this text, would be an impractical task. Two combinations have therefore been chosen that best illustrate the various approaches for point-to-point programming. These are:

1. The full floating zero system with the tab-sequential format and absolute measurement.

2. The fixed zero system with fixed-block format and absolute measurement.

There are also *incremental* measurement point-to-point programming systems, and the incremental concept will be illustrated in a later chapter covering contour programming.

A typical manuscript form is shown in Fig. 5–6(A). In this case, the part is programmed in a *tab sequential format*, and points A and B, as shown in Fig. 5–5, are expressed as *absolute* distances from the X and Y axes. It will be noted that the usual identifications, such as the part name and part number, are shown in the heading of the manuscript. Also note that there is no preparatory or g word used with the particular machine being programmed, nor is there a tape-controlled Z axis. The Z axis, or depth motion, is set by the operator in accordance with the instructions noted by the part programmer in the *Comments* column. Nor is there an f or s word for specifying the feed rate and spindle speed on the tape. These functions are usually found on contouring and point-to-point machines that are more expensive than the type illustrated. The manuscript

Part No. 12345 MANUSCRIPT Date x/x/xx
Part Name FLAT PLATE XXXX MACHINE Prepared By AB
Checked By EF

Sequence No.	TAB or EOB	x Coordinate	TAB or EOB	y Coordinate	TAB or EOB	m Word	TAB or EOB	Comments
RWS								
001	TAB	+1.000	TAB	+1.000	TAB	13	EOB	USE ½" HSS DRILL
								SET FEED DEPTH FOR
								DRILL TO CLEAR
								BOTTOM OF WORKPIECE
								SET SPEED AT
								1040 rpm.
002	TAB	+3.000	TAB	+2.500	EOB			
003	TAB	0.000	TAB	0.000	TAB	30	EOB	MACHINE STOPS AND TAPE AUTOMATICALLY
								REWINDS. REMOVE COMPLETED PART AND
								PUT IN NEW PART.

A

B

Fig. 5–6(A). The manuscript lists the coordinates and operations for drilling the two holes A and B shown in the flat plate in Figs. 5–2 through 5–5. The first block, as it would appear on the tape, is shown in (B). All leading and trailing zeros are shown as well as the plus (+) signs, although in many systems either trailing or leading zeros may be dropped and only the minus (−) signs need be shown. "TAB" characters are noted whenever an additional word or words are to be described in the line. The "EOB" is noted in the TAB or EOB columns when there are no further words to be listed. A TAB or EOB could therefore be listed in any one of the TAB or EOB columns.

described in Fig. 5–6(A) is typical of that used for popular types of lower-cost point-to-point NC drilling and milling machines.

Now refer to the manuscript shown in Fig. 5–6(A) and note the following:

1. The first instruction is the word "RWS" which appears on the first line in the "Sequence No." column. This word means *rewind stop* and is inserted to stop the tape after it has been rewound at the completion of the machining cycle (assuming that the control system has a tape rewind capability). Obviously, the first time the tape is inserted in the tape reader it will not need to be rewound and, hence, at the beginning of the first cycle this instruction will be ignored.

2. The next instruction to appear on the manuscript is the "block" opposite the sequence number 001 and reads:

TAB $+1.000$ TAB $+1.000$ EOB

This means that the drill is to be moved to a point 1.000 inch in the $+x$ direction and 1.000 inch in the $+y$ direction. When the drill reaches this location it moves down and drills the hole. The EOB means End Of

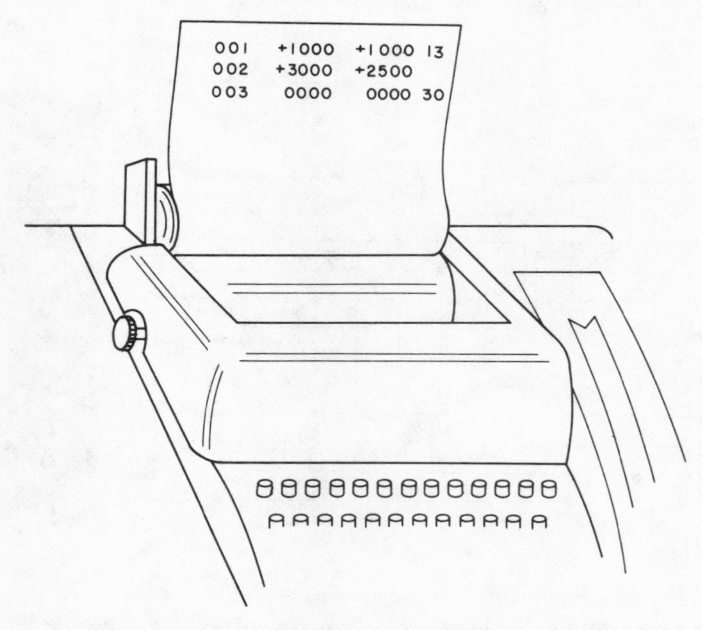

FIG. 5–7. A printout is made at the same time as the tape is punched. A printout of the characters on the tape can also be made by *playing back* the punched tape. Each time the TAB key is struck, a TAB character code is punched on the tape and the carriage of the tape punching machine moves over to a new column position. Striking the EOB (End of Block) key moves the carriage back to the starting point for the typing of a new line across the printout sheet and the punching of a new block on the tape.

Block. The block data for this movement, as it would appear on the tape, is seen in Fig. 5–6(B).

3. When the drill is completely withdrawn from the hole, the block headed by sequence number 002 is read into the control system and the drill moves on to the next point. The coordinates for this point are shown as +3.000 and +2.500. Since the drill is at the point $x = +1.000$ and $y = +1.000$, it needs to move only +2.000 inches in the $+x$ direction to reach $x = +3.000$ and only 1.500 inches in the $+y$ direction to reach $y = +2.500$.

4. After performing the drilling operation at this second point, the drill is instructed to move back to the set-up point in accordance with the coordinates $x = 0.000$ and $y = 0.000$ as shown by the instructions in the block headed by sequence number 003.

It will be noted that each horizontal line on the manuscript is equivalent to a block on the tape. The TAB character code shown on the manuscript has two functions. One is to separate the words on the tape and direct them to their proper registers in the control unit, as discussed in Chapter 4 (Fig. 4–11), and the other is to arrange the words in orderly columns on the printout of the tape punching machine. The TAB function on an NC tape punching machine works very much the same way as a TAB function on a typewriter—that is, to line up data in columns. The printout from the tape punching machine for the program described in Fig. 5–6(A) is shown in Fig. 5–7. If the tab sequential format were not utilized, the

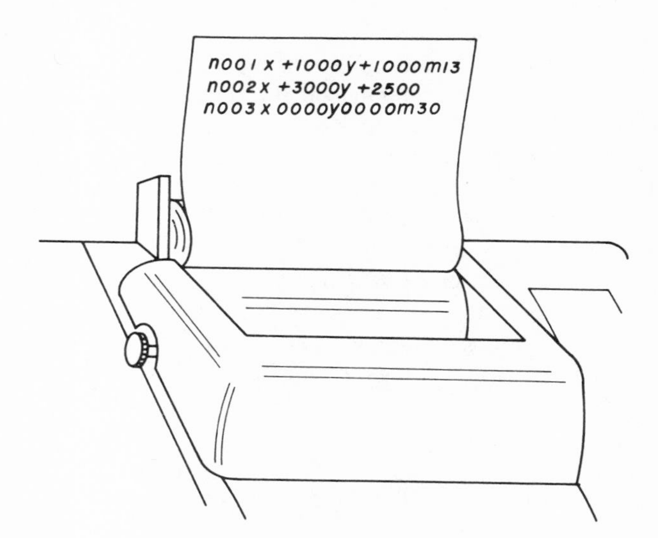

Fig. 5–8. Without the TAB arrangement, the characters for the different words in a block would be pushed together and hard to distinguish.

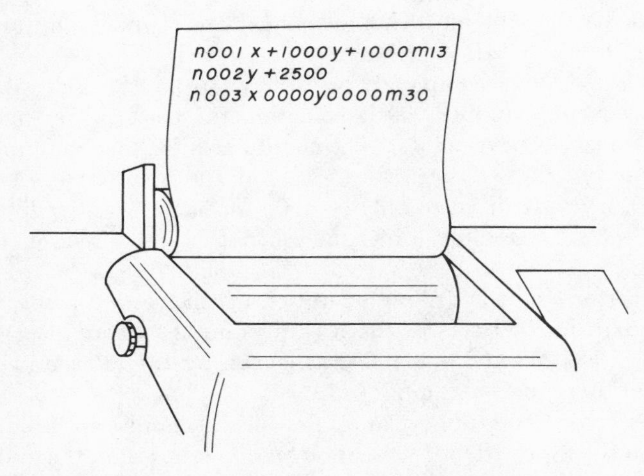

FIG. 5–9. It is difficult to keep track of the words on a printout when they are not in organized columns, particularly with the word address system, as illustrated.

printout would appear as shown in Fig. 5–8, with numbers and signs in a line pushed together and thus difficult to distinguish if the block is lengthy. It will be noted that, in the latter case, it would be necessary to *address* the words with a letter, and this is therefore called the *word address* system. The word address system is especially popular in contour programming, where a computer is frequently used and the coordinates are *automatically* aligned in their proper columns. The tab sequential system is probably the most popular for point-to-point control systems since the majority of tapes are prepared manually in this type of system.

The advantage of the tab sequential format can best be shown by the illustration in Fig. 5–9. If, for example, the x coordinate in the *second* line was +*1.000* instead of +3.000, as shown in Fig. 5–8, it would generally not be necessary to write or type in this coordinate since it did not change from the previous x coordinate, which is shown in the first line and which is also +1.000. If the next figures were then typed-in, the y coordinate, which is +2.500, would fall in line with the x coordinates. In other words, as shown on this printout, things would be pretty mixed up if the TAB character were not used.

PRACTICE EXERCISE NO. 1 CHAPTER 5

1. In addition to performing the calculations necessary to describe the coordinates, what else must the part programmer do?

2. Which is generally more difficult to program: a part requiring point-to-point operations or a part requiring contouring operations?
3. What is the key consideration that an operator must keep in mind when setting a piece of material on the worktable of an NC machine?
4. Referring to the sketch below, what would be the coordinates of points *A*, *B*, *C*, and *D*?

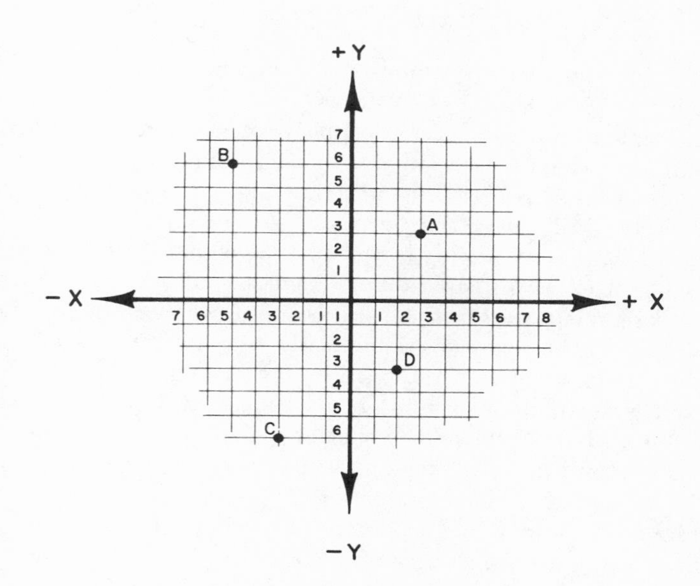

5. What is the significance of a control system having a full floating zero?
6. Is it necessary for the cutting tool to be programmed to move back to the target point after completing the cutting cycle?
7. Prepare a part program for drilling the one hole in the part shown below using the manuscript form shown in Fig. 5–6(A). Before writing the program, prepare a sketch showing the layout of the part as it would appear with respect to the *X* and *Y* axes. (Refer to Fig. 5–5 for type of layout.) The machine to be programmed is of the two-axis type where the depth motion is set manually. The cutter is to be programmed to move to a convenient "park" position.

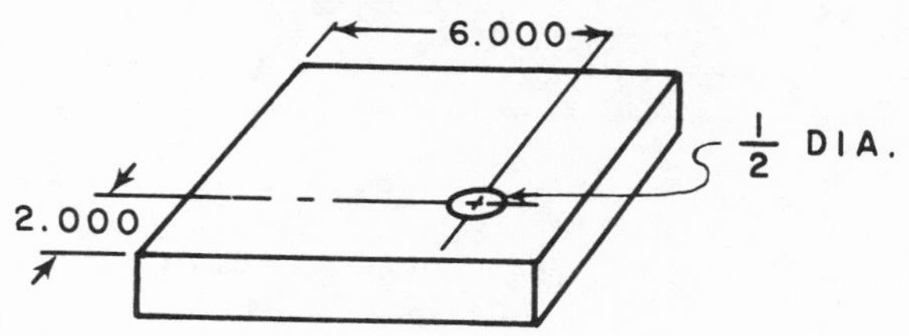

8. What is the advantage of the tab-sequential format with regard to the printout from the tape preparation machine?
9. What is meant by a system having *absolute measurement*?

PROGRAMMING A SIMPLE POINT-TO-POINT PART—System Having Fixed Zero, Fixed Block Format, Absolute Measurement

The point-to-point programming that has been described thus far involves a *free-floating zero* system where the origin, or $x = 0$, $y = 0$ point, can be located anywhere over the full working range of the machine by *manually* adjusting the location of the cutter with respect to the workpiece.

There are also *fixed-zero* systems that allow the operator to *shift* the origin electronically over the *full range* of the NC machine, from an established location, by dialing in the amount of shift in both the X and Y axes. The fixed zero system differs from the full-floating zero system in that the origin is *shifted* from a *home* position, whereas, in the full floating system, there is no home position and the origin is established by moving the spindle manually to a desired location or target point. Thus, for a fixed-zero system, the part must be positioned at a specific location with respect to a fixed origin. Some control systems do allow the origin to be shifted, but only a small amount, say, approximately .050 inch, which usually is sufficient to make minor adjustments.

Figure 5–10 shows a metal plate positioned on the worktable of an NC machine having a fixed-zero system. For purposes of comparison this is the same part configuration that was used to illustrate the full-floating zero program just described. As called for by the manuscript shown in Fig. 5–11, the bottom left-hand corner of the piece is positioned 3.000 inches from the Y axis and 1.500 inches from the X axis ($x = 3.000$,

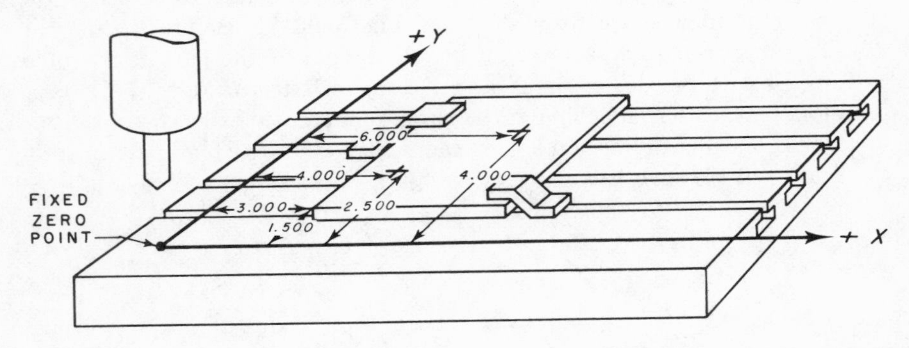

FIG. 5–10. In a fixed zero system the part is positioned with respect to the fixed X and Y axes.

PART NO. 12346 | MANUSCRIPT | DATE: X/X/XX

PART NAME FLAT PLATE | XXXXX MACHINE | PREPARED BY: HK

CHECKED BY: AC

SEQUENCE NO.	X COORDINATE	Y COORDINATE	M WORD	COMMENTS
RWS				
001	4.000	2.500	00	POSITION THE PIECE WITH THE LEFT HAND LOWER CORNER AT X=3.000
				Y=1.500. USE ½ HSS DRILL, SPEED AT 1040 RPM.
002	6.000	4.000	00	
003	0.000	0.000	30	MACHINE STOPS REMOVE COMPLETED PIECE.

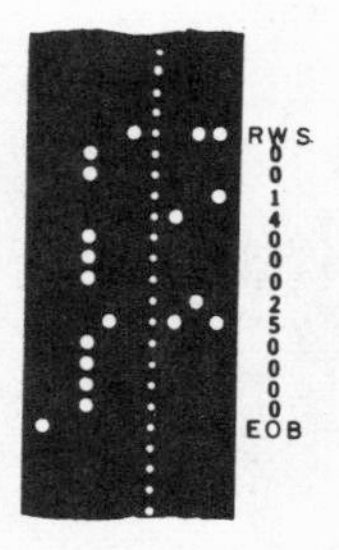

FIG. 5–11. The manuscript and first tape block for a *fixed block format* system are shown above.

$y = 1.500$). Considering that the center of hole A is 1 inch from each edge of the part, the x coordinate for hole A would be 4.000 inches and the y coordinate would be 2.500 inches. The x coordinate for point B would be 6.000 inches, and the y coordinate would be 4.000 inches.[3] Since only the first quadrant is considered with a fixed-zero system, it is not necessary to show the plus or minus signs. It will be noted in Fig. 5–11 that, because of the fixed block format, all zeros must be shown even though there may be no action required of a particular word. This is necessary so that the control system can interpret the proper word in accordance with its position in the block.

[3] If the system had *full zero shift* capability it would be possible to dial in the 3.000-inch distance for the x direction and the 1.500-inch distance for the y direction. This would then put the origin at the bottom left-hand corner of the part, and the condition would be the same as for the full floating zero system in that the coordinates listed on the manuscript would conform to the dimensions shown on the blueprint.

The *motions* of a machine having a fixed-zero control system and one having a free-floating zero control system would be very much the same. In this case the cutter would move from its home position in the left-hand corner, to the first hole, having coordinates $x = 4.000$, $y = 2.500$. After the drilling cycle is complete, the cutter moves on to the second hole which is headed by sequence number 2 and has coordinates $x = 6.000$, $y = 4.000$. And finally, the cutter is directed to its home position by the block headed by sequence number 3 and having coordinates of $x = 0.000$, $y = 0.000$.

The Targeting Procedure

With both the fixed-zero systems having zero shift and the free-floating zero systems, it is generally necessary to *target* the cutter with respect to some point on the workpiece or holding fixture. Where close accuracy is not required, the operator may be able to accomplish this alignment by visual inspection with a drill point or pointed instrument. More often, however, some sort of targeting instrument must be used. After the spindle has been targeted manually by the operator, the targeting device

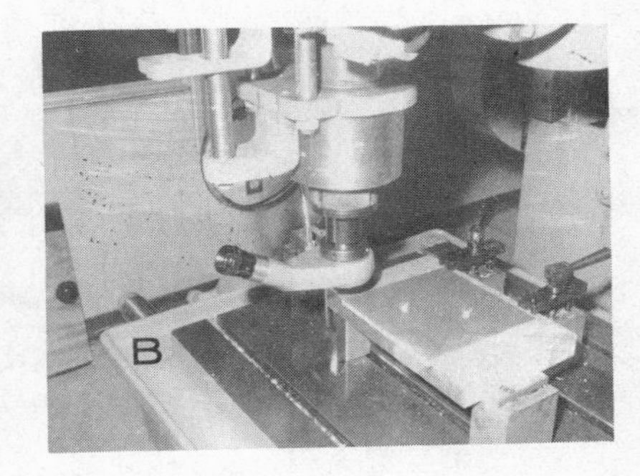

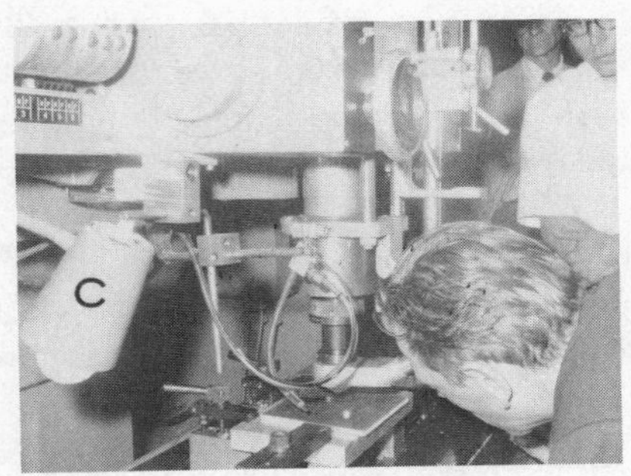

FIG. 5–12. Various ways of targeting: (A) On corner of part with edge finder; (B) on corner of part with scope; (C) on center of hole with scope; (D) on center of hole with dial indicator.

is removed and the cutter inserted. Three popular targeting devices are: edge finder, scope, and dial indicator.

1. An *edge finder* may be used for targeting on the corner of the part, as shown in Fig. 5–12(A). The end of the edge finder can be offset eccentrically so that it wobbles when the spindle rotates. To find an edge, it is deliberately offset. With the spindle rotating, the table is moved under manual control until the end of the edge finder touches the edge of the part. This movement is continued until the end of the edge finder ceases to wobble. At this position the axis of the spindle is offset from the edge a distance equal to one-half of the diameter of the end of the edge finder. When the table is moved toward the edge a distance equal to one-half of the diameter of the edge finder, the axis of the spindle and the edge of the part are aligned. Both edges at the corner must be aligned in this manner.

2. A *scope* is more accurate than the edge finder; however, it is also a more expensive piece of equipment. A scope is shown targeted at the corner of a workpiece in Fig. 5–12(B). The operator aligns cross hairs with the targeting point by positioning the spindle under manual control. See Fig. 5–12(C).

3. A *dial indicator* arrangement is shown in Fig. 5–12(D). In this case the operator rotates the spindle manually, while making X and Y adjustments, so that the center of the spindle is aligned with the target point which, in this instance, is the center of a hole.

PROGRAMMING A MORE COMPLEX POINT-TO-POINT PART

As noted in Chapter 2, most point-to-point machines are capable of milling, tapping, threading, and boring, as well as drilling; and it is for this reason that they are called *machining centers* rather than drilling machines. This capability allows different operations to be performed on a part without having to move the part from one machine to another, thus reducing set-up time considerably. And since most parts require a number of different types of operations, this numerical control feature has proven extremely helpful.

Consider, for example, the part shown in (A) of Fig. 5–13 and the dimensioned blueprint shown below it in (B). This part requires spot drilling, drilling, and boring operations. The entire job can be completed following one set-up of the part on the worktable. The surfaces of the part could also be milled in the same set-up, as will be described later in the chapter.

Assuming that the machine has a full-floating zero, tab-sequential format, and absolute measurement, the manuscript would appear as shown

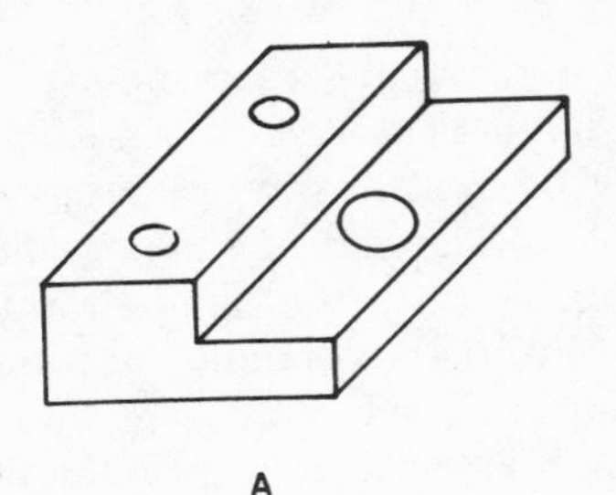

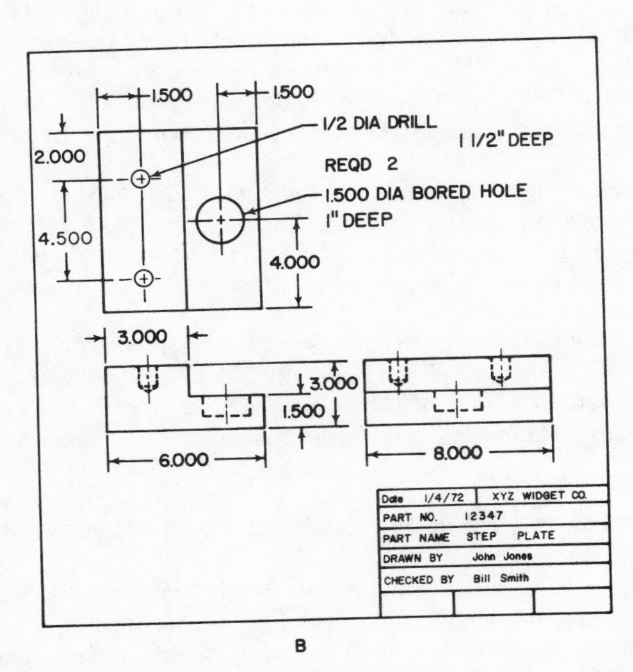

Fig. 5–13 (A) and (B). The part seen in (A) is described on the blueprint or engineering drawing shown in (B). Other illustrations referring to the programming of this part are shown in Figs. 5–14 through 5–16.

in Fig. 5–14(A) or 5–14(B). The layout of the part, with reference to the *X* and *Y* axes, would be as shown in Fig. 5–15. It should be noted that the origin, or target point, is located at the lower left-hand corner of the part. This is not mandatory and any convenient point could have been chosen. However, by choosing the bottom left-hand corner, all coordinate points will be positive, which reduces a potential source of error. On the other hand, sometimes it is more convenient to locate the origin where plus and minus signs may be utilized, and this will be discussed later in the chapter.

A

PART NO. 12347 | PART NAME STEP PLATE

MANUSCRIPT
XXXXX MACHINE

DATE: X/X/XX
PREPARED BY: [illegible]
CHECKED BY: AS

SEQUENCE NO.	TAB or EOB	X COORDINATE	TAB or EOB	Y COORDINATE	TAB or EOB	m WORD	TAB or EOB	COMMENTS
RWS								USE SPOT DRILL & DRILL AT 3 POINTS
001	TAB	+1.500	TAB	+1.500	EOB			
002	TAB	+4.500	TAB	+4.000	EOB			
003	TAB	+1.500	TAB	+6.000	EOB			
004	TAB	-2.500	TAB	+6.000	TAB	02	EOB	SPINDLE MOVES TO CONVENIENT PARK POSITION & STOPS. (REF FIG 5-16)
005	TAB	+1.500	TAB	+1.500	EOB			USE 1/2" HSS DRILL & DRILL 2 HOLES
006	TAB	+1.500	TAB	+6.000	EOB			
007	TAB	-2.500	TAB	+6.000	TAB	02	EOB	SPINDLE MOVES TO PARK POSITION & STOPS.
008	TAB	+4.500	TAB	+4.000	EOB			CHANGE TO 1 7/16" DRILL & DRILL ONE HOLE.
009	TAB	-2.500	TAB	+6.000	TAB	02	EOB	SPINDLE MOVES BACK TO PARK POSITION.
010	TAB	+4.500	TAB	+4.000	TAB		EOB	CHANGE TO BORING TOOL. SET DIAMETER AT 1.500 INCHES & BORE HOLE.
011	TAB	-2.500	TAB	+6.000	TAB	30	EOB	SPINDLE MOVES BACK TO PARK POSITION. TAPE REWINDS. REPLACE DRILL & NEW PIECE OF MATERIAL.

A

PART NO. 12347 | PART NAME STEP PLATE

MANUSCRIPT
XXXXX MACHINE

DATE
PREPARED
CHECKED

SEQUENCE NO.	TAB or EOB	X COORDINATE	TAB or EOB	Y COORDINATE	TAB or EOB	m WORD	TAB or EOB	
RWS								USE SPOT DRILL
001	TAB	+1.500	TAB	+1.500	EOB			
002	TAB	+4.500	TAB	+4.000	EOB			
003	TAB	+1.500	TAB	+6.000	EOB			
004	TAB	-2.500	TAB		TAB	02	EOB	SPINDLE MOVES (REF FIG 5-/
005	TAB	+1.500	TAB	+1.500	EOB			USE 1/2" HSS DR
006	TAB		TAB	+6.000	EOB			
007	TAB	-2.500	TAB		TAB	02	EOB	SPINDLE MOVE
008	TAB	+4.500	TAB	+4.000	EOB			CHANGE TO 1
009	TAB	-2.500	TAB	+6.000	TAB	02	EOB	SPINDLE MO
010	TAB	+4.500	TAB	+4.000	TAB		EOB	CHANGE TO INCHES &
011	TAB	-2.500	TAB	+6.000	TAB	30	EOB	SPINDLE MO REPLACE D

B

FIG. 5–14. (A) The manuscript describes the data to be punched on the tape. Each horizontal line, introduced by a sequence number, represents a block on the tape. Coordinates have been repeated, such as the 1500 in the x coordinate column for sequence 002 and the 6000 for the y coordinate in sequence 003, for clarity. It is not generally necessary to repeat identical figures, and the manuscript could have appeared as shown in (B). Also, it is normally not necessary to show plus (+) signs since the control unit interprets an absence of a sign as a plus (+). All minus (−) signs, however, must be shown. The plus signs have been shown here to give a clearer illustration.

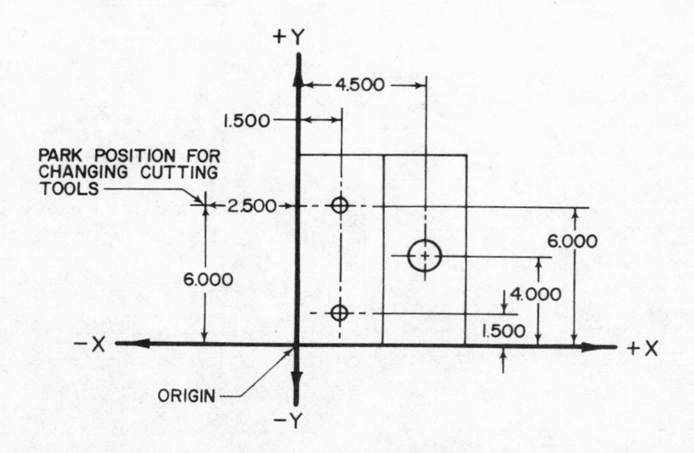

FIG. 5–15. Layout of the part shown in Fig. 5–13 in coordinate axis form.

The sequence of operations is illustrated in Fig. 5–16(A) through (D). In (A) the cutter moves from its targeted position to the coordinates required to spot drill the first hole and then moves on to perform the same operation for the second and third holes. The depth to which the spot drill is to move downward is set manually by the operator prior to the machining cycle since this is a two-axis machine and the *Z* axis, or depth movements, cannot be programmed on tape. Also, as will be noted, it is not necessary for the cutter to return to the target point. A more convenient and accessible point may be selected where it would be easier to change the cutting tools or parts. When the operator pushes the button to start the next cutting sequence, the cutter will move directly to the desired coordinates from its new park location and drill the two $\frac{1}{2}$-inch holes that have been spot drilled, as shown in (B). The drilling and boring of the third hole are illustrated in Fig. 5–16(C) and (D) as a separate operation since a larger drill and a boring tool are used.

After all cutting operations have been performed and the machine stops, the operator may remove the finished piece and position a new piece of material, making sure that the piece is positioned in the same location as the previous part. The cycle is repeated by first re-inserting the spot drill. When the "start" button is pushed, the spot drill will move directly to the point determined by the coordinates of the first hole and will *not* go back to the target point. See Fig. 5–15(E). It should also be noted that direct-line motions to the coordinate locations have been shown in order to simplify the illustrations. Actually the movements would be as shown in Fig. 2–14, Chapter 2.

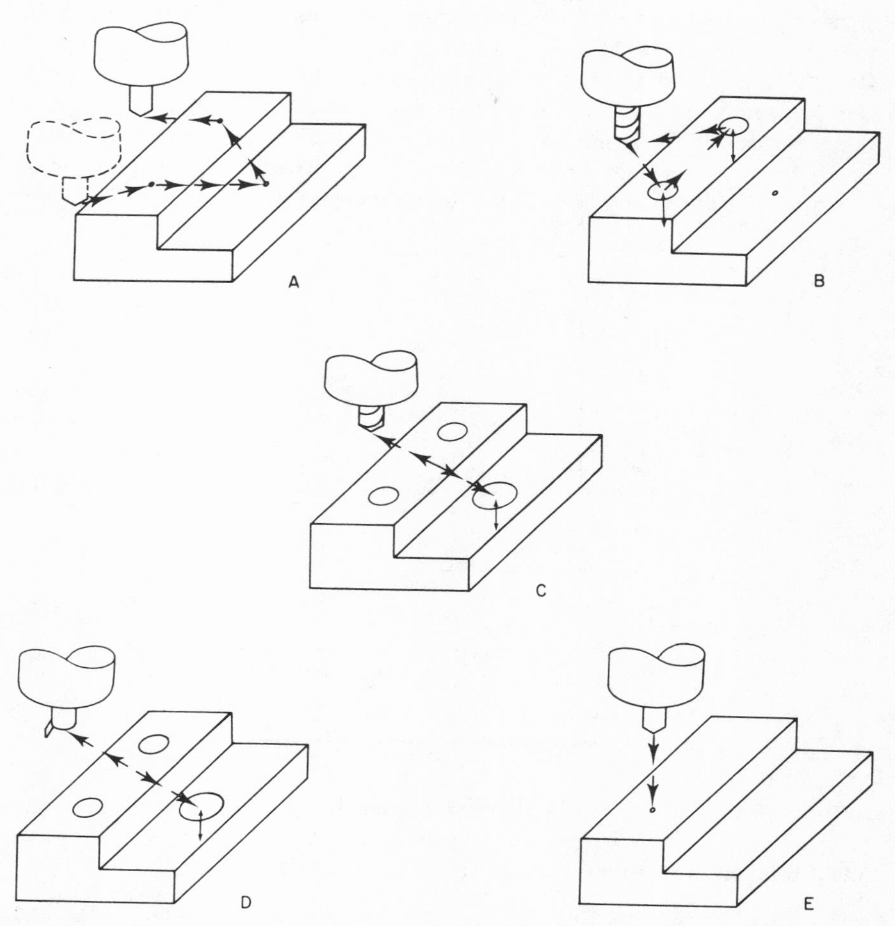

Fig. 5–16 (A, B, C, D, E). The sequence shown above describes point-to-point drilling and boring operations. The part is described in Figs. 5–13 and 5–15, and the manuscript is shown in Fig. 5–14.

PRACTICE EXERCISE NO. 2 CHAPTER 5

1. What is the key difference when positioning a part for a full floating zero system and for a fixed-zero system?
2. Why is it necessary that *all* the words in a block be shown on the manuscript, and punched on the tape, with a fixed-block format?
3. Referring to Fig. 5–16, after the first part has been machined and the operator pushes the *start* button to begin the cycle for machining the

next part, will the cutter move directly to a point above the first hole or to the target point and *then* to the point above the first hole?

4. Referring to the blueprint drawing shown below, draw a sketch as the part would appear with respect to the X and Y axes. Next, prepare a manuscript form similar to that shown in Fig. 5–14 and then write the program for spot drilling *and* drilling the two holes. The machine being considered is a two-axis machining center.

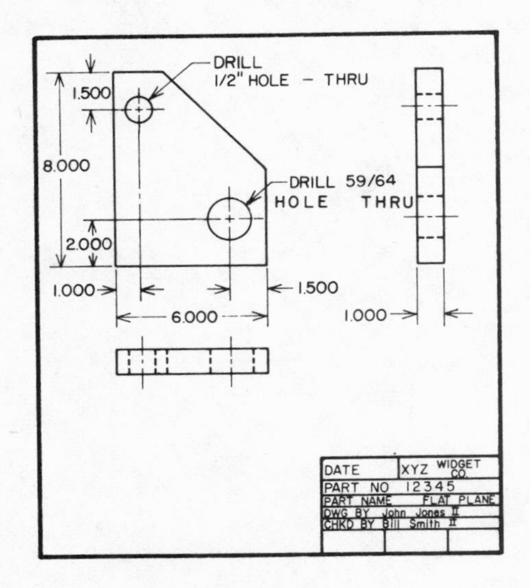

STRAIGHT-CUT MILLING

What Is Meant by Straight Cut Milling?

As noted earlier in this chapter, most point-to-point machines have milling capabilities in addition to drilling, tapping, and boring. However,

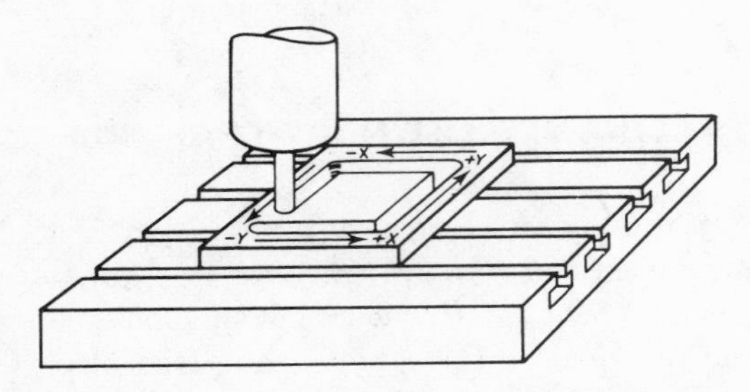

Fig. 5–17. Most NC point-to-point metal-cutting machines have the capability of straight cut milling along any one axis at a time.

the one notable restriction when milling with point-to-point machines is that the milling cuts are confined to straight lines which are parallel to either the X or Y axis. (Some machines, especially those using stepping motor drives, are capable of making accurate 45-degree-angle cuts.) Since the X and Y right-angle cuts can be combined to form rectangles, this type of machining is often referred to as *picture frame milling*. See Fig. 5–17.

Programming a Straight Cut

The chief difference between drilling operations and milling operations in a point-to-point system lies in the feed rate, or speed, that the cutter moves from one point to the next. In a drilling operation the object is to move the drill as quickly as possible from point to point because the tool is moving in air. When cutting metal, however, the cutter moves at a much slower rate and the feed rate must be controlled. The feed rate may be selected in a number of ways. The most common practice with two-axis point-to-point machines is for the operator to dial in the selection (see Fig. 5–18). The coordinate instructions on the tape for moving the cutter

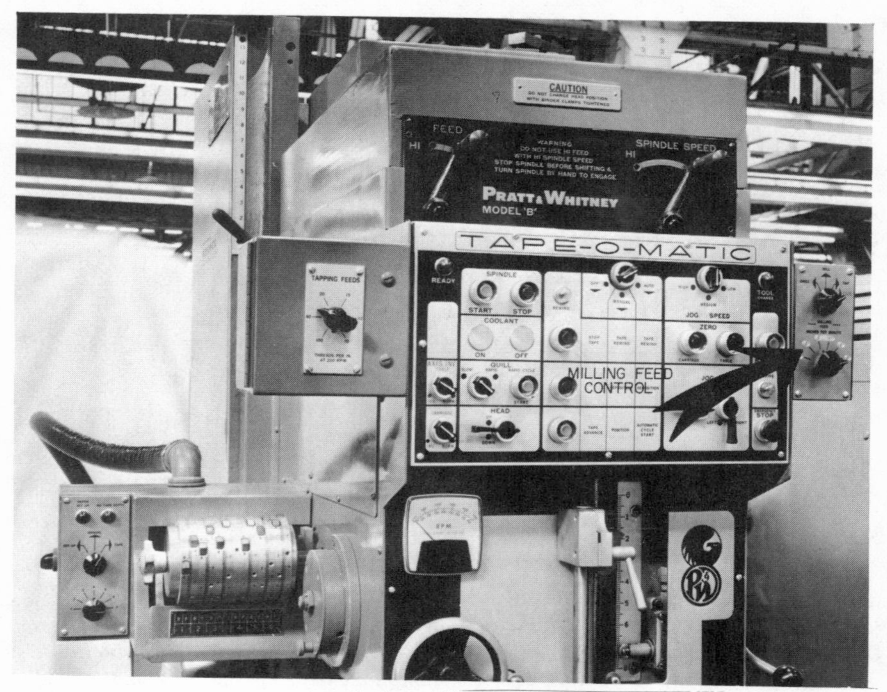

Courtesy of Pratt & Whitney Machine Tool Div. of Colt Industries

FIG. 5–18. A switch may control the milling or drilling mode.

Courtesy of Pratt & Whitney Machine Tool Co.

FIG. 5–19. The straight-line milling cut shown is parallel to the X axis. The part being machined is a major component for an NC machine that is identical to the one machining it. Also refer to Fig. 5–20.

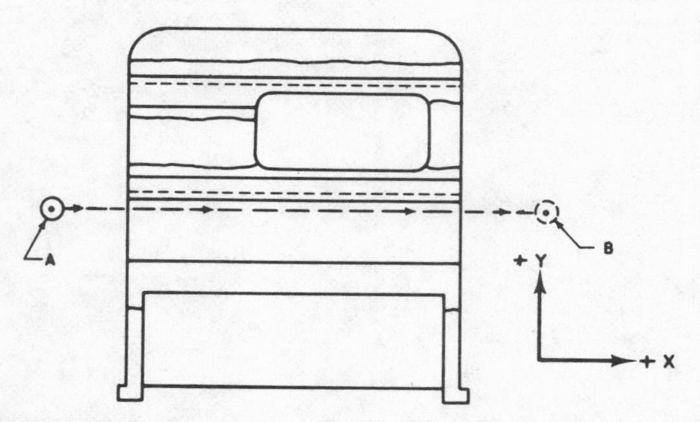

FIG. 5–20. In order for the cutter to move from point A to point B in a milling mode, the coordinates of points A and B must be described. It would also be necessary for the operator to dial in the correct feed rate or for the programmer to specify this for inclusion on the tape. Whether the feed rate is called out on the tape or dialed in would depend on the type of control system.

from the start to the end of its path would be the same as if the machine were moving between two points. With some systems it is necessary for an *m*, or auxiliary word, to be noted on the tape in order to switch from a drilling to a milling mode.

An example of a straight milling cut is shown in Fig. 5–19. The only data required for making the cut as shown would be the coordinates of the two end points *A* and *B*, seen in Fig. 5–20.

Another illustration of straight-line milling is shown in Fig. 5–21, with

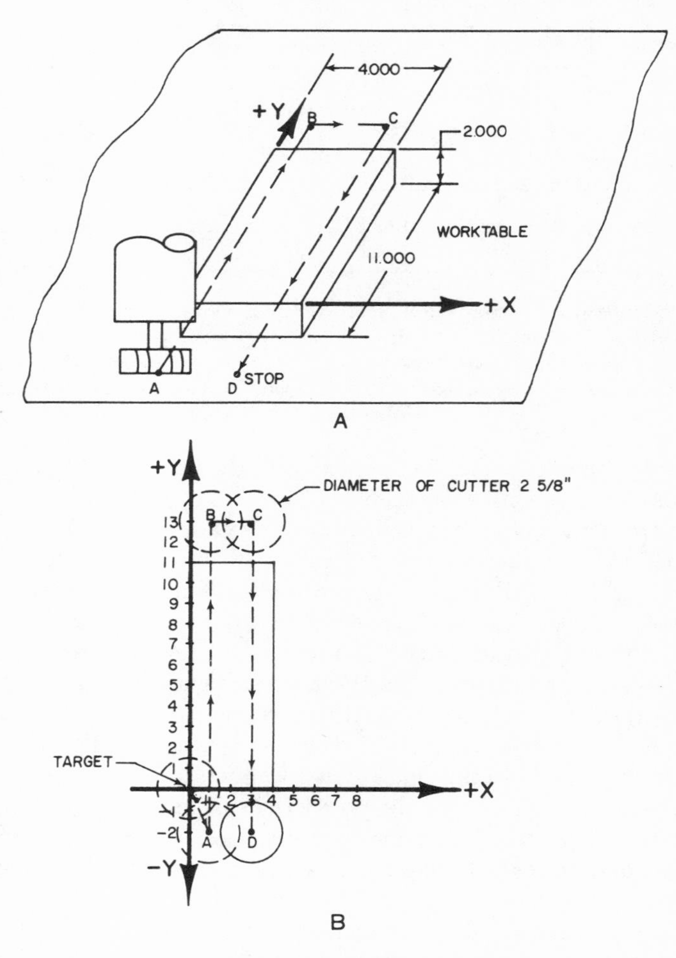

FIG. 5–21. (A) This face mill cutter has been programmed to mill the surface of the piece, in two passes. In (B) the piece is shown as it would appear in the first quadrant of the coordinate system.

PART NO. 12348
PART NAME : MILLED PLATE

MANUSCRIPT
XXXXX MACHINE

DATE : X/X/XX
PREPARED BY :
CHECKED BY :

SEQUENCE NO.	TAB or EOB	x COORDINATE	TAB or EOB	y COORDINATE	TAB or EOB	m WORD	TAB or EOB	COMMENTS
RWS								
001	TAB	+1.000	TAB	-2.000	TAB	02	EOB	TARGET CUTTER AT BOTTOM LEFT HAND CORNER OF PART. CUTTER MOVES TO POINT A AND THEN STOPS WHERE OPERATOR LOWERS SPINDLE TO PROPER DEPTH AND THEN PUSHES START BUTTON.
002	TAB		TAB	+13.000	EOB			CUTTER MILLS TO POINT B
003	TAB	+3.000	EOB					CUTTER MOVES TO POINT C
004	TAB		TAB	-2.000		30	EOB	CUTTER MILLS TO POINT D & STOPS.

Fig. 5–22. Manuscript for the face milling operation shown in Fig. 5–21. After the cutter stops at point *D*—see Fig. 5–21 (B)—and a new piece of material is put into position, the operator pushes the start button and the cutter moves to point *A*. The cutting cycle is repeated by again pushing the start button. The format shown is TAB sequential.

the manuscript shown in Fig. 5–22. The purpose of this machining operation is to mill a flat surface on the top of the piece. This can be accomplished by two passes of a face milling cutter. Note that it is the center or axis of the cutter which is first targeted at the bottom left-hand corner of the part. The cutter center is programmed to move from the target point to point *A*, and stop. The operator adjusts the depth manually and again pushes the start button. The cutter then moves to point *B* in a straight-line motion parallel to the *Y* axis. The coordinates of point *B* are $x = 1.000$, $y = 13.000$. Next the cutter is programmed to move to point *C*, which has coordinates $x = 3.000$, $y = 13.000$. The last move brings the cutter from point *C* back down to point *D*, where the machine is programmed to stop. It will be noted that all motions are parallel to either the *X* or *Y* axis. In machining operations of this type care must be taken to assure that the cuts overlap in order to obtain a smooth surface.

As in the machining of the multioperation part with the two-axis point-to-point type of machine described earlier (manuscript shown in Fig. 5–14), the actual feed rate and cutting speed that are used may, in some instances, be selected by the operator; however, they are usually specified beforehand by the production planner or by the programmer. Many factors determine the best feed and cutting speed to use, such as the type of cutting tool material used, the type and hardness of the work material, the nature of the operation, and the tool life desired. Usually the feed and speed are selected from tabular information which can be found in handbooks, such as *Machinery's Handbook*.[4] Approximate feeds and speeds are given in Appendix A of this book.

Milling a Pocket

Pocket milling, or the cutting out of a rectangular hole in a piece of metal, is a fairly common requirement in machine shop practice. This is also a reasonably simple operation with a point-to-point NC machine having a straight-line milling capability.

As in all point-to-point operations, the programming and operation procedures for milling a pocket would depend on whether a two-axis or a three-axis machine is to be used. As noted earlier, a two-axis machine requires that the operator set the depth movements manually. In a three-axis control system the depth movements can be programmed on the tape, and most of the operations can then be performed automatically from tape instructions.

Figure 5–23(A and B) illustrates a pocket milling operation. First, the cutter is *targeted* at the lower left-hand corner of the part. This targeting location is not mandatory, and any one of the four corners would have been satisfactory. The reason for selecting this point is that all coordinates on the part will then have plus signs, thus reducing the possibility of error due to the programmer's noting an incorrect sign. There are also advantages in using different quadrants that require plus (+) *and* minus (−) signs before the coordinate dimensions, and this will be discussed later in the chapter.

In addition to targeting the cutter with respect to the X and Y axes, the operator must also adjust the Z-axis location to the proper height. This can be accomplished with the aid of a metal block that has been machined to an accurate thickness for insertion between the top of the workpiece and the bottom of the cutter. This piece of measuring equipment is called a *feeler* block. Referring to Fig. 5–23(C), the depth of the 2-inch-diameter cutter can be set by positioning a .500-inch-feeler block on top of the workpiece and then manually moving the cutter down until it just touches the feeler block.[5] The feeler block may then be removed. If a three-axis machine is being considered, and the Z-axis movements

[4] Erik Oberg and F. D. Jones, *Machinery's Handbook* (New York: Industrial Press Inc, 1971).

[5] The operator must be careful not to allow the cutter to strike the feeler block, which might damage the cutting teeth.

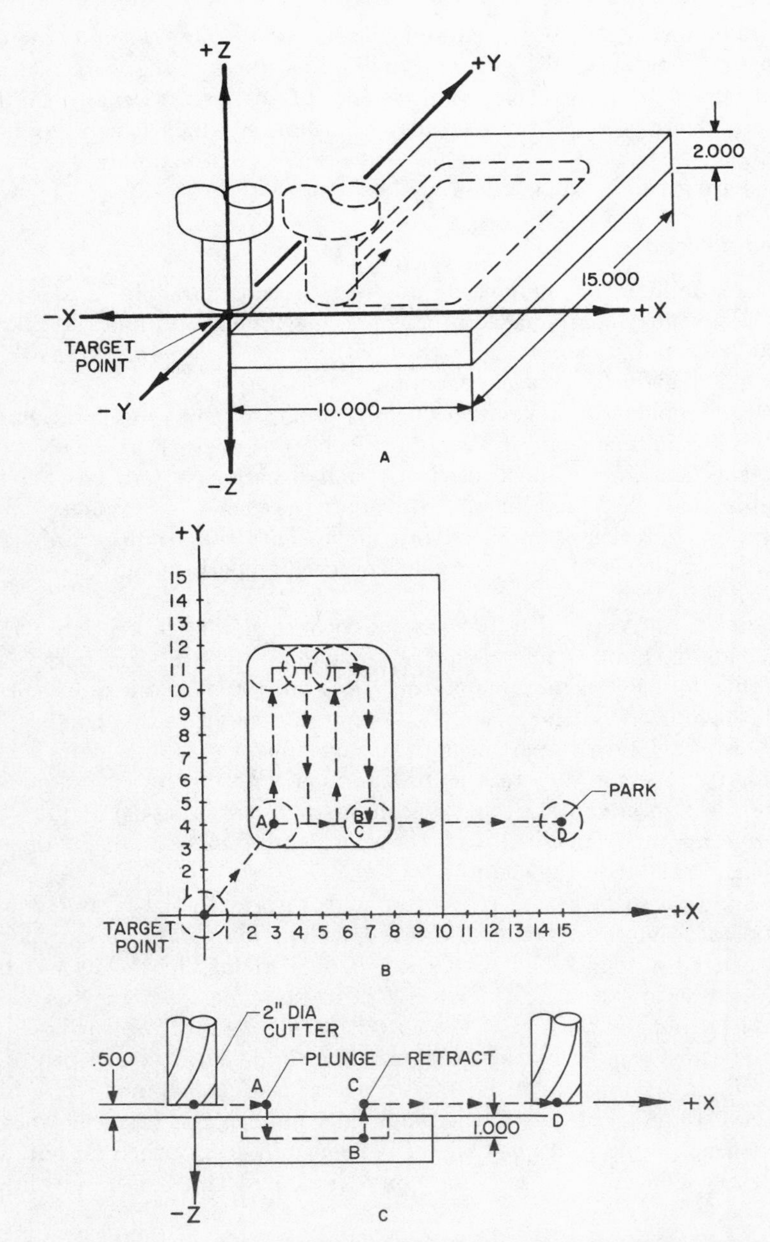

Fig. 5–23. Sketch (A) shows the position of the cutter after it has been targeted manually by the operator. In (B) the cutter moves to point *A*, then on to cut out the pocket. A side view is shown in (C). Following the pocket cutting operation the cutter is programmed to move to a convenient park location, such as the point *D*.

can be programmed on the tape, the only requirement left in this case would be for the operator to push the *start* button. The cutter, which would be of a center-cutting type, would move to point A, plunge, and then proceed to cut out the pocket.

With a two-axis machine, where the operator is required to set the *Z*-depth movements, the operation is not quite so simple. In this case, after the cutter has been targeted and the depth has been set by using the feeler block, the machine is set for a *drilling* cycle.

The start button is pushed, and the milling cutter moves to point A, where it performs a plunge milling operation. After the cutter has made its plunge cut, it retracts and is programmed to stop. (The cutter will automatically retract since this is part of the machine's normal drilling cycle.) The operator may next move the milling cutter manually back down to the depth just cut and turn the switch to the milling mode.

For most two-axis machines it is also necessary for the operator to dial in the desired feed rate and set the speed. After the start button has been

PART NO. 12349
PART NAME: POCKET PLATE

MANUSCRIPT
XXXX MACHINE

DATE X/XX/XX
PREPARED BY ABC
CHECKED BY DFT

SEQUENCE NO.	TAB or EOB	x COORDINATE	TAB or EOB	y COORDINATE	TAB or EOB	m WORD	TAB or EOB	COMMENTS
RWS								
001	TAB	+ 3.000	TAB	+ 4.000	TAB	30	EOB	USE FEELER BLOCK AND SET DEPTH OF 2" DIA CUTTER @ .500 FROM SURFACE. SET DEPTH FOR POINT A AT 1.000 FROM TOP SURFACE OF PART. MACHINE WILL STOP AFTER DRILLING CYCLE. RE-SET DEPTH MANUALLY TO BOTTOM OF HOLE RESULTING FROM PLUNGE CUT. CHANGE TO MILLING CYCLE AND SET MILLING FEEDRATE.
002	TAB		TAB	+11.000	EOB			START MILLING CYCLE.
003	TAB	+ 4.500	EOB					
004	TAB		TAB	+ 4.000	EOB			
005	TAB	+ 5.500	EOB					
006	TAB		TAB	+ 11.000	EOB			
007	TAB	+ 7.000	EOB					
008	TAB		EOB	+ 4.000	EOB			MANUALLY WITHDRAW CUTTER FROM POCKET AND SET SWITCH TO "DRILL" CYCLE. PUSH START BUTTON AND CUTTER WILL MOVE AS SHOWN IN SEQUENCE NO. 009.
009	TAB	+15.000	TAB		TAB	30	EOB	CUTTER MOVES TO PARK POSITION AND STOPS. TAKE OUT FINISHED PART AND PUT IN NEW WORKPIECE.

FIG. 5–24. Manuscript for cutting pocket shown in Fig. 5–23 on a two-axis machine where the depth or *Z* motion is set manually.

PART NO. 12349 MANUSCRIPT DATE: x/x/xx

PART NAME POCKET PLATE XXXXX MACHINE PREPARED BY: HJ

CHECKED BY: AS

SEQUENCE NO.	TAB or EOB	X COORDINATE	TAB or EOB	Y COORDINATE	TAB or EOB	Z COORDINATE	TAB or EOB	f WORD	TAB or EOB	m WORD	TAB or EOB	COMMENTS
RWS												
001	TAB	+3.000	TAB	+4.000	TAB	0.000	TAB	1500	EOB			USE FEELER BLOCK & SET DEPTH AT .500" FROM SURFACE.
002	TAB		TAB		TAB	-1.500	TAB	20	EOB			CUTTER PLUNGES IN -Z DIRECTION @ 2 IPM
003	TAB	+3.000	TAB	+11.000	TAB		TAB	50	EOB			START OF POCKET.
004	TAB	+4.500	EOB									CUT
005	TAB		TAB	+4.000	EOB							
006	TAB	+5.500	EOB									
007	TAB		TAB	+11.000	EOB							
008	TAB	+7.000	EOB									
009	TAB		TAB	+4.000	EOB							
010	TAB		TAB		TAB	0.000	TAB	1500	EOB			CUTTER MOVES UP & OUT OF POCKET.
011	TAB	+15.000	TAB		TAB		TAB		TAB	30	EOB	CUTTER MOVES TO PARK POSITION AT POINT D.

FIG. 5–25. Manuscript for cutting pocket in Fig. 5–22 on a NC machine having three axes of control. The feed rate remains the same for subsequent blocks unless changed and need not be shown for each block. Thus the movement noted for sequences 004 through 009 would be at 5.0 ipm.

pushed, the cutter moves about its milling cycle just as it would with the three-dimensional machine, except that for the final retraction, the operator must move the cutter manually up and out of the pocket. The manuscript for the type of machine having two-axes of control is shown in Fig. 5–24. Figure 5–25 illustrates the manuscript for the three-axis type of machine. In the manuscript for the three-axis machine the Z-axis motions as well as the feed rate for the milling operation are programmed on the tape, thus making the operation completely automatic. This is shown in the additional *z coordinate* and *f* word columns of Fig. 5–25. The *f* word in this case is programmed directly in inches per minute (please refer to Chapter 4). The *f* word shown in the first line is 1500, and means that the cutter is to move along its path at a feed rate of 150.0 inches per minute. This is considered a *rapid traverse* feed rate, to be used when the cutter moves in air—such as when it goes from one point to another without cutting metal.

Again, referring to Fig. 5–25, the line beginning with sequence number

001 would move the cutter from the target point having coordinates $x = 0$, $y = 0$, $z = 0$, to point A, having coordinates $x = 3.000$, $y = 4.000$, $z = 0.000$. Next, the cutter would plunge, or move down, into the workpiece, at a feed rate of 2.0 inches per minute. The command for this move is shown on the second line of the manuscript (sequence number 002). The x and y coordinates are the same as for sequence number 001, and therefore nothing need be shown in the *x coordinate* and *y coordinate* columns. The z coordinate changes, however, and the cutter is commanded to move to a point having a z coordinate of -1.500 inches. Another way of looking at this point is to consider it as lying below a plane formed by the X and Y axes. (Please refer to Chapter 2.) The concept of a point lying below a plane formed by the X and Y axes is shown in Fig. 5–26. Referring again to Fig. 5–25, after the milling cutter has moved to a depth of 1.000 inch into the workpiece, it then automatically proceeds to move to the coordinate described in the third line (sequence number 003). Lines beginning with sequence numbers 004 through 009 describe the motions to be followed in cutting out the pocket. Sequence number 010 moves the cutter back up and out of the pocket to point C at a rapid traverse of 150.0 inches per minute. The last instruction (sequence number 011) calls for the cutter to move away from the part and to a convenient park position, which is at point D.

It might be pointed out that the part programmer is not restricted to the pocket cutting pattern shown in Fig. 5–23. Thus, a number of alternate paths may be selected, such as the two shown in Fig. 5–27.

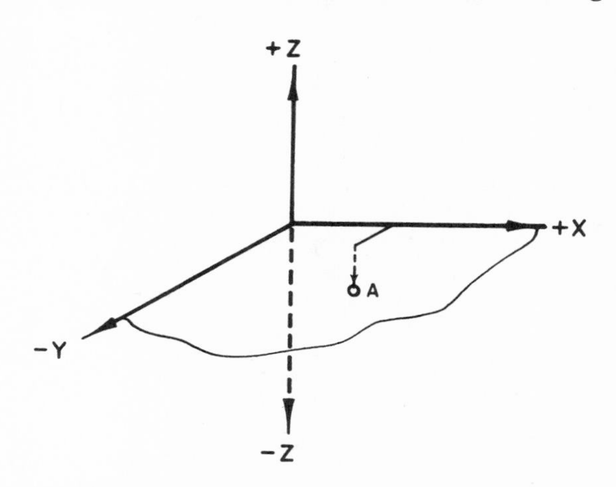

FIG. 5–26. If the plane formed by the X and Y axes were to be considered as a pane of glass, we could see point A lying below it and the sign of the z coordinate would be minus (−). The further down along the Z axis, the greater the minus value.

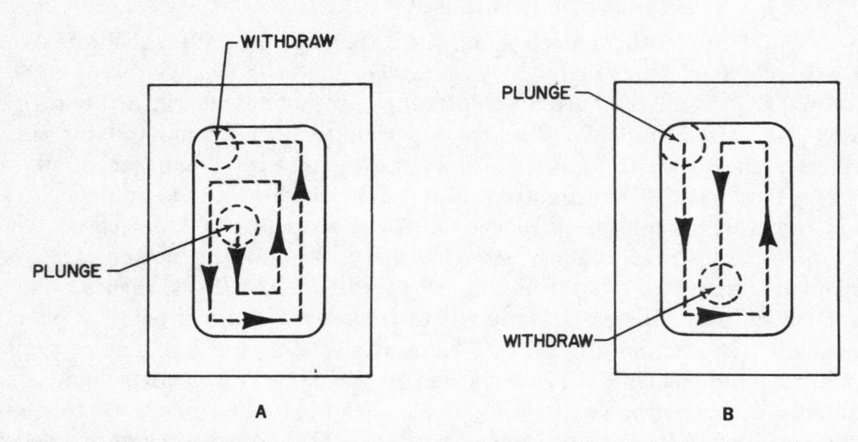

Fig. 5–27. Pocket cuts may be made in a spiral fashion as shown in (A) and (B).

Using Plus and Minus Programming

Although most of the examples thus far have utilized the first quadrant only—where both the x and y coordinates are plus, it is often convenient to use the other three quadrants—where signs may be either plus or minus. By way of review, consider Fig. 5–28. The coordinates of point A are $x = +3.000$, $y = +3.000$. The coordinates of point B are $x = -5.000$, $y = +4.000$; the coordinates of point C are $x = -3.000$, $y = -4.000$; and of point D, $x = +2.000$, $y = -3.000$.

With a full-floating zero system, where the target point, or origin, can

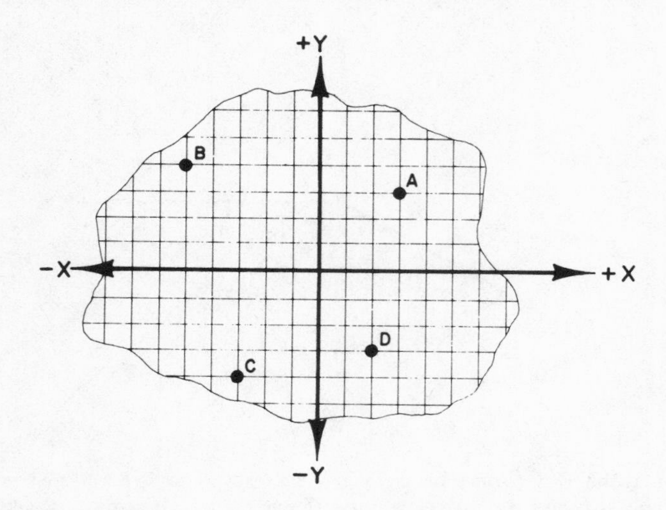

FIG. 5–28. The coordinates of points A, B, C, and D are described as plus (+) and minus (−) values.

be positioned anywhere over the full working range of the machine and all four quadrants can be used, it is possible to reduce the number of calculations by carefully selecting the location of the target point. This would usually be the origin where the coordinates of x and y are equal to zero, although there are some control systems that allow for the target point to be at a different location from the origin.

For example, if the origin, or point, where both the x and y coordinates are equal to zero, is set at the center of the part shown in Fig. 5–29, then only two points have to be calculated. The numerical value of the other points would be the same except for their signs. Consider the points A, B, C, and D in Fig. 5–29. If the distances from the X and Y axes for points A and B are the same, respectively, as for points C and D, then the x and y coordinates for points A and B would be the same as for points C and D, except for the *signs*; and points A, B, C, and D may be said to be symmetrical about the Y axis. This would also apply to points E, F, G, and H. If points E and F, for example, were the same respective distances from the X and Y axes as points C and D, then points E and F would be symmetrical with points C and D about the X axis. The same relationships would apply to points G and H with respect to points A and B and with respect to points E and F. The only difference, therefore, between the coordinates of points A and B and the six other respective points would be the signs.

Another example where consideration for placement of the target point

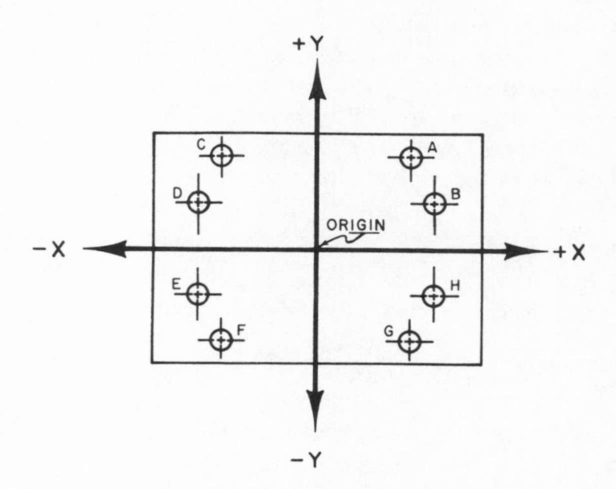

FIG. 5–29. If the target point, or origin, is put at the center of the pattern, then the x and y coordinate values would be respectively the same for C and D, E and F, and G and H, as for A and B, except for the signs. This reduces the amount of calculation required for determining the coordinates of the various points.

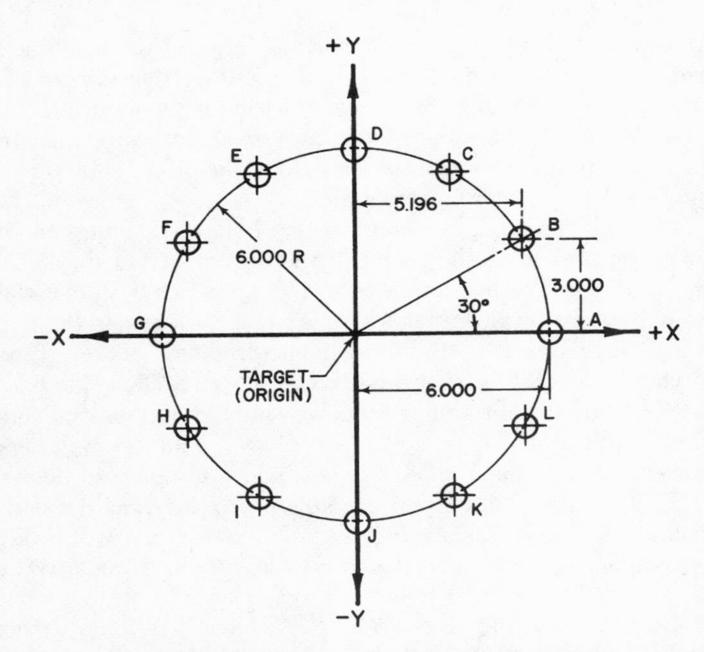

FIG. 5–30. By placing the target, or origin, at the center of the circular pattern it is possible to reduce the calculations required.

would prove helpful is in Fig. 5–30. In this case there are twelve equally spaced holes around the circumference of a circle, for which the radius is 6.000 inches. If the origin is placed at the center of the circle, points A, D, G, and J can be described without any calculation. The coordinates of point A would be $x = +6.000$, $y = 0.000$. The coordinates of point D would be $x = 0.000$, $y = +6.000$. For point G the coordinates would be $x = -6.000$, $y = 0.000$; and for point J the coordinates would be $x = 0.000$, $y = -6.000$. In order to find the coordinates of points B and C, some simple trigonometry is required. The x and y coordinates of point B would be calculated by:

$$\begin{aligned} x &= 6.000 \cos 30^\circ \\ &= 6.000 \times .8660 \\ &= 5.196 \\ y &= 6.000 \sin 30^\circ \\ &= 6.000 \times .5000 \\ &= 3.000 \end{aligned}$$

The coordinates for point C would be calculated in a similar way. The coordinates of points F and E would be the same as for points B and C except that the sign of the x coordinate would be minus $(-)$ since these points lie in the second quadrant. The x coordinate of point F would therefore be -5.196 and the y coordinate would still be plus (+), or $+3.000$. The coordinates for all of the other points are as follows:

Point C:	$x = +3.000$	Point H:	$x = -5.196$
	$y = +5.196$		$y = -3.000$
Point D:	$x = 0.000$	Point I:	$x = -3.000$
	$y = +6.000$		$y = -5.196$
Point E:	$x = -3.000$	Point J:	$x = 0.000$
	$y = +5.196$		$y = -6.000$
Point F:	$x = -5.196$	Point K:	$x = +3.000$
	$y = +3.000$		$y = -5.196$
Point G:	$x = -6.000$	Point L:	$x = +5.196$
	$y = 0.000$		$y = -3.000$

If the entire circular hole pattern were placed in the first quadrant, as shown in Fig. 5–31, it would be necessary to perform a greater number of calculations since the part programmer must consider the coordinates of

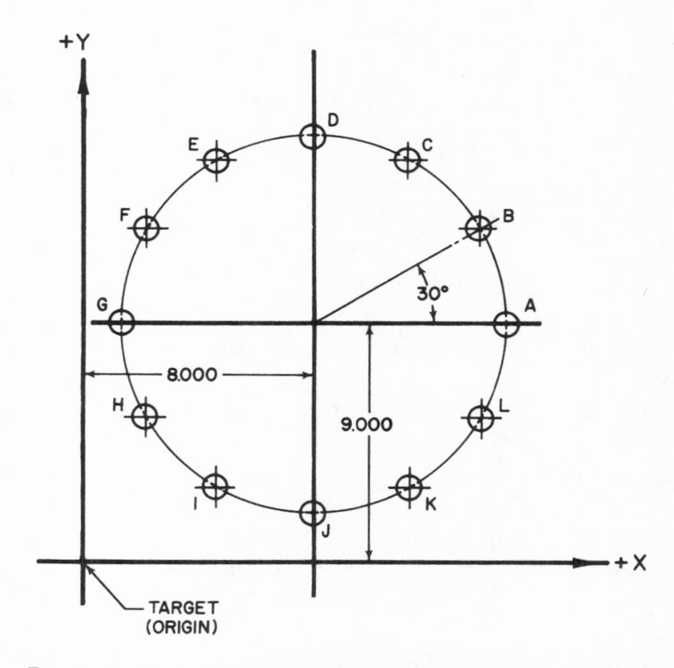

FIG. 5–31. By placing the entire circular pattern in the first quadrant, all of the coordinate values will have a plus (+) sign. The x coordinate of point B, for example, would be equal to $8 + 5.196 = 13.96$. The x coordinate of point F would be $8 - 5.196 = 2.804$.

the center of the pattern in every one of his calculations. The coordinates for the points in this case would be as follows:

Point A:	$x = +14.000$	Point G:	$x = +\ 2.000$
	$y = +\ 9.000$		$y = +\ 9.000$
Point B:	$x = +13.196$	Point H:	$x = +\ 2.804$
	$y = +12.000$		$y = +\ 6.000$
Point C:	$x = +11.000$	Point I:	$x = +\ 5.000$
	$y = +14.196$		$y = +\ 3.804$
Point D:	$x = +\ 8.000$	Point J:	$x = +\ 8.000$
	$y = +15.000$		$y = +\ 3.000$
Point E:	$x = +\ 5.000$	Point K:	$x = +11.000$
	$y = +14.196$		$y = +\ 3.804$
Point F:	$x = +\ 2.804$	Point L:	$x = +13.196$
	$y = +12.000$		$y = +\ 6.000$

Whether the target is placed at the center of the circular pattern and all four quadrants are used, or whether all calculations are performed in the first quadrant, would depend on individual choice and the configuration of the part to be machined. In the former case the plus and minus signs must be watched very carefully and, in the latter case, the calculations are greater. These calculations, however, generally involve no more than additions and subtractions. Still, care must be taken to avoid error in the calculations.

PRACTICE EXERCISE NO. 3 CHAPTER 5

1. What is the key advantage of having all points located in the first quadrant?
2. Why is it necessary to adjust the depth motions manually on a two-axis machine?
3. What is the standard word for stopping the movement of an NC machine? What is the standard word for stopping the machine *and* rewinding the tape?
4. What is the significant difference in programming a straight milling cut and in programming a movement between two points?
5. With respect to the provision for operating the machine at the selected feed rate, what difference is often found on two-axis versus three-axis machines?
6. The part shown below is to be programmed for a three-axis machine. There are two holes that require drilling and one pocket. Draw a sketch as the part would appear in the first quadrant and then prepare the program on a manuscript form as shown in Fig. 5–25. The target

point and the origin are to be set at the lower left-hand corner and .500 inch above the surface of the part. Use 150 ipm for rapid traverse, 2 ipm for drill feed rate, and 5 ipm for mill feed rate. Note: The material is soft aluminum and the pocket may be cut in one pass, and at full depth.

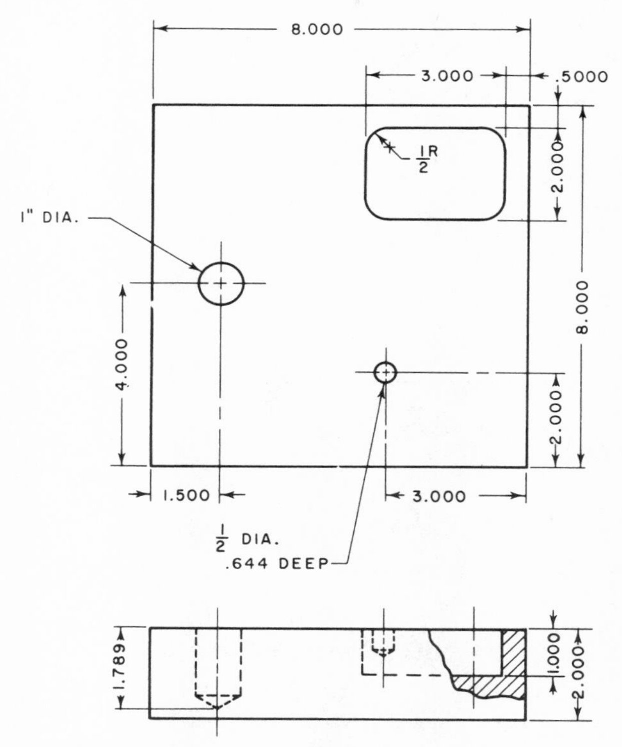

7. In the sketch shown below, point *B* is symmetrical with point *A*, about the *Y* axis. What are the coordinates of point *B*?

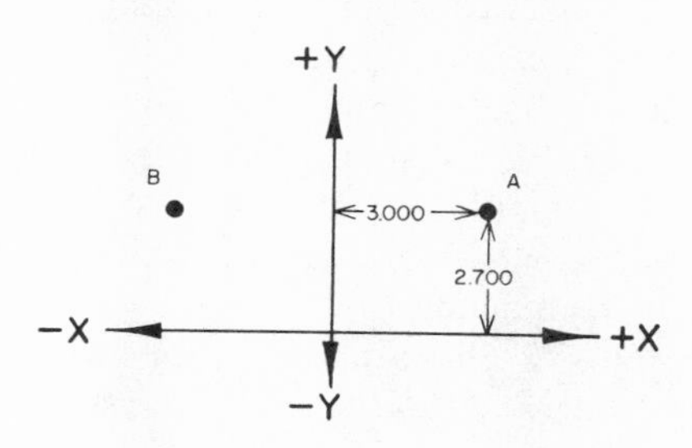

8. In the sketch of the part shown below, what would be the most logical

location for the origin, considering that the system has a free floating zero?

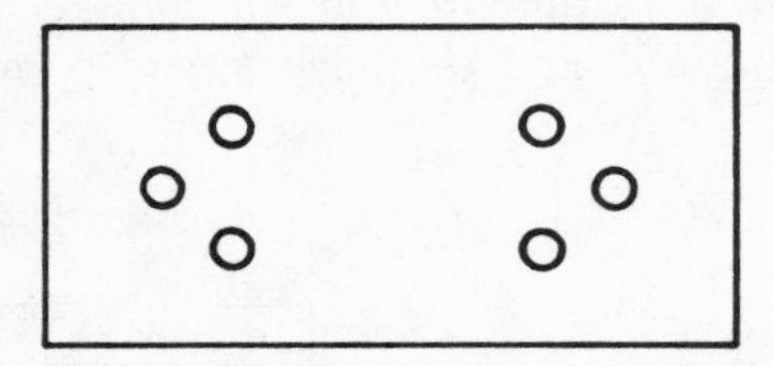

9. Referring to the sketch below, what are the x and y coordinates of points A, B, C, and D?

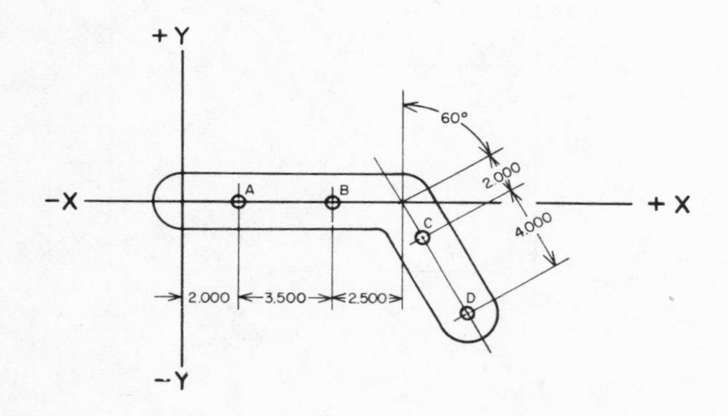

QUESTIONS CHAPTER 5

1. What are the two types of part programming?
2. What is the first word to be listed on a line across the manuscript? What is the location of this word in a block?
3. What does RWS stand for, and what is its function?
4. What are the three operations of a machining center?
5. What is the key advantage of a machining center?
6. What is the chief restriction of the straight-cut milling feature on a point-to-point machining center? Because of this restriction what designation has been applied to this type of milling?
7. Who is generally responsible for selecting the proper speeds and feed rates on a three-axis machine?
8. When programming a three-dimension machine, does the negative z value increase or decrease as the cutter moves into the workpiece?
9. When programming the path of a facemill over a surface, what must the programmer be particularly concerned about?

10. Prepare a manuscript for spot drilling and drilling the two holes shown in the part below. The machine being considered is of the two-axis type with full floating zero and tab sequential format.

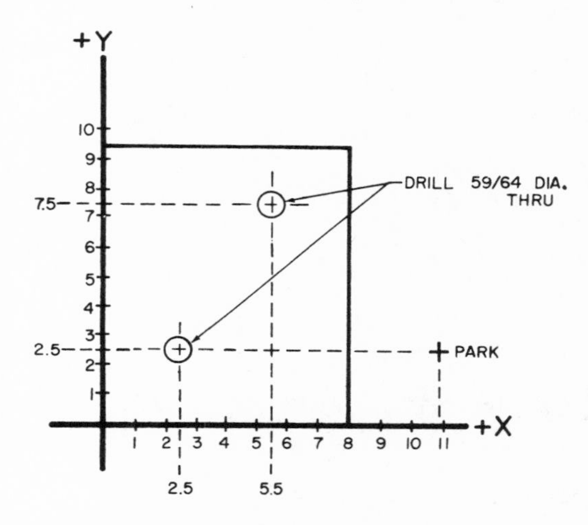

11. Locate the most convenient origin for the pattern below. Then calculate and list the coordinates for the holes *A* through *J*. The machine has a free floating zero.

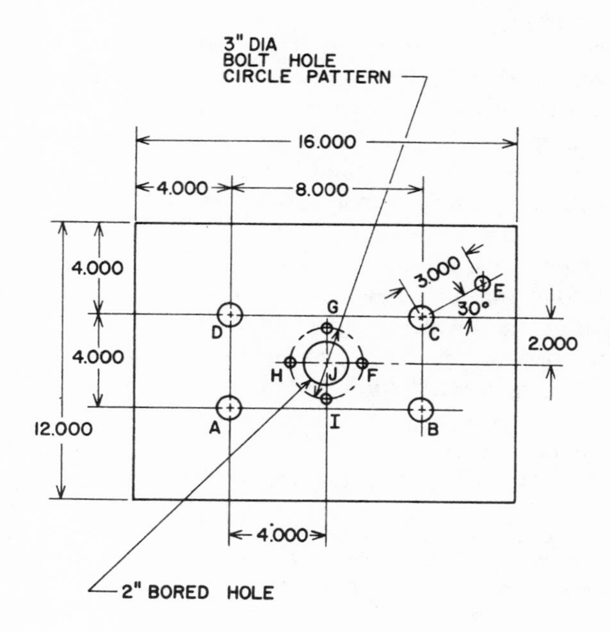

CHAPTER 6

Contour Programming—Manual Method

Point-to-Point Programming Versus Contour Programming

As described in Chapter 2, there are two types of numerical control: point-to-point and contouring. It was also pointed out in Chapter 2 that, while the point-to-point and contouring *machine tools* might look somewhat alike, the *electronic control* systems are quite different. The contouring control system is more complex and usually more expensive than the point-to-point system, although often this might not seem to be the case from appearances. At first glance the relatively low-cost *contouring* milling machine shown in Fig. 6–1 seems very similar to a point-to-point installation. On closer examination, however, one will find that the dials, knobs, and switches are different and that the machine tool itself is more heavily constructed. More important, the electronic components and wiring inside the control cabinet are much more elaborate. It should be noted that a contouring machine can usually perform all of the functions of a point-to-point machine, whereas a point-to-point machine is limited to straight-line motions which are parallel to or at 45 degrees to the axes.[1]

As far as the part programmer is concerned, the major difference is that contour machining generally requires more calculations to prepare a manuscript since more complex parts are involved. Most often the mathematical requirements are also of a higher order, and the calculation problems more difficult. With point-to-point machining, it is necessary only

[1] While point-to-point machines have been programmed to perform contouring operations by describing closely spaced coordinate points, this method is usually not too practical because of excessive calculation, long tapes, and the reduced feed rates resulting from relatively slow reader speeds and no buffer storage in the control system.

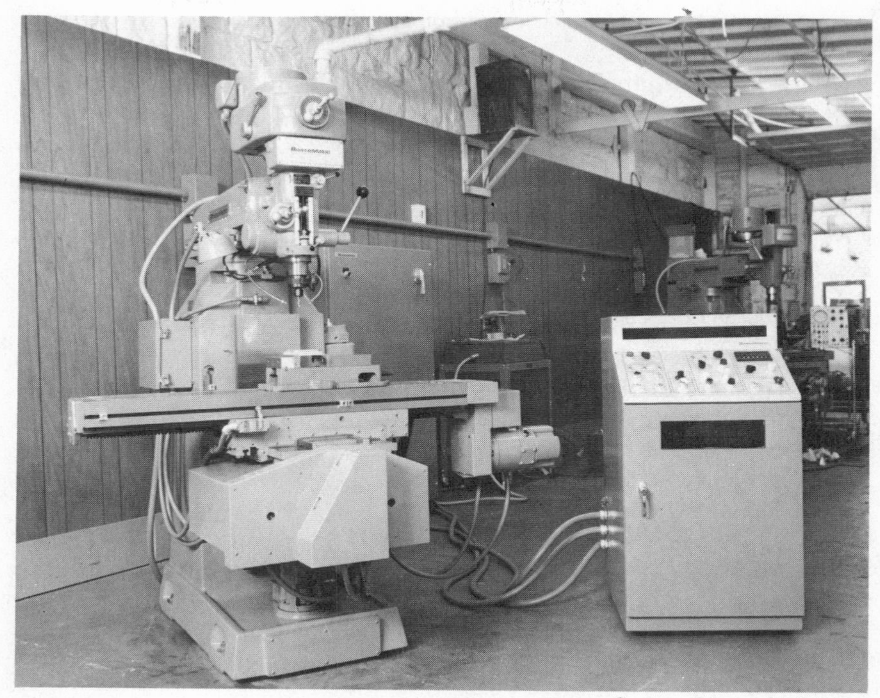

Courtesy of Boston Digital Corp.

Fig. 6–1. This relatively low-cost contour milling machine is not too unlike a point-to-point NC installation in *appearance*. Electronic components, located inside the control cabinet, account for the main difference. An NC milling machine is also usually more heavily constructed than a point-to-point drilling machine or machine center.

that the point of operation, such as the center location of a hole, be noted on the manuscript. Because the cutter is traveling in air when moving from one point to the next, the part programmer need not be concerned with its path as long as the cutting tool does not hit a portion of the part or an obstruction, such as a clamp. In the contouring system the cutter moves through metal, and thus the accuracy of its path will determine the accuracy of the part. Even with non-metal cutting NC contouring applications, such as drafting and plotting machines, the path of the moving elements must be accurately controlled.

Chapter 5 describes the milling capabilities of most point-to-point machines, where the cutter is controlled along straight-line paths that are parallel to the axes. Also, there are NC point-to-point machines that can machine accurate 45-degree-angle cuts. These machines are, therefore, normally limited to rectangular or diamond-shaped cuts, and while they can machine other than 45-degree angles and even arcs if a series of closely spaced points are calculated, this is often impractical, especially without a computer.

Many parts require cuts other than those parallel and at 45-degree angles to the axes, such as the part found in Fig. 6–2. This part is one of the major wing components of the Air Force fighter bomber shown in Fig. 6–3. Contour machining is widely used in the aircraft industry, where many of the parts must conform to the contoured surface of an airplane. Two such contoured parts, belonging to the airplane shown in Fig. 6–3, are shown in Fig. 6–4.

In addition to NC contour milling machines, most NC lathes are equipped with contour control systems to guide the single-point cutting tool as it feeds across the workpiece. The key difference is that a lathe cutting tool is confined to two axes of motion since the workpiece is rotating, whereas the milling cutter may be programmed to move in three axes, or more, at the same time. An example of a contoured lathe part is shown in Fig. 6–5.

Courtesy of Republic Aviation Div. of Fairchild Hiller

Fig. 6–2. Wing structure for an Air Force fighter bomber. The part is approximately 6 feet long and requires a series of machining cuts running at different angles with the X and Y axes. The part is shown positioned on the vertical worktable of a large NC contour milling machine.

Courtesy of Republic Aviation Div. of Fairchild Hiller

FIG. 6–3. This Air Force bomber contains many contoured machined parts.

Approximating a Curve with Straight Lines

As already stated, the main difference between a point-to-point and a contouring machine is that the contouring machine can move at varying angles to the axes, but the point-to-point machine tool must move in lines parallel and at 45 degrees to the axes. The question might rightly be asked then as to how *contour curves* can be produced with solely *straight-line* motions. The answer is that the straight lines are calculated to be of such a length that they *fit*, or *approximate* the curved line to within the tolerances required. This is called *linear interpolation,* an example of which is shown in Fig. 6–6. The length of each straight line is calculated so that it approximates the curved line to within the allowable tolerance. However, this is a *calculation* tolerance, which must be *less* than the allowable tolerance for the finished part since there will be a normal addition error due to the machining of the part. It can be seen in Fig. 6–6 that

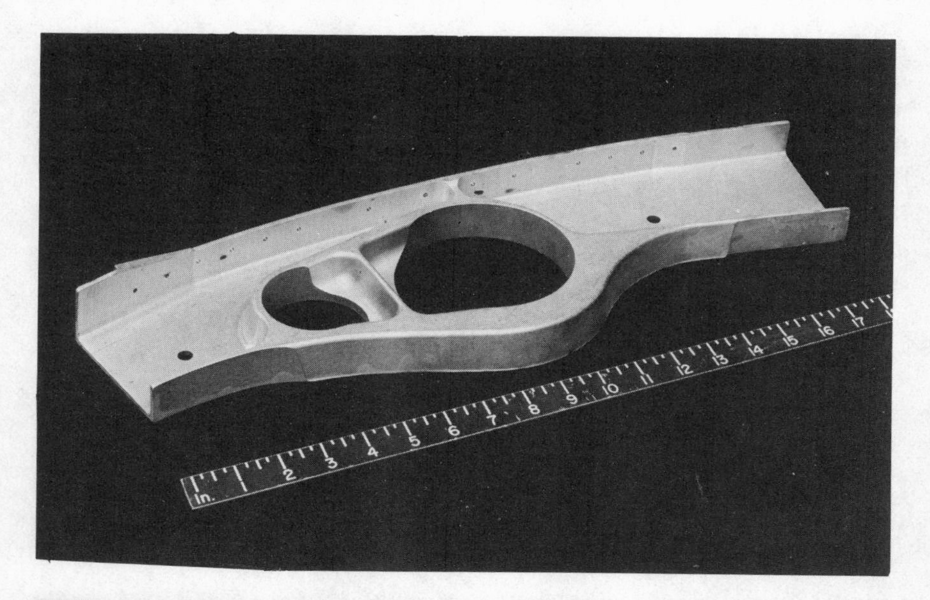

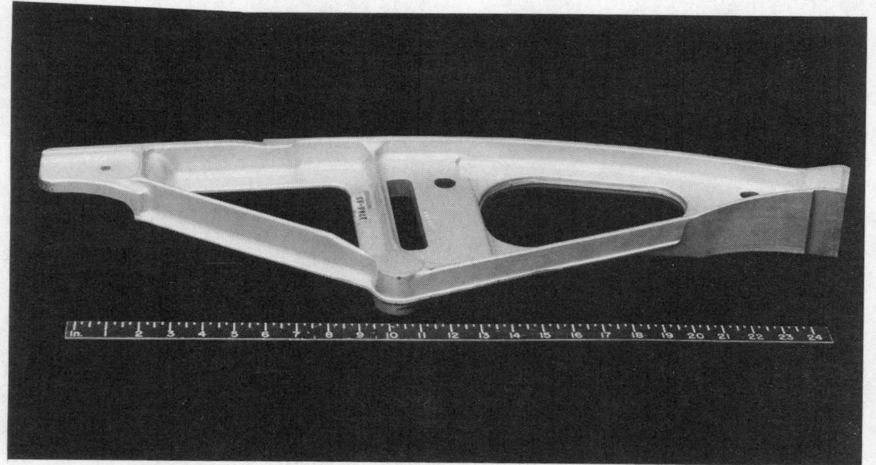

Courtesy of Republic Aviation Div. of Fairchild Hiller

FIG. 6–4. These contoured parts help to make up the airplane shown in Fig. 6–3.

the straight lines approximating the curve also pass *through* the curve, and that the allowable tolerance is applied to *both* the inside and outside of the curve. Curves may also be approximated by straight lines lying either inside or outside of the curve, as shown in Fig. 6–7.

As the calculations for determining the length of the straight lines could become quite complex, a computer is generally used for programming the

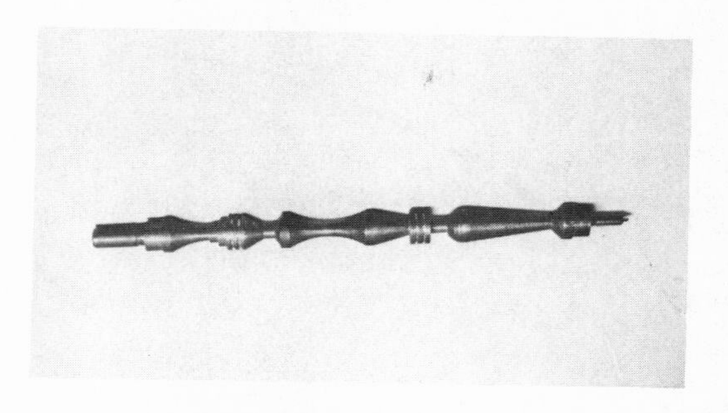

Courtesy of Teledyne Taber

FIG. 6–5. This part was machined on an NC contour lathe. Most NC lathes are of the contouring type.

cutter movements along a path of this type. Another means is to plot the curve graphically, at from 10 to 40 times the normal size, and then to draw the straight lines to fit the curve so that they stay within the allowable tolerance limits. By using a 40 to 1 scale on a grid paper having ten divisions per inch, for example, it is possible to approximate the curve to within $\frac{1}{25}$th of an inch, which is equal to a calculated accuracy of .001 inch with respect to the part. The whole part need not be plotted; only those areas where coordinate information is required. The graph method, however, is generally far more time consuming than the computer method and

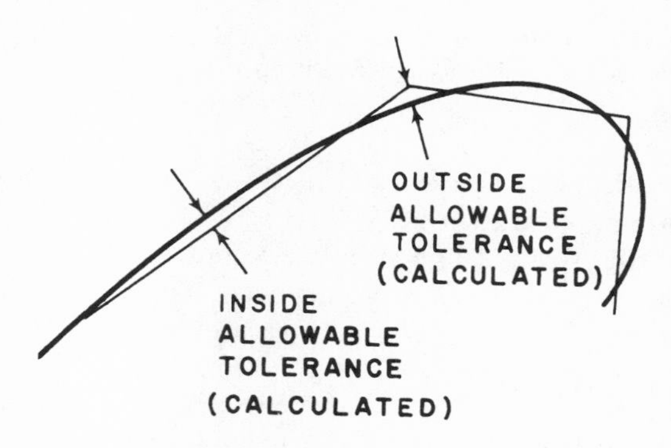

FIG. 6–6. The lengths of the straight lines shown have been calculated so that they fit the curve and do not exceed the allowable tolerance. The lengths of the lines have been exaggerated to give a clearer illustration.

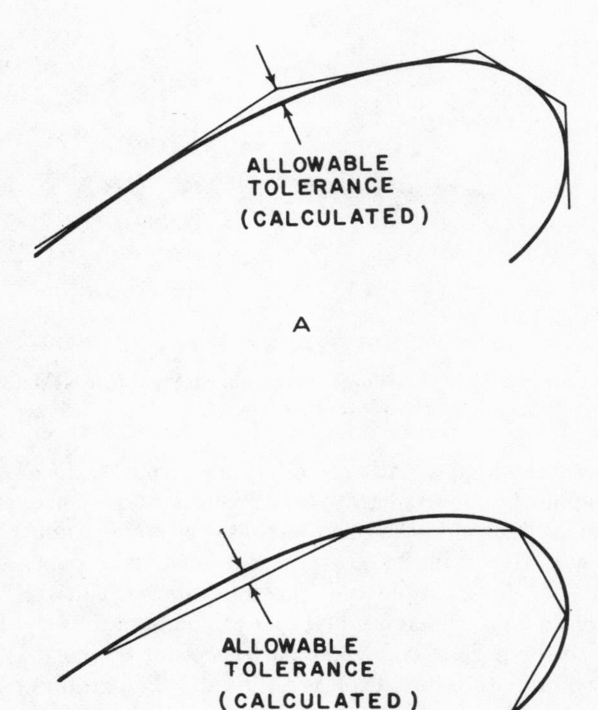

FIG. 6–7. (A) Approximating a curve using straight lines on the *outside* of the curve. (B) Approximating a curve using straight lines on the *inside* of the curve.

would be impractical for large curved sections. The computer method is covered in Chapters 7 through 10.

While calculating straight-line approximations for complex curves without the aid of a computer is laborious, and sometimes just plain impractical, approximations for *circular arcs* are considerably more useful, although the calculations may still be time-consuming. Figure 6–8 shows two straight lines, *AB* and *BC*, that approximate a circular arc. The lines are of the same length, as all the lines approximating this circular arc would be. The outside tolerance limit is .005 inch and the inside tolerance limit is also .005 inch. Any point on the straight lines, therefore, may not exceed the total tolerance limit of .010 inch. Figure 6–8 also shows two arcs which are drawn parallel, i.e., concentric with the given arc and at a distance of .005 inch on either side of it. These two arcs represent the limits of the tolerance zone on or within which any lines approximating the given arc must lie. With the use of some fairly simple trigonometry it is possible to calculate the length of the approximating straight line. If, for example, the radius of the circular arc is 10.000 inches, a line extending from the center of the circle to the outside tolerance arc would be equal to the radius of the circle plus .005 inch; a line extending from the center of the circle to the inside tolerance arc would be equal to the radius *minus* .005 inch. The first step in calculating the length of the line BC which approximates the arc is to calculate one-half of the line by means of plane trigonometry. With the adjacent side and the hypotenuse of the right triangle shown in Fig. 6–8 known, the included angle can be

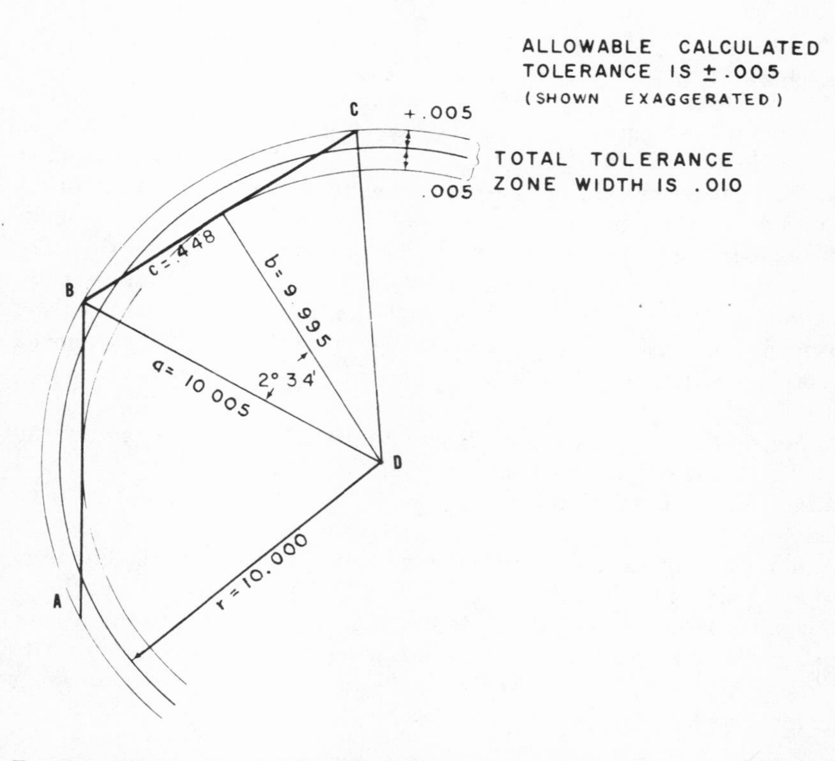

FIG. 6–8. The length of the straight line that approximates the circular arc may be calculated using simple trigonometry. The length of the approximation lines will be the same around the entire circle. For the sake of illustration the lines and angles have not been shown as being proportionate.

calculated. If θ is the included angle, then:

$$\cos\theta = \frac{9.995}{10.005} = .99900$$

$$\theta = 2°34'$$

$$\begin{aligned} c = \text{length of half the line} &= 10.005 \times \sin\theta \\ &= 10.005 \times \sin 2°34' \\ &= 10.005 \times .04478 \\ &= .448 \text{ inch} \end{aligned}$$

The full length of the line approximating the circular arc is therefore $2 \times .448$ or 0.896 inch. It should be pointed out that the lengths of the lines and the angles in Fig. 6–8 have been exaggerated for clarity. It will also be noted that *all* of the lines that approximate the circular arc are of the same length. This is not true of the lines approximating the irregular curve shown in Figs. 6–6 and 6–7.

After the length of the straight line has been calculated, the next step is to calculate the lengths of the x and y *incremental* distances. Incremental dimensioning, discussed in Chapter 3 (see Fig. 3–10), is the most common way of describing movements in contouring control systems. This differs from the absolute dimensioning approach, which is the most common with point-to-point control systems. The programmer, rather than describing the coordinates of the end points of the straight-line movements, as is done in an absolute dimensioning system, describes the *length* of the movement that is parallel to the X axis and then the *length* of the movement that is parallel to the Y axis. These are called *delta* movements, designated by the Greek letter Δ.

In Fig. 6–9, which shows the calculated line BC of Fig. 6–8, Δx and Δy form the two legs of a triangle and are parallel, respectively, to the X and Y axes. The lengths of Δx and Δy would be the x and y words to be noted on the manuscript if the cutter were to move from point B to point C, or along the hypotenuse of the triangle. In this instance, the line from A to B is parallel to the Y axis and the line DE, which passes through the center of the circular arc and bisects the line AB, is parallel to the X axis.

Before the delta movements can be calculated, it will be necessary to determine all of the angles. This can be done by recalling two axioms from geometry.

1. Like angles of similar and congruent triangles are equal.
   Thus the angle between BD and DE is equal to the angle between DF and BD since triangles BDE and BDF are congruent triangles.
2. The alternate interior angles formed by a line intersecting two parallel lines are equal.

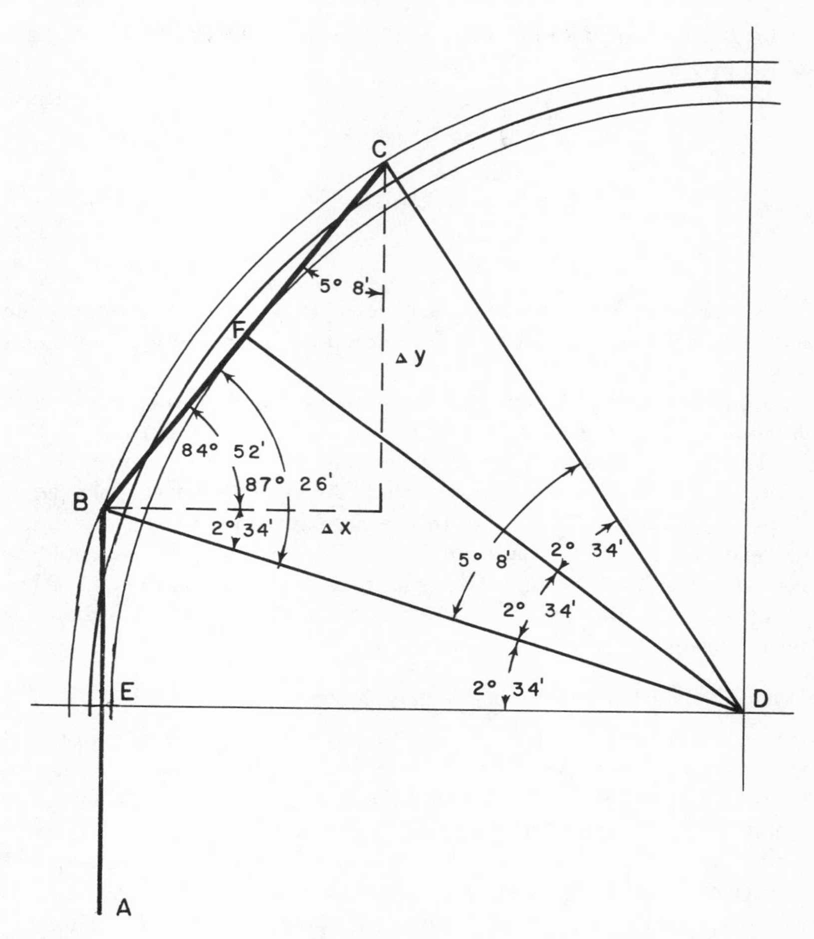

FIG. 6–9. The Δx and Δy incremental movements must be described on the manuscript as x and y words. In order to describe the principles of the Δx and Δy movements the sides are not shown as being proportionate; nor are the angles.

Thus the angle formed by BD and ΔX, Fig. 6–9, must be equal to 2°34′; because the angle formed by BD and DE is 2°34′.

3. The sum of the two acute, or smaller, angles in a right triangle must equal 90 degrees.
   Thus the angle formed by BF and BD in triangle BFD must be $90° - 2°34' = 87°26'$; because $87°26' + 2°34' = 90°$.

Knowing the value of these two angles (2°34′ and 87°26′), the angle between BC and ΔX can be found by the following subtraction: $87°26' - 2°34' = 84°52'$. It is now possible to calculate ΔX and ΔY (Fig. 6–9) since the length of BC is known (.872 inch) and one angle in the right triangle (84°52′) is also known.

$$\begin{aligned}\Delta x &= .872 \times \cos 84°52' \\ &= .872 \times .08947 \\ &= .078 \\ \Delta y &= .872 \times \sin 84°52' \\ &= .872 \times .99599 \\ &= .868\end{aligned}$$

Next, the Δx and Δy movements are calculated for each straight line segment, and then these movements are noted on the manuscript as x and y words.

As another example, consider the incremental movements required to travel from A to B, and then from B to C, and then from C to D, in Fig. 6–10. The manuscript data for these movements are also shown in Fig. 6–10. It should be especially noted that the Δy movement between C and D is minus $(-)$. This is due to the negative direction of travel and has no relationship to the quadrant that the movement is in, as would be the case with absolute coordinate dimensioning (refer to Chapter 3). Also, note that the tape format is *word address*, the most popular format for contouring systems. (Refer to Chapter 4.)

Simplifying Circular Motions—Circular Interpolation

Because of the demand for the machining of circular cuts, many contouring control systems have a feature called *circular interpolation* in addition to linear interpolation. In contrast to a system having only linear interpolation, a system with this feature can be programmed to move about a circular arc with one block of data. This eliminates the requirement for calculating the straight-line segments, as described earlier. Not all contouring control systems have circular interpolation, since it is generally furnished by the machine tool builder as an option, and at additional cost.

Because any control system with circular interpolation will also have linear interpolation, it is necessary to note, on the manuscript, whether the block on the tape is to describe a *linear* or *circular* motion. This is accomplished by using a preparatory, or *g* word. (Refer to Chapter 4.) With a system having both linear and circular interpolation, the linear interpolation block is described by a *g01* word and a circular interpolation block by either a *g02* or *g03* word. The *g02* word is used for a clockwise motion of the cutter, and the *g03* word for a counterclockwise motion of the cutter. The proper *g* word need only be noted in the first block, as the following blocks will conform to the same *g* word until it is changed by another *g* word.

As an example, consider the motion shown in Fig. 6–11. The cutter is

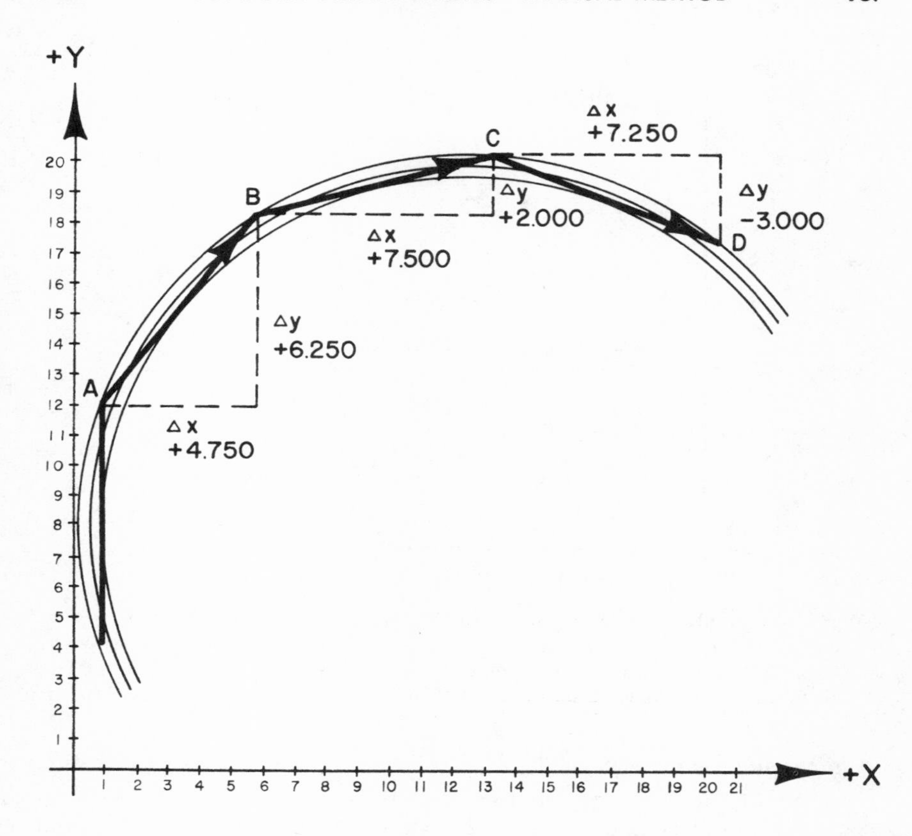

PART NO. 12354 PART NAME..........................			MANUSCRIPT XXXXX MACHINE			DATE : PREPARED BY : CHECKED BY :
SEQUENCE NO.	x WORD	y WORD	f WORD	m WORD	EOB	COMMENTS
RWS						
001	X+4.750	y+6.250	f50		EOB	MOVEMENT FROM A TO B
002	X+7.500	y+2.000			EOB	MOVEMENT FROM B TO C
003	X+7.250	y-3.000			EOB	MOVEMENT FROM C TO D
↓	↓	↓	↓	↓	↓	

Fig. 6–10. (Upper) The incremental movements from A to B, B to C, and C to D are shown. (Lower) The instructions on the tape describing the Δx and Δy components of these movements are shown on the manuscript.

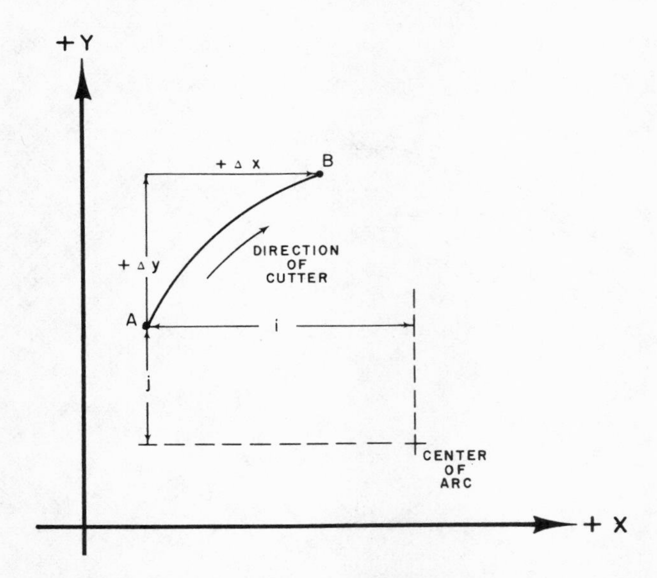

FIG. 6–11. A single block instruction for moving around an arc is possible with systems having a *circular interpolation* feature. The words that have to be described involve the Δx and Δy incremental distances and the i and j distances, which are equal to the horizontal and vertical distances from the center of the arc to the point at the start of the arc.

to move along an arc from point A to point B in a clockwise direction. The words required to be listed on the manuscript, in addition to the n (sequence number) and m (auxiliary function) words, normally found, would be:

g02, which notes circular interpolation and a clockwise motion of the cutter.

x, which describes the x incremental distance (Δx). The direction sign must also be noted.

y, which describes the y incremental distance (Δy) and the direction.

i, which denotes the distance from the center of the arc to point A, in the X direction.

j, which denotes the distance from the center of the arc to point A, in the Y direction.

f, which is the feed-rate word. It may be expressed directly in inches per minute or in accordance with the formula:

$$f = \frac{ipm}{r}$$

where ipm stands for inches per minute and r equals the radius of the arc. The means of expressing the *f* word will depend on the particular control system.

It will be noted that, except for the calculation of the *f* word, the radius of the arc is not used.

An important consideration in circular interpolation programming is that a separate block is required for each section of the arc lying within each of four right-angle (90-degree) segments oriented with the *X* and *Y* axes. In Fig. 6–12 three blocks would be required: one for the arc from *A* to *B*; a second for the arc from *B* to *C*; and a third for the arc from *C* to *D*. Each of these sections would be treated as separate arcs and would follow the programming rules described with Fig. 6–11.

Referring again to Fig. 6–12, the movement from *A* to *B* would require that the values $+\Delta x_1$ and $+\Delta y_1$ be listed in the x and y columns of the manuscript and the i and j values be listed in their appropriate i and j columns. In moving from *B* to *C*, $+\Delta x_2$ and $-\Delta y_2$ would be noted; as would be the j value. The i value in this case would be equal to zero and therefore need not be listed. In moving from *C* to *D* the $-\Delta x$, and the $-\Delta y$, values would be listed in addition to the i value. In this instance the j value would be equal to zero.

Both linear and circular interpolation are described in Fig. 6–13. In this example a cutter is to move from point *A* to point *B* along a straight line that is parallel to the *X* axis. Then the cutter will move around an arc to point *C*; and then along the straight line from *C* to *D*, which is

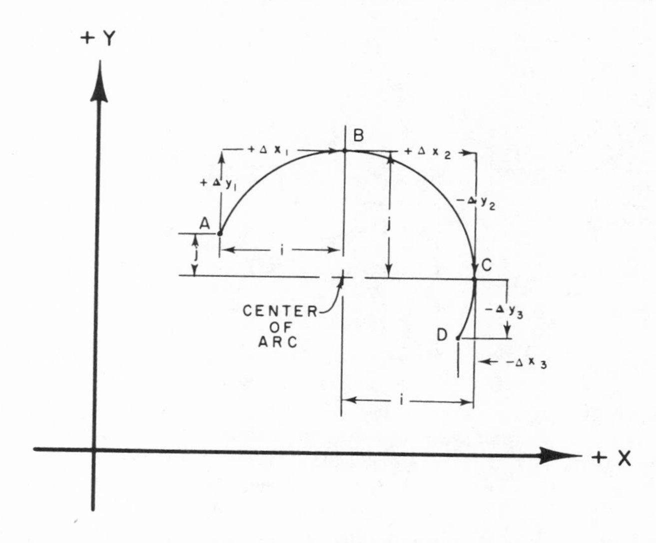

FIG. 6–12. Three blocks would be required to move from point *A* to point *D*, when utilizing the circular interpolation feature.

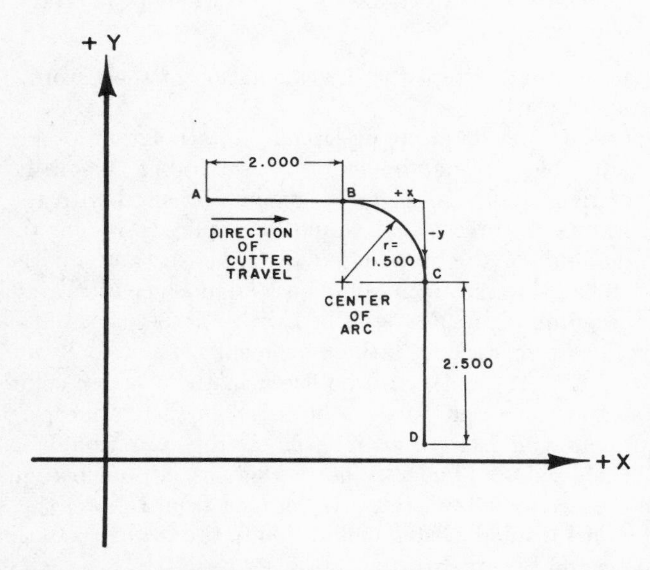

FIG. 6–13. Three blocks are required to move from point A to B, and then to point C and on to point D. The movement involves two linear blocks and one circular interpolation block.

MANUSCRIPT

PART NO:
PART NAME:

CXXXXX MACHINE

DATE:
PREPARED BY:
CHECKED BY:

SEQUENCE NO.	g WORD	x WORD	y WORD	i WORD	j WORD	f WORD	m WORD	EOB	COMMENTS
↓	↓	↓	↓	↓	↓	↓	↓	↓	↓
M016		X+2.000						EOB	STRAIGHT LINE MOVEMENT FROM POINT A TO POINT B
M017	g02	X+1.500	y-1.500		j1.500			EOB	CIRCULAR MOVEMENT FROM POINT B TO POINT C
M018	g01		y-2.500					EOB	STRAIGHT LINE MOVEMENT FROM POINT C TO POINT D
↓	↓	↓	↓	↓	↓	↓	↓	↓	

FIG. 6–14. A manuscript for a circular interpolation system is more complex, and has more columns than one for a system having only linear interpolation. The arrows indicate that the three lines (M016, M017 and M018) are *part* of a program and that there are blocks that precede and follow those that are shown. It should be pointed out that an x and/or y word is shown only when there is motion in the particular axis. This concept is similar to the point-to-point absolute dimensioning system in which no coordinate need be shown if it does not change.

parallel to the Y axis. The manuscript in Fig. 6–14 describes the program for these motions. On the first line the arrows mean that these three moves are a *part* of a continuing program and that there were other instructions prior to these. Sequence number *n016* consists of an incremental movement only, and therefore only an x word is shown. Sequence number *n017* changes the mode from linear interpolation to circular interpolation via the preparatory word *g02*. The Δx word, which is $x+1500$, denotes the incremental distance that the cutter is to move in the plus (+) X direction, and the Δy word denotes the incremental distance that the cutter is to move in the minus (−) Y direction. Also, since point B, which is the start of the arc, lies directly over the center of the arc, there is no i distance in the Y direction. There is, however, a j distance in the Y direction; and, in this case, since point B lies directly over the center of the arc, the offset is equal to the radius of the arc, or 1.500 inches, and this is noted as a j word. On the next line the *g01* word changes the mode back to linear interpolation, and the y word describes the incremental movement in that direction, which is −2.500 inches. There is no incremental x motion from C to D.

PRACTICE EXERCISE NO. 1 CHAPTER 6

1. Why are the programming calculations generally more difficult and time-consuming for a contouring system than for a point-to-point system?
2. What is meant by *linear interpolation*?
3. What two factors govern the length of the linear interpolation lines?
4. In the sketch below, calculate the length of the line AB that approximates the circular arc. (The dimensions are not shown to scale in order to provide a clearer illustration.) Note that, in this instance, the outside tolerance and inside tolerance differ.

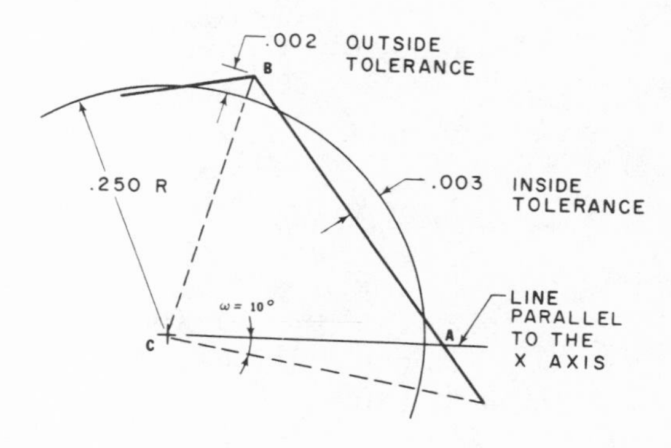

5. From the sketch given in Question 4, calculate the lengths of Δx and Δy.
6. What preparatory code would be required on the tape to change from a linear interpolation mode to a circular interpolation mode with the motion of the path in a clockwise direction?
7. Referring to the following sketch, how many circular interpolation blocks would be needed to move a cutter from point A to point B?

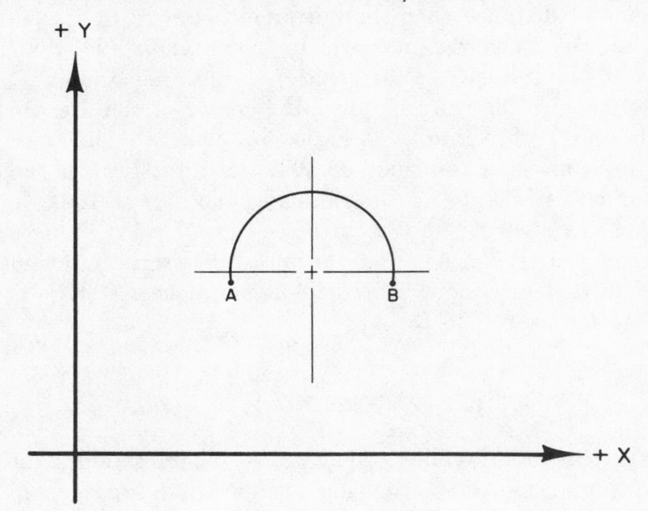

8. From the information given below, prepare a manuscript for moving a cutter from point A, through points B, C, D to E. Use the manuscript form shown in Fig. 6–14. The coordinates of the points are:

Point A: $x = 1.000$
$y = 4.000$
Point B: $x = 4.000$
$y = 4.000$
Point C: $x = 8.000$
$y = 7.200$
Point D: $x = 9.000$
$y = 7.500$
Point E: $x = 10.300$
$y = 6.200$
Point F: $x = 10.300$
$y = -2.000$
Point G: $x = 9.000$
(Center of arc)
$y = 6.200$

Cutter Offset

As has been noted earlier, the Δx and Δy movements, described in the manuscript as x and y words, refer to the movements of the center of the cutter. And this center-line motion may normally be obtained without too much calculation from the blueprint dimensions when the cutter moves in air, or performs such operations as cutting grooves, or face milling. There are many occasions, however, when it is required to cut around the perimeter of a part. Since the blueprint, or engineering drawing, describes the dimensions of the *part*, it is the part programmer's responsibility to calculate the incremental movements that make up the continuous path of the *center of the cutter* in order to obtain the required part dimensions. The distance of this center-line path to the perimeter of the part at any given point is equal to the radius of the cutter, as shown in Fig. 6–15. On rectangular type parts, where the sides are parallel to the X and Y axes, the calculations are relatively simple. In Fig. 6–16, for example, the x coordinate of point B would be less than the x coordinate of point A by an amount equal to the radius of the cutter. The same consideration would apply to the y coordinate of point B, which would be *greater* than the y coordinate of point A by an amount equal to the radius of the cutter. Moving about the circular arc would require little additional calculation, *providing* the system has circular interpolation.

On the other hand consider the pattern shown in Fig. 6–17 in which the lines are not parallel to the axes. In order to find the Δx and Δy movements of the center of the cutter it is necessary to find the coordinates of G, D, and H. From the blueprint information we are given:

Coordinates of A, B, and C
The angles α and β
The radius of the cutter

To find the x and y coordinates of G we first draw triangle GJA with

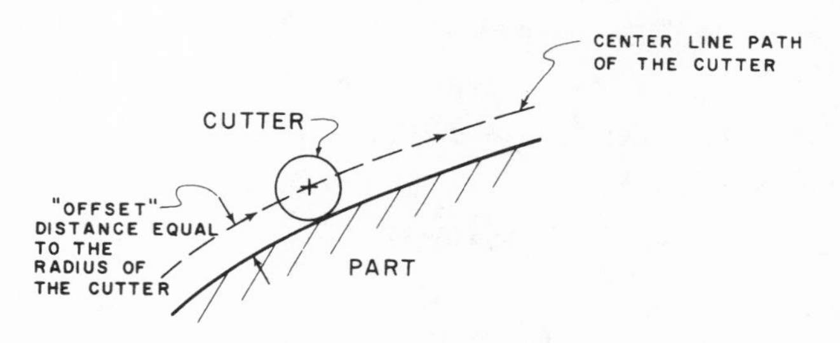

FIG. 6–15. The incremental movements of the center-line path of the cutter are listed on the manuscript.

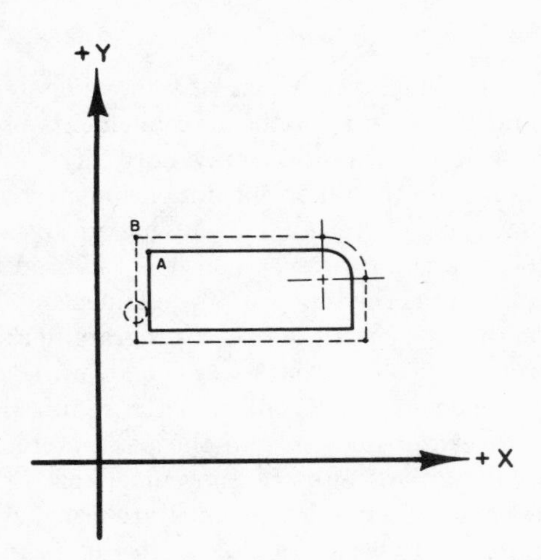

Fig. 6–16. Calculating the cente: path of a perimeter cut is relatively easy when the part is of a rectangular shape and the sides are parallel to X and Y axes.

side JA parallel to the X-axis and side JG parallel to the y-axis. Note that angle JGA is equal to angle α since its respective sides are at right angles to the sides of that angle.

Hence x coordinate of G = x coordinate of $A - JA$
or x coordinate of G = x coordinate of $A - GA \sin \alpha$
and y coordinate of G = y coordinate of $A + JG$
or y coordinate of G = y coordinate of $A + GA \cos \alpha$
since $GA = r$, the radius of the cutter
then x coordinate of G = x coordinate of $A - r \sin \alpha$
and y coordinate of G = y coordinate of $A + r \cos \alpha$

In order to find the vertical (Y) distance from B to D and therefore the y coordinate of D it is necessary to find the distance from K to D, which is the same. The steps are as follows:

1) Construct line EBF parallel to the X axis.
2) Calculate the length of line EBF, as follows:

$$EB = \frac{r}{\sin \alpha}$$

$$BF = \frac{r}{\sin \beta}$$

$$EBF = EB + BF$$

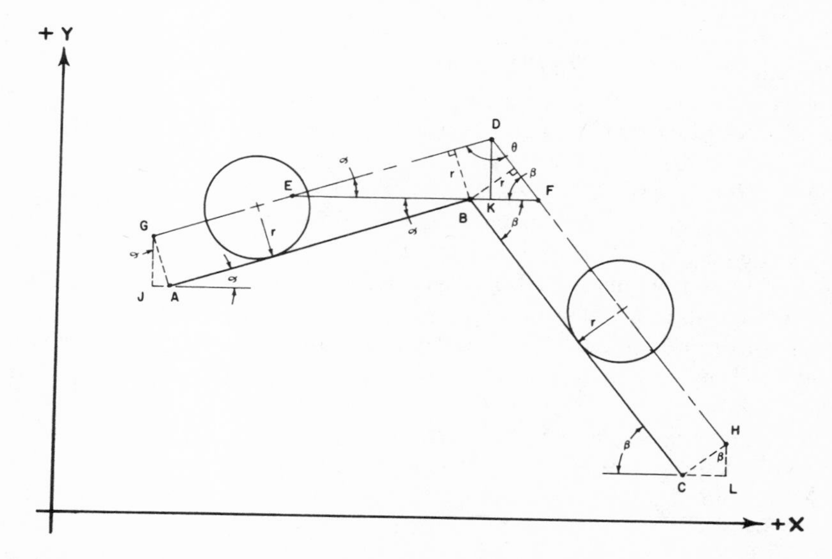

FIG. 6–17. Calculating the center-line path of a cutter can be time-consuming and complex, especially when movements are not parallel to the X and Y axes, as shown.

3) Knowing the distance EBF, calculate ED, as follows:

$$\frac{ED}{\sin \beta} = \frac{EBF}{\sin(180° - \alpha - \beta)}$$

(*Note:* This is the law of sines that states that one side of a triangle divided by the sine of the opposite angle is equal to another side divided by the sine of its opposite angle.)

$$ED = \frac{EBF \times \sin \beta}{\sin(180 - \alpha - \beta)}$$

4) Knowing ED, we can now calculate DK, as follows:

$$DK = ED \times \sin \alpha$$

The distance DK may now be added to the y coordinate of B in order to get the y coordinate of D.

The x coordinate of D may be found as follows:

1) x coordinate of D = x coordinate of $B + BK$

2) $BK = EK - EB = ED \times \cos \alpha - \dfrac{r}{\sin \alpha}$

3) x coordinate of D = x coordinate of $B + ED \times \cos \alpha - \dfrac{r}{\sin \alpha}$

To find the x and y coordinates of H we construct triangle CHL with CL parallel to the X-axis and LH parallel to the Y-axis.

Then x coordinate of H = x coordinate of $C + CL$
and y coordinate of H = y coordinate of $C + LH$
but $CL = CH \sin\beta = r \sin\beta$
and $LH = CH \cos\beta = r \cos\beta$
hence x coordinate of H = x coordinate of $C + r \sin\beta$
and y coordinate of H = y coordinate of $C + r \cos\beta$.

Note that the calculations described for Fig. 6–17 involve only two axes (X and Y). If the cutter were programmed to perform three-, four- or even five-axis motions at the same time, the calculations would become considerably more complex. A three-axis motion would be a combination of movements in the X, Y, and Z axes. A four-axis motion would be one in which movements in the X, Y, and Z axes plus a rotary motion were carried on at the same time. And a five-axis motion would be one in which the X, Y, and Z motions were combined with *two* rotary motions—all to

Courtesy of Sundstrand Machine Tool

Fig. 6–18. This milling machine can move in five axes at the same time. The arrows show the direction of the movements. The "drum"-type device shown at the left stores cutting tools that are changed automatically from commands on the tape.

Courtesy of U.S. Naval Ship Research and Development Center

FIG. 6–19. The machining of the propeller shown on the rotary table requires multi-axis movements.

be performed at the same time. The rotary motions would normally be at right angles to one another. Figure 6–18 shows a machine having five-axis motion capability. Although the requirement for moving all five axes at the same time is rare, design engineers are beginning to take advantage more often of the multiaxis capability of numerical control.

For example, more parts like the rocket engine impeller shown in Fig. 6–19 are being designed. In view of the calculations required for contour programming, it is not surprising that the computer has become a very handy assistant to the part programmer. Computer-assisted part programming for contouring operations is covered in Chapter 8.

Programming Contour Lathes

The most popular contouring machine is the NC lathe, of which there are several types. Figure 6–20 illustrates a NC horizontal *turret* lathe. This machine is particularly well suited for machining relatively short "chunky" parts, such as castings or forgings. Longer, shaft-type work is normally machined on *engine* lathes, which are also capable of handling shorter parts. Figure 6–21 shows an example of an NC engine lathe. (Another example of an NC engine lathe is illustrated in Chapter 1, Fig. 1–5. Since the engine lathe illustrated in Fig. 6–21 is the most common type, the programming procedures described here will be centered around this type of lathe. The programming concepts for an NC turret lathe are very similar to those of an NC engine lathe; the key difference being in the type of parts handled.

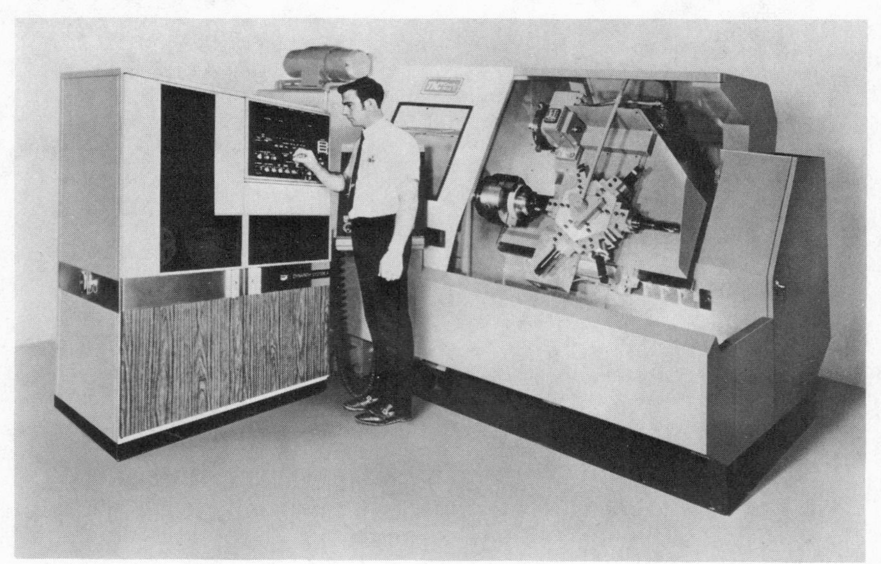

Courtesy of Jones and Lamson

FIG. 6–20. This turret lathe is usually reserved for machining relatively short parts. In addition to the large hexagonal turret, which is well suited for internal boring work, the machine shown is also equipped with a second, and smaller, rotary turret that is used for outside shaft-type work.

Courtesy of South Bend Lathe

FIG. 6–21. This is a popular variety of a medium size engine lathe. The arrows illustrate the two axes of motion.

Referring again to Fig. 6–21, note that the motion of the carriage is designated as a Z axis and that the motion of the turret, which rides on the carriage and moves perpendicular to it, is designated as the X axis. This conforms with the *right-hand rule* already explained in Chapter 2. (See Fig. 2–8.) While most NC engine lathes are equipped with turrets, these are generally of a smaller type and hold fewer cutting tools than those of a turret lathe. NC lathes also contain several tape-controlled features usually not found on milling machines or on the lower cost machining

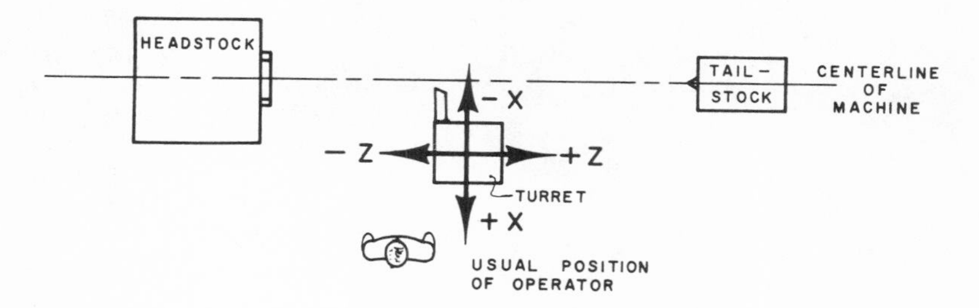

FIG. 6–22. Most lathes operate on an incremental two-axis basis. In accordance with standard practice, the two axes used are X and Z.

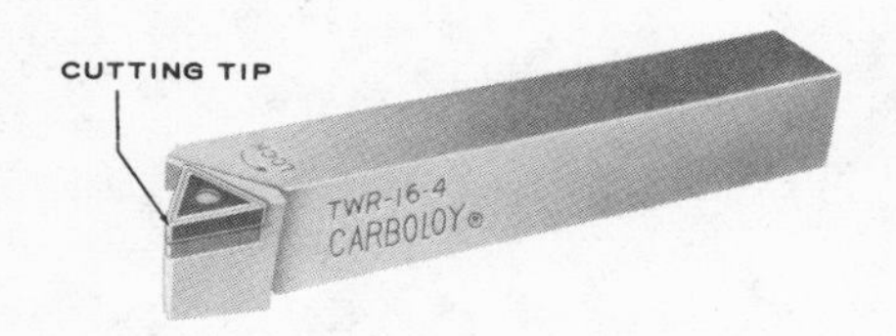

Courtesy of Metal Products Dept., General Electric Co.

FIG. 6–23. Single-point cutting tool and holder.

centers. One is that the spindle *speeds* of the lathes are normally controlled by an s word, which is used only for this function. Another feature is a *t* word, which is used exclusively to command the indexing turret, thereby selecting the cutting tool to be used. A preparatory word, or *g* word (normally *g04*), may be used to initiate the dwell cycle. This is a safety measure that stops axis motions while the turret is being auto-

Courtesy of Lodge and Shipley

FIG. 6–24. Close-up of the turret arrangement on an NC engine lathe. The calculations required for programming the path of the single-point cutting tool are very similar to those for programming the center path of a milling cutter.

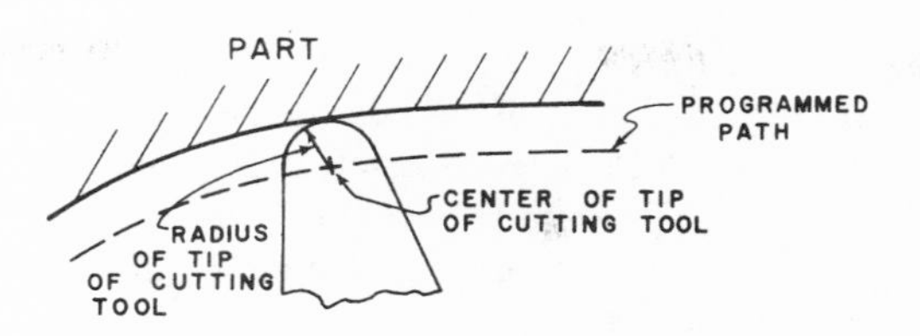

FIG. 6–25. The part programmer must calculate the path of the center of the tool tip.

matically indexed or while spindle speeds are being automatically changed, via the *s* and *t* words. Also of interest is that practically all NC lathes now being produced are equipped with *contouring* systems; in addition, most of these systems include the circular interpolation feature mentioned earlier in this chapter.

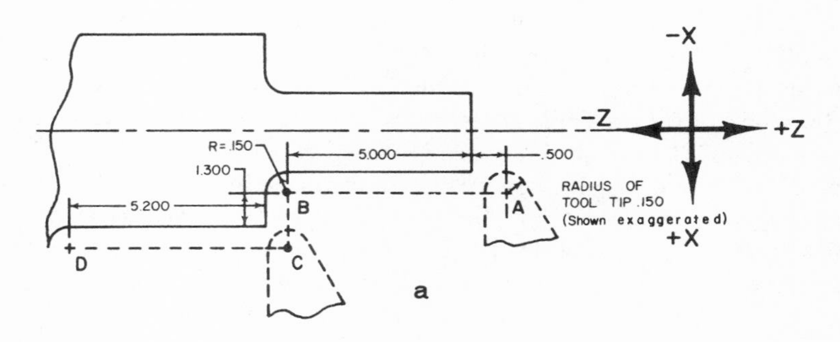

MANUSCRIPT

PART NO. XXXXX MACHINE DATE :
PART NAME : PREPARED BY :
CHECKED BY :

SEQUENCE NO.	g WORD	x WORD	z WORD	i WORD	k WORD	f WORD	s WORD	t WORD	m WORD	EOB	COMMENTS
RWS											
001	01		Z-5.500			50	360	01		EOB	MOVE FROM POINT A TO POINT B
002		X+1.450								EOB	MOVE FROM POINT B TO POINT C
003			Z-5.350							EOB	MOVE FROM POINT C TO POINT D

b

FIG. 6–26(a and b). Turning a shaft requiring straight line motions parallel to the Z and X axes. The shaft and cutting tool are shown in (a); the manuscript in (b).

Programming the cutter movements of an NC lathe is very much the same as programming the path of a milling cutter on a contour milling machine, except for the NC lathe characteristics just discussed, such as the *s*, *t*, and *g* words. And while the calculations for the tool movements are most often not too complex, generally being limited to straight lines and circular arcs, the programmer must exercise care in calling for the proper tape instructions, at the proper time.

To appreciate the axis arrangement of an NC lathe, see the bird's eye view, looking down on the headstock, tailstock, and turret, given in Fig. 6–22. Movement of the turret to the right, i.e., toward the tailstock, would be a plus (+) Z movement. Movement toward the headstock would be a minus (−) Z movement. A movement away from the center line of the machine, and toward the operator, would be a plus (+) X movement. Movement in toward the center line of the machine would be a minus (−) X movement.

Aside from a difference in axis designations, the part programming considerations that apply to contour milling machines also apply to lathes. In fact, except for special lathe features, the electronic control systems are very similar. The main difference is that the programmer must calculate the path of a single-point tool rather than the path of a circular rotating cutting tool. One very notable similarity is that *cutter offset* must be considered with the single-point tool of the lathe just as it is with the rotating tool of the milling machine. The reason is that most lathe cutting tools have a rounded nose, even though the radius of the nose may be quite small. A single-point cutting tool is shown in Fig. 6–23. Figure

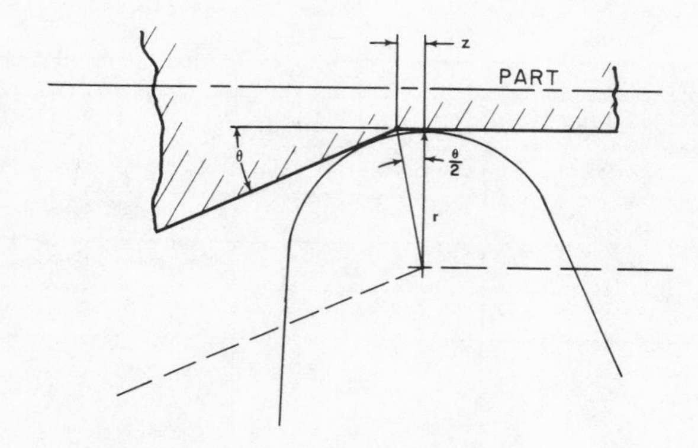

Fig. 6–27. The difference between the z coordinate on the part and the z coordinate that has to be described on the tape is equal to $r \tan \theta/2$ where r equals the radius of the tip of the cutter. This is shown as z, above.

Courtesy of American Edelstaal, Inc.

FIG. 6–28. A convenient way of checking a tape without cutting metal is to plot the path of the cutter by inserting a pen or pencil into the spindle, as shown above. The machine, in this instance, includes a special plotting function to accommodate motions in the XY and XZ planes as well as the plotting of lathe parts since the machine is a combination milling machine and lathe. See Fig. 1–7(A).

6–24 illustrates a four station turret with single-point cutting tools in place.

Figure 6–25 contains a close-up sketch showing the radius of a single-point cutter and its relationship to the edge of a part. Just as the path of the center of the circular cutter on a milling machine must be programmed, so must the center of the arc at the end of a single-point cutting tool on a lathe be programmed. As in programming the path of a milling cutter, the calculations are relatively simple when the movements are parallel to either the X or the Z axis. Circular interpolation also reduces the calculations for movements about an arc. An example in which the machining of a portion of a shaft requiring straight-line motions that are parallel to the axes is shown in Fig. 6–26(a). The manuscript shown in Fig. 6–26(b) includes a column for the *t* word, which calls out the side of the turret and cutting tool to be brought into play, and a column for the *s* word, which calls out the spindle speed in revolutions per minute. The *j* word, used with the circular interpolation feature for milling in the *X* and *Y* axis on a milling machine, is now a *k* word on a lathe manuscript. The reason for this is that, according to the standards, the *j* word, used with circular interpolation on a milling machine, is associated with the *Y* axis; and the *k* word, used with circular interpolation on a lathe, is associated with the *Z* axis of the lathe, which is normally the center line of the lathe.

Courtesy of Tektronix, Inc.

Fig. 6–29. Tapes may also be checked out by having the tool path plotted on a cathode ray tube, as shown above. In this instance the tapes may be corrected by typing in the necessary changes and then by having the corrected tape automatically typed.

Referring again to the sketch in Fig. 6–26(a), the movement from point A to point B is equal to the 5.000-inch dimension plus .500 inch. The movement from point B to point C is equal to the 1.300-inch dimension plus the radius of the cutter tip. And the movement from point C to point D is equal to the 5.200-inch dimension plus the radius of the cutter tip.

Programming a part that contains lines not parallel to the axes generally requires trigonometry calculations. In the example shown in Fig. 6–27 the distance z noted represents the difference between the z coordinate on the part and the z coordinate of the center of the tip of the cutter. This difference is calculated as $R \tan \theta/2$; where R is equal to the radius of the tip of the cutter.

It must be pointed out that the lathe cuts described in the preceding figures show only one pass. This has been done to simplify the illustrations and still describe the essential procedures. In actual practice several passes are usually required before the final cut, to blueprint dimensions, is made.

Verifying Tapes

It is not necessary to machine a part in order to determine whether a tape is correct. Other checks may be made prior to machining. And although these checks cannot be 100 percent conclusive, they can help reduce time in checking out a tape and the amount of scrap metal caused by machining while following incorrect tapes. One way of checking a tape without cutting metal is to insert a pen or pencil into the spindle of the machine tool and then to let the machine plot the path that the cutter would take. An example of this is shown in Fig. 6–28.

Figure 6–29 illustrates an arrangement using a cathode ray tube that shows the path of the cutter. This arrangement also allows the operator to edit the tape. What occurs is that the data on the punched tape is read into the system, and then the part programmer can make any corrections via a keyboard. When the necessary corrections have been made, a new tape is automatically punched.

PRACTICE EXERCISE NO. 2 CHAPTER 6

1. What is the amount of the cutter offset equal to?
2. Referring to the sketch below, what would the x and y coordinates be for points A, B, C, and D? The sides of the rectangle are parallel to the axes. The radius of each corner is 1.000 inch.

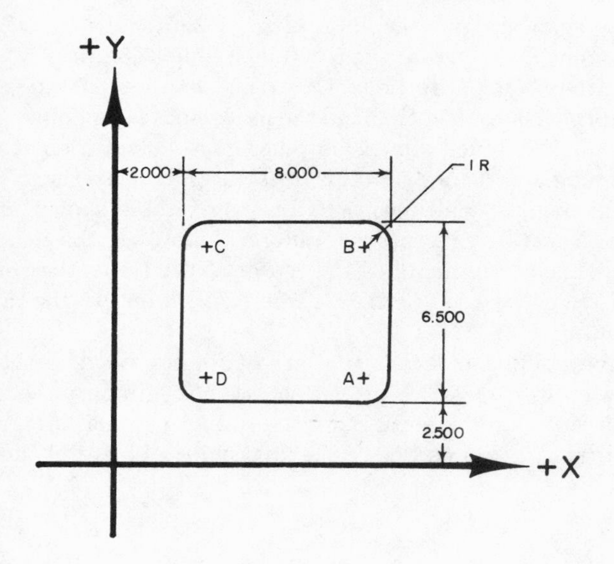

3. Give the incremental movements (Δx and Δy) for the center of a cutter when it moves from A to B; B to C; and C to D. The radius of the cutter is .500 inch.

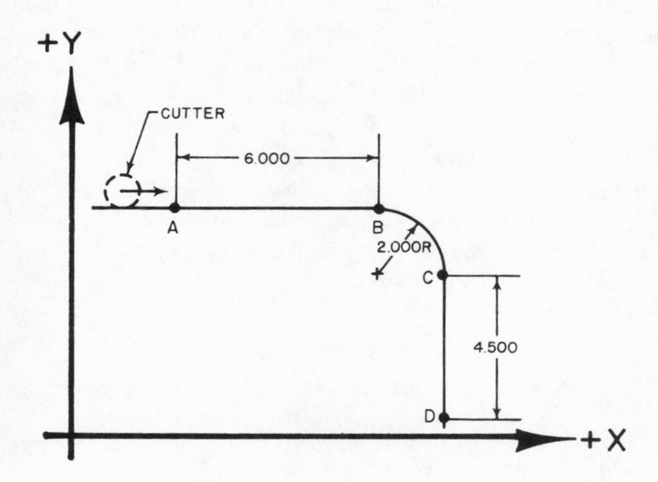

4. Calculate the Δx and Δy movements for the center of a cutter when moving from point A to point B, and then to point F. The radius of the cutter is 1.500 inches.

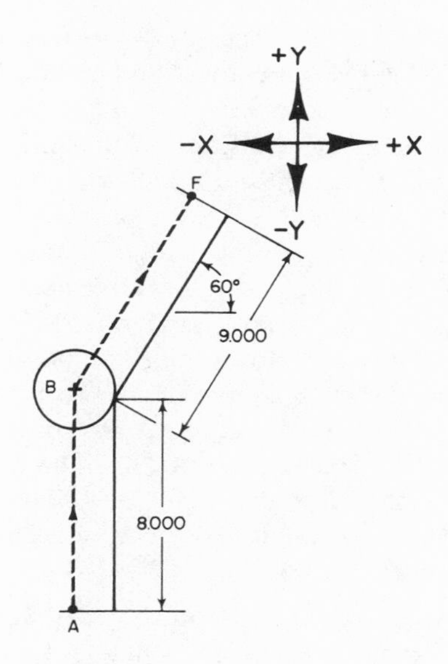

Answer to problem 4:

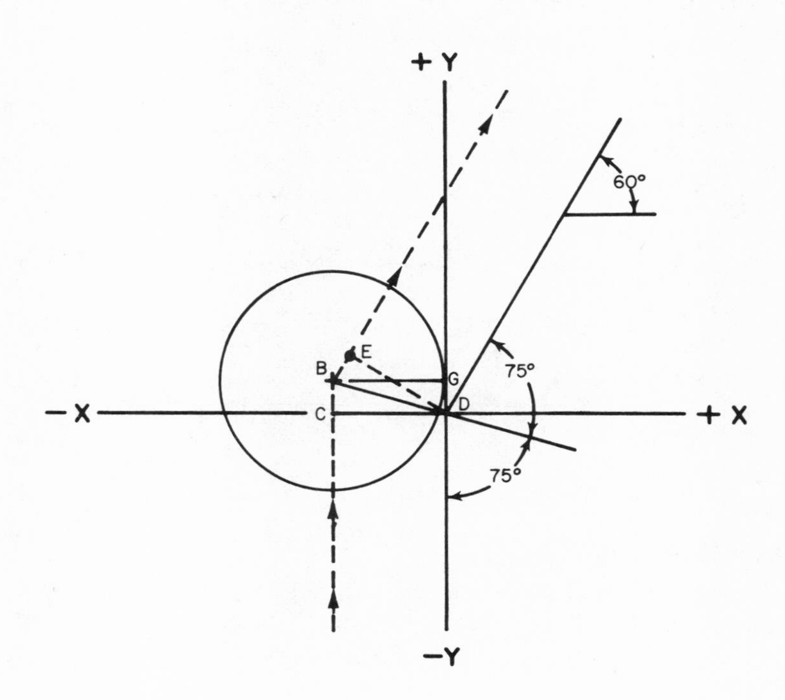

In the sketch at the bottom of page 157 it is noted that to find the y distance of the movement from point A to point B it is necessary to find the distance from point C to point B, and to add it on to the 8.000-inch vertical dimension. The distance BC, which is the same as the distance GD, is calculated as: Since angle $BDG = 75°$, $GB/GD = \tan 75°$, $GD = GB/\tan 75°$.

Since GB is equal to the radius of the cutter, GD then is equal to: $GD = 1.500/3.7320 = .402''$, the y movement is then equal to: $8.000 + .402 = 8.402''$; the x movement would be 0.000.
The distance from point B to point F would be equal to the 9.000 dimension plus the distance from point B to point E. The distance from point B to point E is the same as from point B to point C, which has been calculated and is .402″. The distance from point B to point F would therefore be $9.000 + .402 = 9.402''$. Δx for the move from point B to point F would be: $9.402 \cos 60° = 4.701''$, and Δy would be: $9.402 \sin 60° = 8.142''$.

5. The sketch below describes the outline of a part on an NC lathe. The path of the center of the tip of the single-point cutting tool runs from point A through point H, and then back to point A. Prepare the part program according to the manuscript form shown in Fig. 6–26. (Note: The ½-inch radius cutter is shown unusually large for illustrative purposes.)

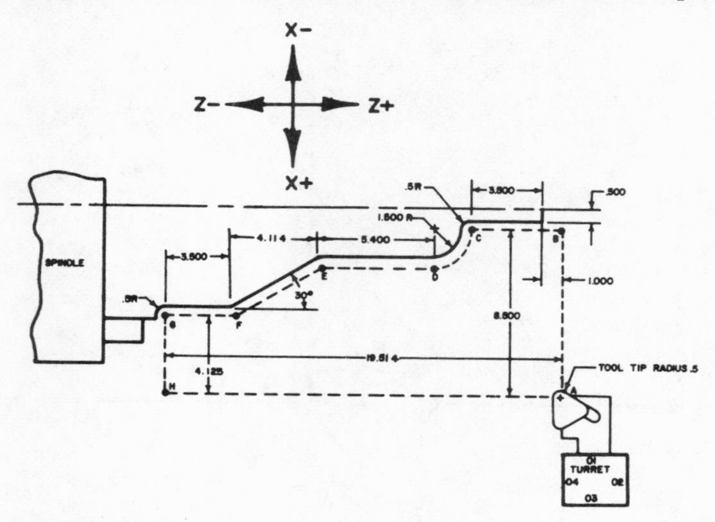

Answer to problem 5:

The movement from the starting point A to point B, and then on to points C and D, is described in the lines headed by sequence numbers

001, 002, and 003 on the manuscript shown below. A close-up of the movement from point C to point D is illustrated as follows:

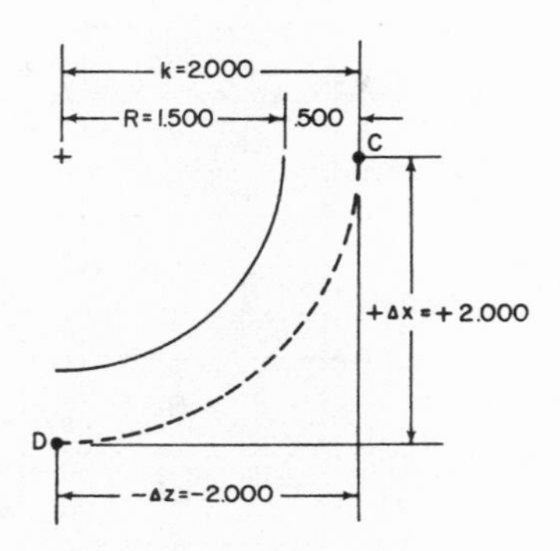

The $\Delta - Z$ movement from point D to point E would be equal to the 5.400 less an amount due to the tool tip offset. This is illustrated below. The amount of offset in the Z axis would be the distance from point I to point J, which is equal to $r/\tan 75° = .500/3.7320 = .134$. The $\Delta - Z$ distance from point D to point E would then be $5.400 - .134 = 5.266$.

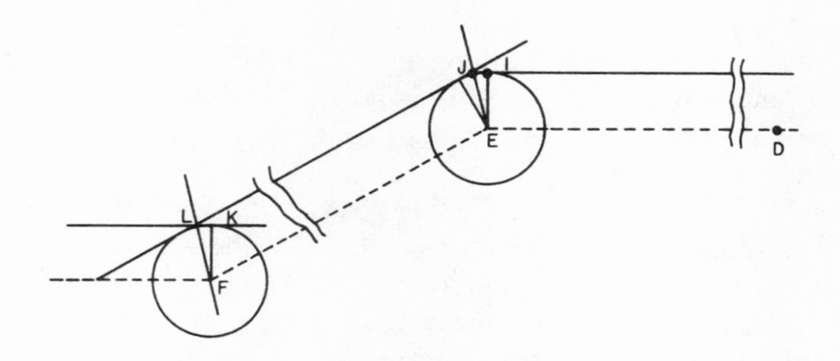

Next, since the distance IJ is the same as the distance KL, the distance from point E to point F would be the same as J to L. In going from E to F, the $-z$ movement would be 4.114, and the $+x$ movement would be $4.114 \tan 30° = 2.375$. The movement from point F to point G would be $3.500 + LK$ which is equal to JI or $3.500'' = .134'' = 3.634''$. These movements from point D to point G are shown as sequence numbers 004, 005, and 006.

PART NO. | MANUSCRIPT | DATE :
PART NAME | XXXXX MACHINE | PREPARED BY :
CHECKED BY :

SEQUENCE NO.	g WORD	x WORD	z WORD	i WORD	k WORD	f WORD	s WORD	t WORD	m WORD	EOB	
RWS											
001	g01	X-8.500				f1.500	S8.600	t01	m03	EOB	MOVE CENTER OF TOOL TIP FROM POINT A TO POINT B AT A RAPID
											TRAVERSE OF 150 IPM. AT THE SAME TIME THE TURRET IS
											AUTOMATICALLY INDEXED TO POSITION 01 AND THE SPINDLE SPEED IS SET
											860 RPM. THE M03 WORD TURNS THE SPINDLE ON AND IN A CLOCKWISE
											DIRECTION LOOKING OUT FROM THE SPINDLE FACE.
002			z-4.500							EOB	MOVE FROM POINT B TO POINT C
003	g02	X+2.000	z-2.000		k2.000					EOB	MOVE FROM POINT C TO POINT D
004	g01		z-5.266							EOB	MOVE FROM POINT D TO POINT E
005		X+2.375	z-4.114							EOB	MOVE FROM POINT E TO POINT F
006			z-3.634							EOB	MOVE FROM POINT F TO POINT G
007		X+4.800								EOB	MOVE FROM POINT G TO POINT H
008			z+19.514						m30	EOB	MOVE TO POINT A AND STOP

QUESTIONS CHAPTER 6

1. From a hardware standpoint, what is the key difference between a point-to-point and a contouring system?
2. What is the significant difference, in capability, between the motions of a point-to-point machine and those of a contouring machine?
3. What are the three ways of approximating a curve with linear interpolation?
4. Would the allowable *calculation* tolerance be greater or smaller than the final tolerance required for the part? Where would the calculation tolerance be noted on the blueprint?
5. In addition to performing linear interpolation calculations for irregular curves, which can be time-consuming and complex, what other two means can the part programmer use to determine the length of straight-line segments?
6. Calculate the Δx and Δy distances for the line segment AB shown below.
7. Which feature allows the part programmer to move a cutter about an arc in a true circular path?
8. How many blocks would be required to move a cutter around a 360° circular arc?
9. What are the two ways of expressing the *f* word in a circular interpolation block? Which factor determines the *f* word to be used?

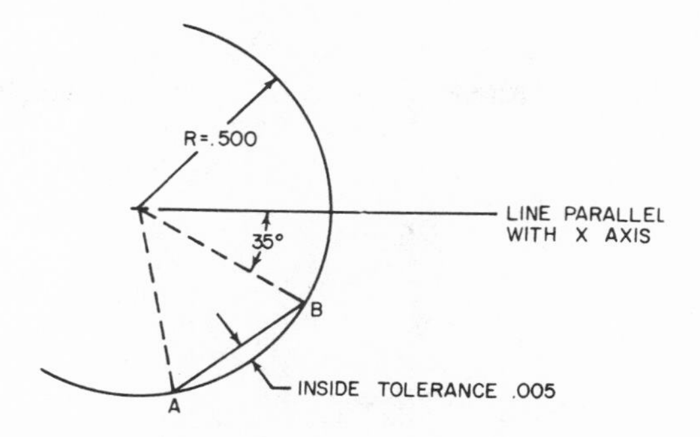

10. Using the manuscript form found in Fig. 6–14, program the groove in the part which is described on the blueprint shown below. The machine tool being used is a three-axis contouring machine having circular interpolation capability and a free-floating zero. Therefore a column for the Z coordinate would be added to the form shown in Fig. 6–14. Before writing the program, draw a sketch describing the part as it would appear in the first quadrant of the coordinate system. The target point and the origin are to be at the bottom left-hand corner, 1 inch above the surface of the part.

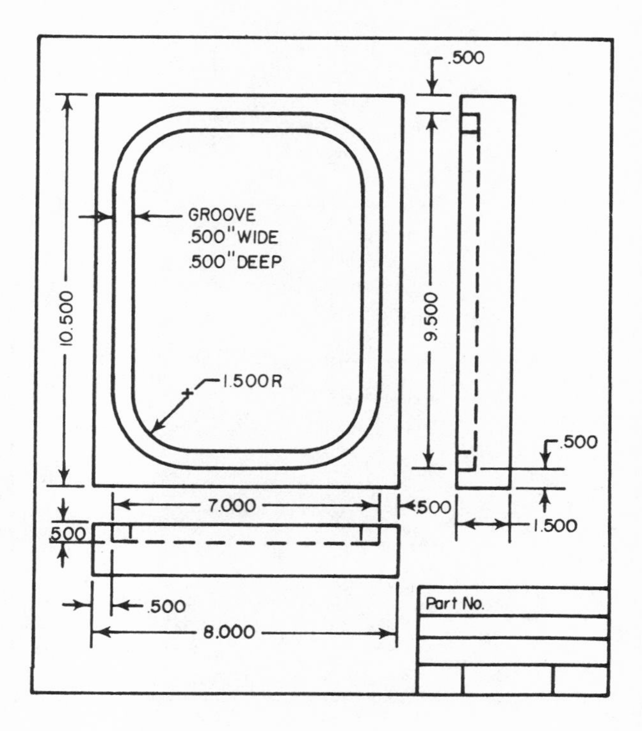

11. In the sketch below what are the Δx and Δy movements for the path of the center of the cutter? The diameter of the cutter is 1.000 inch.

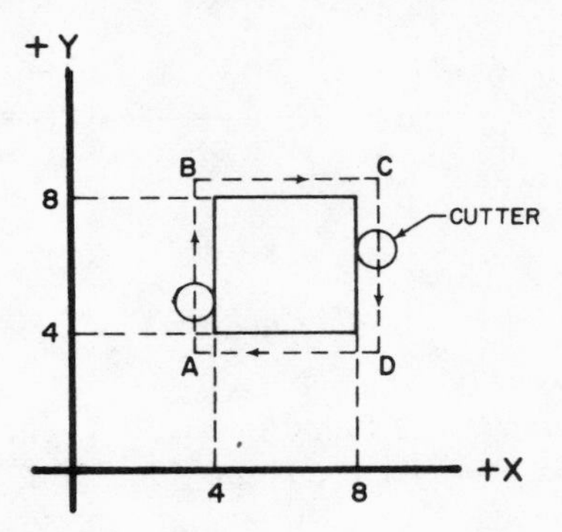

12. Using the manuscript form shown in Fig. 6–14, and adding a column for a z word (third axis), prepare a part program for milling the perimeter of the part shown on the blueprint below. The part, as it

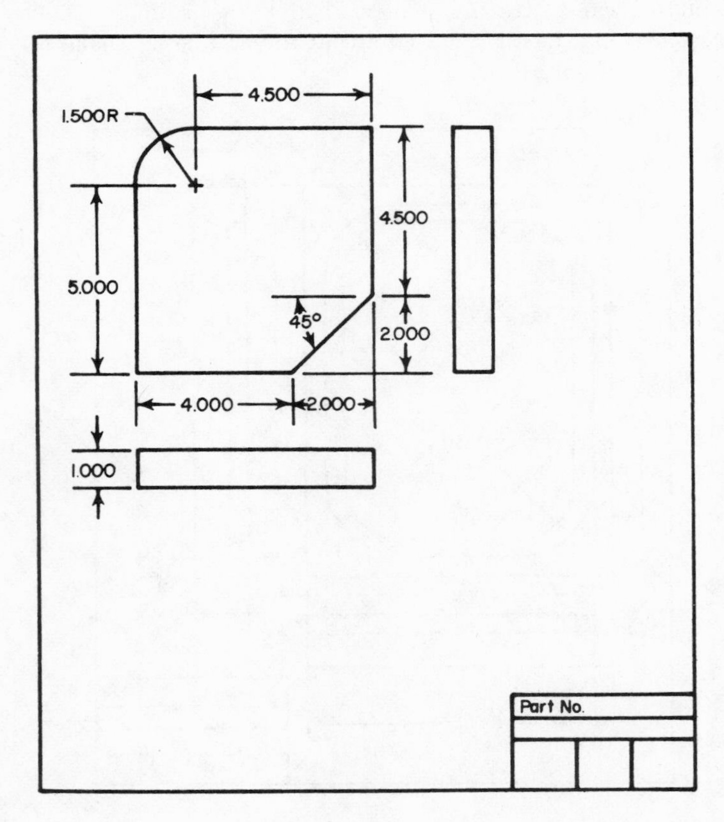

would appear located with respect to the X, Y, and Z axes, is also shown below. The target point has coordinates of $x = -.250$; $y = -1.000$; $z = +1.500$. This is the point where the bottom of the cutter would be positioned by the operator. The cutter to be used has a diameter of .500 inch. The control system is of the three-axis type and has circular interpolation. The cutter is to move around the part in a clockwise direction and then return to the target point after cutting the part.

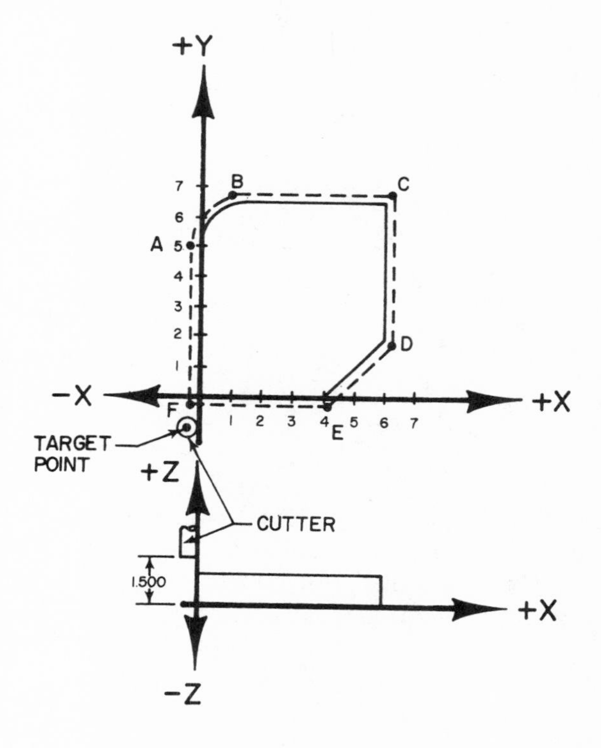

CHAPTER 7

Numerical Control and the Computer

Why the Computer for NC?

Although there were only a handful of numerical control machines in the field in 1956, it became evident that the calculations required to prepare a tape, particularly for contour programming, could be extremely time consuming. Consequently, the U.S. Air Force, which had sponsored the development of the first numerical control machine at the Massachusetts Institute of Technology (MIT), requested MIT to follow up this machine tool project with a computer program that would significantly reduce the number of calculations required of the part programmer. The program developed by MIT became known as the Automatically Programmed Tool (APT) program, and its uses will be discussed in detail in a later chapter.

The part programmer may use the computer as a tool, much as he might use any other tool in the shop. In addition to performing calculations at far greater speeds than a human can, the computer can perform these calculations with greater accuracy and reliability. The computer rarely makes a mistake and never gets tired. The computer system, by means of a tape punching device, also has the capability for automatically punching the tape in the exact format for any NC machine tool.

In order to appreciate the value of the computer as an aid in part programming, consider the example shown in Fig. 7–1. The curve AB represents the perimeter of a part that must be cut. Since the cutting tool moves along a straight line, the path of the cutting tool must be broken down into straight line segments. These straight lines must be of such lengths that they approximate the curve within the allowable tolerance.

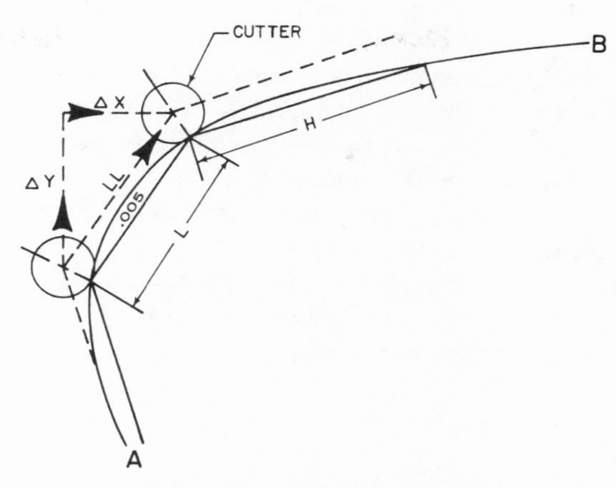

FIG. 7–1. In order to cut around the curve *AB* it is necessary to calculate straight line segments that approximate the curve. Two of the straight line segments, *L* and *H*, are shown (exaggerated). First it is necessary to calculate the length of segment *L*. Then it is necessary to calculate the distance *LL* that the center of the cutter moves. Finally, the x and y components of *LL*, Δx and Δy, must be calculated. (Refer to Chapter 6 for further details).

(The lengths of the straight lines shown in Fig. 7–1 have been exaggerated to illustrate the approximation.) The lengths of each of these straight lines must be calculated, and each one will be different if the curve is irregular as shown in Fig. 7–1. If only the line marked *L* is considered, the center of the cutter would have to move along the line marked *LL*. This would require calculations to determine the lengths of both *L* and *LL*. Also, because *LL* is at an angle with respect to the coordinate table movements of the machine, it is necessary to calculate the distances Δx and Δy since this is the data required on the tape. Considering that these calculations must be performed for every straight line segment, the total amount of calculation could become almost overwhelming. Details of this type of calculation have already been described in Chapter 6.

Using a computer programming language such as APT, if we wanted the cutter to move around the outside of the curve, it would merely be necessary to note the diameter of the cutter; the maximum distance on the inside of the curve that the straight line segments are to be from the curve (see .005 inch maximum distance shown in Fig. 7–1); and the direction that the cutter is traveling. The exact statements to the computer would be as follows:

```
INTØL/.005
CUTTER/1
GØFWD/CURVE
```

The first line states that the greatest distance the straight line approximation can be from the curve is .005 inch on the *inside* of the curve. The second line states that the cutter is to be 1 inch in diameter. This figure is necessary for the computer to calculate the center line path of the cutter. The last statement instructs the cutter to "go forward" around the curve which has previously been described to the computer. In this case *no* calculations are required of the part programmer when using the computer as an assist. This example is just one of many that could be described in which the computer can reduce the amount of calculation required of the part programmer. Computer-assisted part programming is covered, in detail, in a later chapter.

What Is a Computer?

The electronic computer, as we know it, is a relatively new development. The first commercial unit was announced approximately 20 years ago. Since its appearance many science fiction writers have had a field day accusing it of all sorts of weird and miraculous actions—even to the extent of overtaking man and controlling his world.

What is a computer? Can it actually *think*? The answer to this last question is *no*, the computer cannot think, at least not in the human sense. It can recognize only two signals, either *on* or *off*. However, it can combine and operate on these two electronic signals with such speed that even the most complex problems can be solved in a fraction of the time a human being would need to solve them. In less than 10 seconds, for example, a medium size computer can add a million four-digit numbers—and without error. To put it another way, it would take over 100 days of manual calculation to equal the results of 10 seconds of computer operation; and this is based on a 24-hour working day, without eating or sleeping. Even if it were possible to stay awake 100 days and nights, the chances would be pretty slim that anyone could go the stretch without a sizable pile of errors.

On the other hand, the computer requires accurate and carefully planned information, for its performance is only as good as the information and data that are fed to it. Thus, there is a term used among computer programmers called *GIGO* which stands for *G*arbage *I*n, *G*arbage *O*ut. The programmer—that is, the person who writes the detailed instructions for the computer—must thoroughly understand the problem that he is working with and exercise extreme care and accuracy.

Computers are grouped into different types. There are *special purpose* and *general purpose* computers. The special purpose computer is designed to perform one task, such as navigation aboard a ship or airplane. The general purpose computer, on the other hand, can handle a multitude of different types of problems, depending on how the programmer sets it up. Then there are *business*-type and *scientific*-type computers. There are also *analog* and *digital* type computers. The analog-type computer offers "approximate" answers which are still generally well within the accuracy requirements of the problem. The digital computer offers exact answers to whatever degree it has been designed for. It has this capability since it "counts" separate pulses, or *on* and *off* signals.

Computers vary considerably in size and cost, ranging from the digital desk-top mini-computer shown in Fig. 7–2 that has a price tag of approximately $15,000 to the giant model shown in Fig. 7–3 that costs up to several million dollars. The key differences, with regard to performance,

Courtesy of Honeywell

Fig. 7–2. The size of the computer varies considerably, ranging from the one shown above to the one shown in Fig. 7–3.

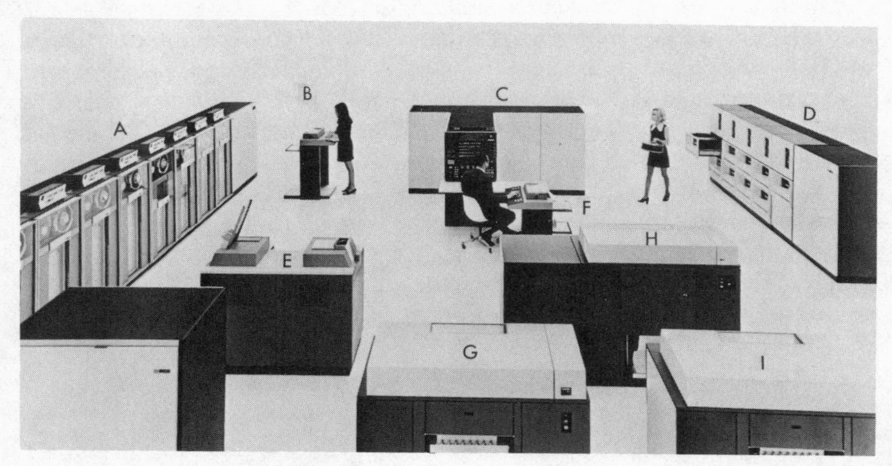

Courtesy of IBM

FIG. 7–3. A large computing system is comprised of a number of units. The speed with which this computer can perform calculations is measured in billionths of a second. The units involved are (A) storage units, (B) typewriter input, (C) central processing unit, (D) storage units, (E) card input unit, (F) operator's console, (G) printer, (H) printer, and (I) printer.

lie in the amount of data that the computer can store and work on, and the speed with which it can perform calculations.

The computers used with numerical control calculations are classed as *general purpose, scientific, digital-type* computers. Also, because numerical control problems usually require a good deal of storage capacity they are usually, but not necessarily, of the larger type.

Time-Share Terminals

It is not necessary to own or rent a computer to be able to use one. Because of a relatively recent development it is possible to communicate directly with a computer over conventional telephone lines. The programmer simply dials a local phone number and is connected directly to the computer. Instead of using voice communication, the connection is made through a teletypewriter terminal as shown in Fig. 7–4. The machine tool tape is automatically punched at the *terminal* by signals sent back from the computer to the terminal. The large, general-purpose, computers that the teletypewriter communicates with are usually located in major cities, and the customer pays for the amount of time that he uses the computer.

One of the key developments that has made this type of computer communication possible is called *time-sharing.* Because of the tremendous speed and capacity of the large computers it is possible for a number of programmers in different locations to use the computer at the same time. Actually no two programmers are using the computer at *exactly* the same time; but since the computer can switch from one program to another with such speed, it seems as if all the programs are being worked on simultaneously. Figure 7–5 illustrates a remote time-sharing arrangement. The terminals may be located hundreds, or even thousands, of miles from the computer. Because of the increasing popularity of time-share terminals and the convenience of the telephone, these terminals are being made so

Courtesy of General Electric Co.

FIG. 7–4. The part programmer, shown in the foreground, is communicating directly with a computer that may be hundreds, or even thousands, of miles away. The terminal is hooked directly to a telephone and the machine tool tape is prepared automatically at the terminal by a tape punching device. All data and instructions are transmitted over conventional telephone lines.

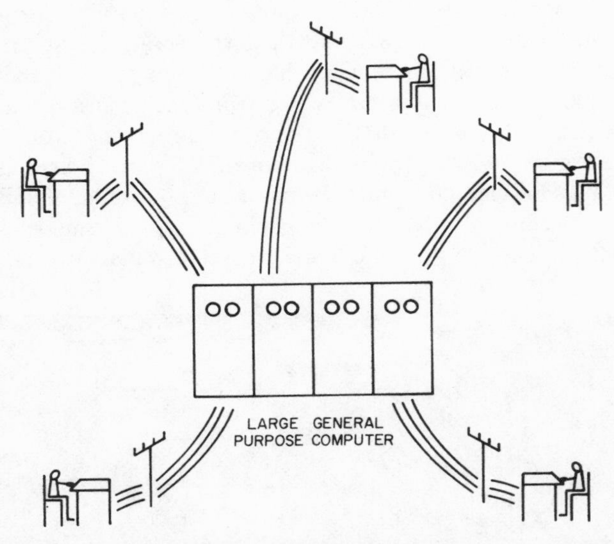

FIG. 7–5. The part programmer can, from a time-share terminal, communicate directly with the computer over conventional telephone lines. The machine control tape is also punched at the terminal. As many as 40 time-share terminals may be operated from different locations at the same time.

Courtesy of The National Cash Register Co.

FIG. 7–6. Terminals are becoming smaller and more portable. The one shown above is smaller than an average size typewriter.

that they can be carried around much like a portable typewriter. See Fig. 7–6.

HOW DOES A COMPUTER OPERATE?

The Input Language

As noted earlier, the digital computer can recognize only two signals—either *on* or *off*. However, it can perform this task with lightning speed (in billionths of a second). It can also *store* a tremendous amount of information and perform calculations far more *accurately* than a human being. It also never tires.

Since the computer recognizes only two signals, all of the instructions and data fed to it must be coded in a form based upon only two signals. The requirement is the same as that for the format used to describe numbers on a numerical control tape. Only, in this case, it is the electronic signals that may be arranged as either "on" or "off" combinations instead of a "hole" or "no hole" arrangement as on the tape. The number *5*, for example, would be expressed on tape as follows:

87654 321 ROW NUMBER
EBXOP8 421 ROW VALUE

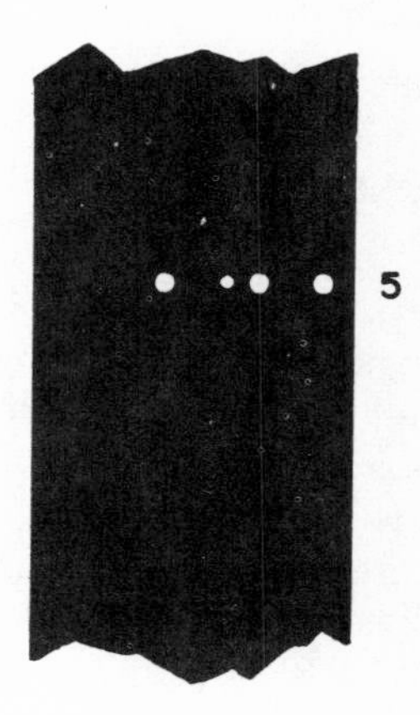

If we were to express this same number to the computer we would use a combination of on-off circuits such as illustrated below.

The numerical value for the closed circuits, in this instance, would amount to $4 + 1 = 5$, just as holes in the columns on the tape valued at 4 and 1

would equal the numerical value 5. The difference between the binary code used for the numerical control tape and that used in the computer is that—whereas the tape has a limit of four on-off values (namely 8, 4, 2, 1), which can be combined to form numbers from 1 through 9, the number of on-off values of the computer can be limitless and therefore so can the number formed by the combination of these on-off circuits. For example, the values of a series of on-off computer circuits could be as illustrated below.

In this case the combination of closed circuits adds up to the total numerical value of *115* (64 + 32 + 16 + 2 + 1). The code used with the computer is called a *straight binary code,* whereas the code on the numerical control tape is called *binary coded decimal,* and sometimes referred to as *BCD.*

The values 1, 2, 4, 16, 32, etc., are all powers of 2 as shown in the following:

$$2^0 = 1$$
$$2^1 = 2$$
$$2^2 = 4$$
$$2^3 = 8$$
$$2^4 = 16$$
$$2^5 = 32$$
$$2^6 = 64$$
etc.

Rather than address the computer in *on* and *off* terms, it has been found much more convenient to use the number *1* and *0.* Thus, *1* would mean *on,* and *0* would mean *off.* The number 114, instead of being expressed as on-on-on-off-off-on would be expressed as 111001.

Alphabetical letters and instructions must also be fed to the computer, as well as numbers, and have therefore been assigned special codes which are a combination of 1's and 0's. The code to instruct the computer to *add* two numbers, for example, might be 1000001. The code for the letter *M* might be 10101010. This kind of a coding system is called *machine* language, with the computer being the machine referred to.

It was soon recognized that instructing a computer in machine language was a cumbersome and difficult task since all numbers, letters, and instructions had to be sets of *on* and *off* signals, as expressed by 1's and 0's. The answer was to develop a special computer program that would automatically translate a language that would be easy for the programmer to use into a machine language that the computer would "understand."

For example, to add two numbers, instead of noting 1000001, we could write the word *ADD*. The special computer program that had already been prepared and fed into the computer would translate the word *ADD* into the machine instruction 1000001. The same would apply to numbers and letters as well as instructions. If we wanted the computer to accept the number *114*, we would merely write the number *114*, rather than 1110010. The program that performs this kind of a translation is called an *assembly* program.

Another type of program, which goes farther than an assembly program, is called a *compiler*. In addition to direct translation a compiler can also perform mathematical functions such as multiplying 25 by 25 or calculating the square root of 625; 25×25 would be expressed as 25*25 and $\sqrt{625}$ would be expressed as SQRTF(625). Trigonometric functions may also be handled. The sine of 45°30′, for example, would be expressed as SINF(45.5).

The most popular compiler program for engineering and scientific problems is called FORTRAN (from *for*mula *tran*slation) and is the basis for the majority of NC computer programs. The automatically programmed tool (APT) system, which will be reviewed in detail later in the text, relies on the FORTRAN compiler. APT is a special computer program designed for numerical control part programming, whereas FORTRAN is a general-purpose scientific type program. The relationship of the *APT*, *FORTRAN*, and *machine language* programs is shown in Fig. 7–7.

Major Elements of a Computer

When many people think of a computer they imagine that everything is contained in a single box. And this is true to a certain extent since the primary calculations may be handled by a single unit (the central processing unit or CPU). However, for a CPU to operate there must be some means of getting the data and instructions into it. Also, when the computer has completed its calculations, there must be some means of listing,

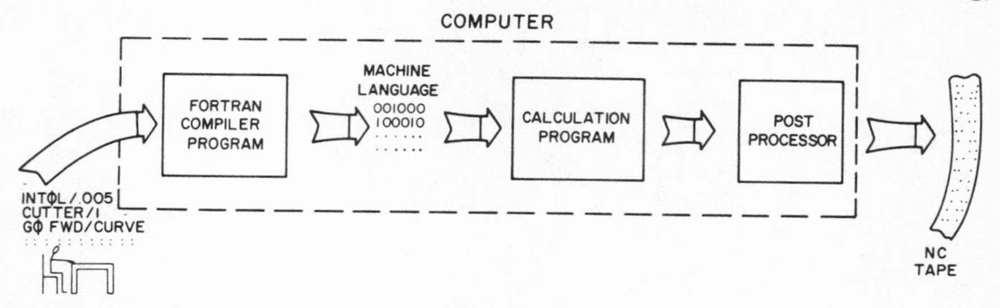

FIG. 7–7. Major steps involved with the preparation of a tape via the APT programming system.

Courtesy of IBM

Courtesy of IBM

Courtesy of Rocketdyne, a Division of North American Rockwell Corp.

FIG. 7–8. Three different means of entering data and information into a computer: (upper) a keypunch, on which the operator punches computer cards, (middle) a card reader, which reads the punched cards into the computer, (lower) a cathode ray tube, by means of which the programmer feeds data into the computer.

or printing out the information. In addition, many computers require additional storage capacity for excess data that cannot be handled in the central computer so that other cabinets of electronic equipment must be added to the system. Figure 7–3 illustrates a large computer system, major sections of which have been labeled.

In addition to the teletypewriter unit shown in the foreground of Fig. 7–2, a number of other on-site devices may be used to feed information and data to the computer. The most common is a *punched card reader.* The punched cards are prepared on a *keypunch* unit by a keypunch operator. In numerical control, the data on the cards would be the part program, which is prepared by the part programmer. Instructions such as INTØL/.005, or CUTTER/1, or GØFWD/CURVE would be punched

```
CUTTER/   0.5000
 INTOL/   0.0010    0.0010    0.0010
OUTTOL/   0.0050    0.0050    0.0050
 FEDRAT  /    10.0000
 FROM /SETPT
                          X                  Y                  Z
                      2.0000000          2.0000000          0.0000000
DS IS/L1
                          X                  Y                  Z
                      2.0000000          4.7500000          0.0000000
DS IS/L1
                          X                  Y                  Z
                     20.0000000          4.7500000          0.0000000
C1     (     0) = CIRCLE/  20.0000     7.0000    0.0000    2.0000
DS IS/C1
                          X                  Y                  Z
                     20.2313615          4.7569002          0.0000000
                     20.5453143          4.8119284          0.0000000
                     20.8483723          4.9106725          0.0000000
                     21.1344804          5.0511595          0.0000000
                     21.3979226          5.2305828          0.0000000
                     21.6334355          5.4453575          0.0000000
                     21.8363138          5.6911926          0.0000000
                     22.0025040          5.9631766          0.0000000
                     22.1286859          6.2558755          0.0000000
                     22.2123385          6.5634413          0.0000000
                     22.2517904          6.8797292          0.0000000
                     22.2462534          7.1984201          0.0000000
                     22.1958382          7.5131466          0.0000000
                     22.1015520          7.8176209          0.0000000
                     21.9652786          8.1057599          0.0000000
                     21.7897406          8.3718067          0.0000000
                     21.5784451          8.6104459          0.0000000
                     21.3356136          8.8169099          0.0000000
                     21.0637339          8.9826675          0.0000000
DS IS/L2
                          X                  Y                  Z
                      8.0637339         15.9573823          0.0000000
```

Fig. 7–9. A portion of a computer listing for a part program.

directly onto the cards. After the cards are punched, the remaining operation is completely automatic, including the preparation of the machine tool tape. Another means of input to the computer is *punched tape*. In this case, the tape format is not the same as that on the numerical control tape for the machine tool although its dimensions are. Still another, more recent, input device is a cathode ray tube, which is similar to a television tube. In this case the operator may draw on the face of the tube with what is called a light pen, and the image is transmitted to the computer. Close-ups of the keypunch, card reader, and cathode ray tube are shown in Fig. 7–8.

Computer *output* devices also vary. The most common is called a *printer*, which lists the information and data prepared by the computer. This output is referred to as the *computer listing*. If we want to run a numerical control machine we can obtain a punched tape as output. In most cases the part programmer will want *both* the punched tape *and* the computer listing, which may describe the data on the tape or the path that the cutting tool will take. He may use the computer listing to check his program. Figure 7–9 shows an example of a computer listing. The

Courtesy of Universal Drafting Machine Co.

FIG. 7–10. The punched tape that has been prepared by the computer is used to operate the drafting machine shown.

cathode ray tube and the teletypewriter may also be used as *output* devices as well as input devices. Quite often the output information from the computer will be put on magnetic or punched tape, and the tape data may then be read by another device and displayed. Figure 7–10 shows a numerical control drafting machine being operated by punched tape that has been prepared by a computer.

Referring again to Fig. 7–3, it will be noted that there are four major units that comprise a computer system. These are:

1. An *input device* to transmit the information into the computer.
2. A *central processing unit* to perform the calculations.
3. *Storage units* to store data that can be worked upon by the *CPU*. In addition to performing calculations, the CPU also monitors and controls the selection and utilization of the data in the storage units. Data may be stored on magnetic tape, as shown in the left portion of Fig. 7–3, or on discs as shown to the right of the photo in Fig. 7–3. Data are also stored in the CPU in a *magnetic core*. The advantage

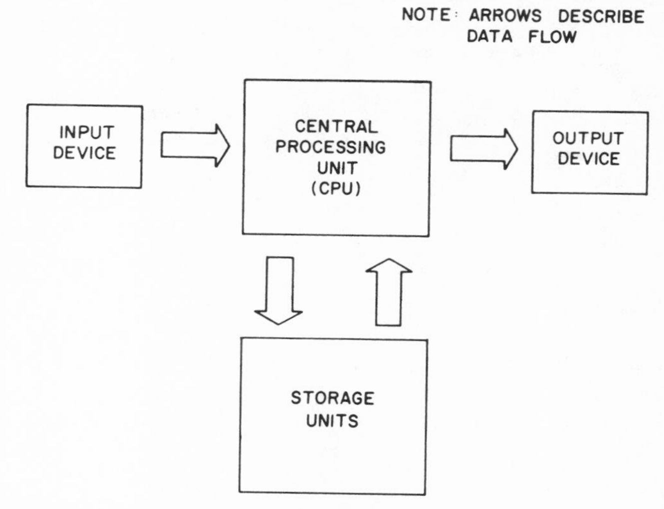

Fig. 7–11. The computer system consists of four major units or devices. The *input* device feeds the data and instructions *into* the main portion of the computer, which is called the *central processing unit*, or CPU. The CPU performs the calculations and controls the operation with the assistance of the storage units. The *storage units* store data and information which can be utilized by the CPU. The ability to store tremendous amounts of data that can be retrieved and utilized and then restored, is one of the significant features of the computer along with its speed and accuracy. The function of the *output* device is to prepare and present the data in a permanent form, such as a computer listing or printout.

of magnetic core storage is that information can be retrieved and worked upon more quickly than it can be from either magnetic tape or drum storage. However, magnetic core storage is more expensive.

4. An *output device* to deliver the output of the computer in understandable form, such as a printout.

A diagram of the data flow into and out of a computer is shown in Fig. 7–11.

The APT System

While there are a number of part programming languages available for numerical control, the most popular is the automatically programmed tool or APT system, which was developed to relieve the mathematical burdens of numerical control part programming. Developed by MIT under U.S. Air Force sponsorship, APT is a very powerful programming language that is capable of handling programs for the most complex machines. It has sometimes been accused of being *too* powerful for handling the programs of simpler machines such as two-axis drills or two-axis milling machines. A number of other programming languages have therefore been developed to handle simpler programming requirements.[1] Simpler versions of APT have also been developed for the simpler NC machines such as two-axis drilling and milling machines and lathes. These programs can usually be run on the smaller size computers and are considered sub-sets of APT since the language is the same. One such subset is ADAPT, which, like APT, was also developed under government sponsorship.

Since the APT language can be used for the widest variety of work—from simple two-axis point-to-point drilling to the most complex five-axis contour milling—and can be applied to the largest number and greatest variety of machine tools, this text will concentrate on this programming language.

As noted earlier, the purpose of the APT programming system is to reduce the amount of calculation required and, at the same time, to reduce the number of errors likely to develop as the part programmer prepares his program. As with most other computer programming systems, APT relies heavily on the English language. If we wish to describe a line to the computer, we spell out the word LINE. A point would be described as PØINT, and a circle would also be described to the computer just as it is spelled in English, in other words, as CIRCLE. It should be noted that only capital letters are used in the APT language. If we want the computer to prepare a tape that would move a cutting tool to a specific

[1] Other popular programming languages include: SPLIT, AUTOSPOT and COMPACT 11

point that is marked P1, the programming instruction would be GØTØ/P1. *Note that slash (/) marks have been put through the O's here as well as in the word PØINT just cited.* This is to help the keypunch operator, or whoever is preparing the input to the computer, to distinguish between a zero and the letter *O*. The remainder of the text will follow this practice. It should also be pointed out that there is a clear distinction between a *part program* and a *computer program.* A part program is prepared by a part programmer and describes the geometry and machining operations of a part. A computer program, which has been previously prepared by a computer programmer, processes the part-programming information within the computer.

Steps in Preparing a Machine Tool Tape via the APT System

The step-by-step procedure for preparing a tape via the computer and the APT programming system is shown in Fig. 7–12.

Step 1. The part programmer prepares a manuscript from the blueprint information and his knowledge of machining practice. This handwritten manuscript is unlike that used for manual programming, which lists the step-by-step movements of the cutting tool, in that the statements appear in English-like words. There is also far less arithmetic required in preparing a manuscript for a computer program than for manual part programming since the computer performs the bulk of the mathematical calculations. The part programmer must be precise, however, and all words and notations must be exactly as called for by the APT programming system. An example of a handwritten APT manuscript is shown in Fig. 7–13.

Step 2. Computer cards are prepared by a keypunch operator. The information for the cards is taken directly from the manuscript.

Step 3. The card information is then automatically fed into the com-

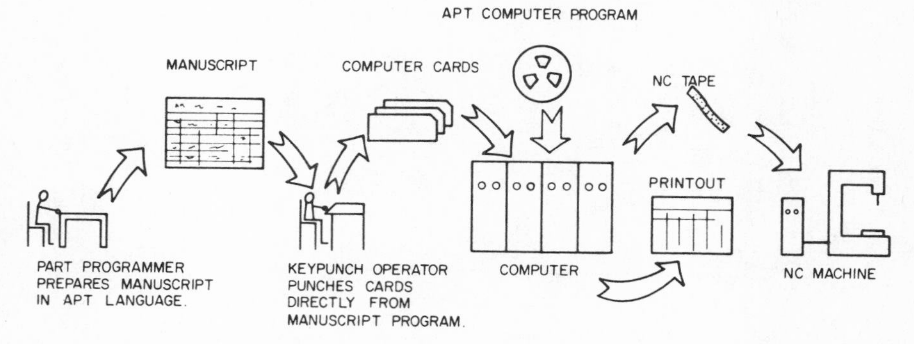

FIG. 7–12. The flow process for preparing a tape utilizing an on-site computer.

JAMES J. CHILDS ASSOCIATES
P. O. BOX 3086 ALEXANDRIA. VA. 22302
AREA CODE (703) 549-4592

PART NO. 12 T 94862
COMPANY ABC
POST PROC ABC QT

DATE xx/x/xx
PREPARED BY J. J. C.
SHEET 1 OF 2

SYMBOL		APT STATEMENT	SEQ. No.
PARTNØ		SØLID BLØCK 12T94862 JØHN SMITH JR.	10
		NØPØST	20
		CLPRNT	30
		CUTTER/.500	40
		INTØL/.001	50
		ØUTTØL/.005	60
		FEDRAT/10	70
SETPT	=	PØINT/2, 2	80
P1	=	PØINT/5, 5	90
P2	=	PØINT/20, 6	100
P3	=	PØINT/0, 20	110
L1	=	LINE/P1, ATANGL, 0	120
C1	=	CIRCLE/CENTER, P2, TANTØ, L1	130
L2	=	LINE/P3, LEFT, TANTØ, C1	140
L3	=	LINE/P1, PERPTØ, L1	150
C2	=	CIRCLE/XLARGE, L3, YSMALL, L2, RADIUS, 2	160
		FRØM/SETPT	170
		GØ/TØ, L1	180
		TLRGT, GØRGT/L1	190
		GØFWD/C1	200
		GØFWD/C2	220
		GØFWD/L3, PAST, L1	230
		GØTØ/SETPT	240
		FINI	250

FIG. 7–13. Example of a manuscript written in the APT language.

puter where the manuscript data is processed by the APT *computer program.* The computer program is normally stored on a roll of magnetic tape, and the data and instructions on the tape must be fed into the computer *before* the APT *part* program, which describes the particular part to be machined in the APT part-programming language, is fed into the computer. This is done because it is the APT *computer* program that processes the APT *part* program.

Step 4. The output of the computer is the punched tape that will run the NC machine tool. In certain instances it may be necessary to convert from either magnetic tape or punched cards to the punched tape by using special output equipment; however, what is significant is that the process is entirely automatic after the punched cards have been prepared.

In addition to the punched tape, the computer also puts out a printed

Courtesy of United Computing Corp.

FIG. 7–14. Both the input to the computer, which is the program listed on the part programmer's manuscript, and the output, which consists of the calculated information that is punched on the machine control tape, are handled via the teletypewriter.

PART PROGRAM
INTϕL/.005
CUTTER/1
GϕFWD/CURVE

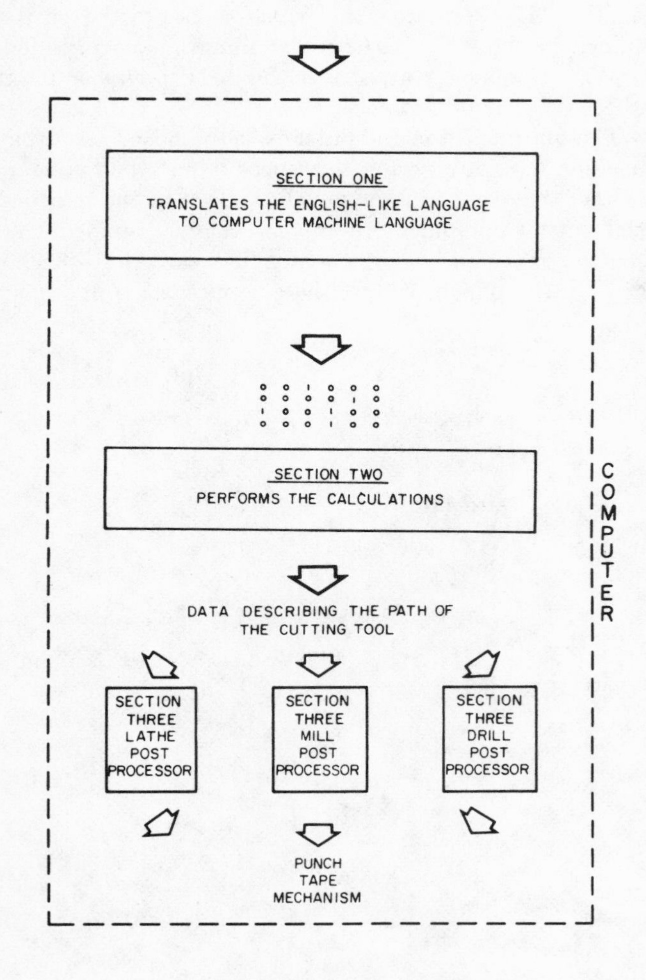

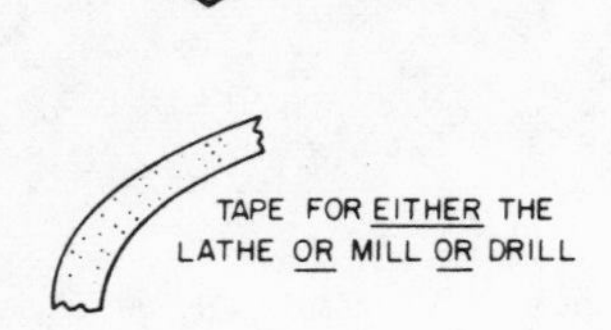

listing describing the movements of the cutting tool. This listing is used by the part programmer in order to determine that his program is correct. (Refer to Fig. 7–9.)

When using a small on-site mini-computer, as shown in Fig. 7–2, or a time-share terminal as shown in Fig. 7–4, access to the computer is much more direct. In this case the machine control tape is produced by the same teletypewriter unit that the part programmer uses to type his program into the computer. There need be only two major units for the on-site arrangement—namely, the computer and the teletypewriter, as illustrated in Fig. 7–14. This arrangement is considerably less expensive than that shown in Fig. 7–3.

The Post Processor

In Fig. 7–15 it will be noted that the APT computer program is broken down into three major sections in the computer. These sections contain information and instructions necessary for processing the part program that the part programmer writes and that is fed into the computer.

Section one translates the part programmer's English-like language into the computer machine language. Actually section one is a special FORTRAN compiler.

Section two is the part of the APT computer program that performs the calculations. The output of section two is a series of points, in coordinate form, that describe the path of the cutting tool. The cutting tool may be attached to any kind of numerical control machine since the *path* of the cutting tool is independent of the type of machine. This section is therefore concerned with the geometry of the motions.

Section three, called the *post processor* section of the computer program, takes the geometry path information from section two and arranges it in the proper format for the numerical control tape. Such tape words as the *sequence number,* the *g* words, and the *m* words, as well as the *feed rate word,* are handled by the post processor section of the APT computer program. Since the exact tape format for different machines will be different, each type of numerical control machine must have its own unique post processor. Referring to Fig. 7–15, it will be noted that the part program for any type of machine, whether it be a mill, drill, or lathe, will be proc-

---

Fig. 7–15. The part program, which has been punched on computer cards, is translated into computer machine language in *section one.* Next, *section two* performs the calculations that describe the path of the cutter. *Section three* consists of the *post processor.* In the illustration shown there are three post processors, each applying to three different machine tools. All three post processors may be stored in the computer at the same time. The part programmer notes in *which* of the three post processors the output of section two is to go. This would depend on which machine the part program is intended for.

essed by sections *one* and *two*. The output of section two will then be fed to the particular post processor part of the program that applies to the specific machine and control system being considered. Figure 7–15 shows three types of post processors, namely one for a lathe; one for a mill, and one for a drill. Actually the post processor would apply to a *specific* lathe and control system combination, or a *specific* mill and control system combination, or a *specific* drill and control system combination. Generally speaking, every different machine tool and control system combination requires its own special post processor. This would seem like a lot of work to prepare all of the necessary post processors, and it is. However, since many numerical control machine tools are similar and the control systems conform to standards, many of the post processors can be similar.

Operating Machine Tools Directly from a Computer (DNC and CNC)

Direct numerical control (DNC) is a relatively new development that refers to the automatic operation of a machine from instructions fed directly from the computer. In other words, there is no tape. Instead of the computer's preparing instructions and then punching these on tape, the instructions are transmitted from the computer directly to the drive motors of the machine tool. And since the computer can be programmed

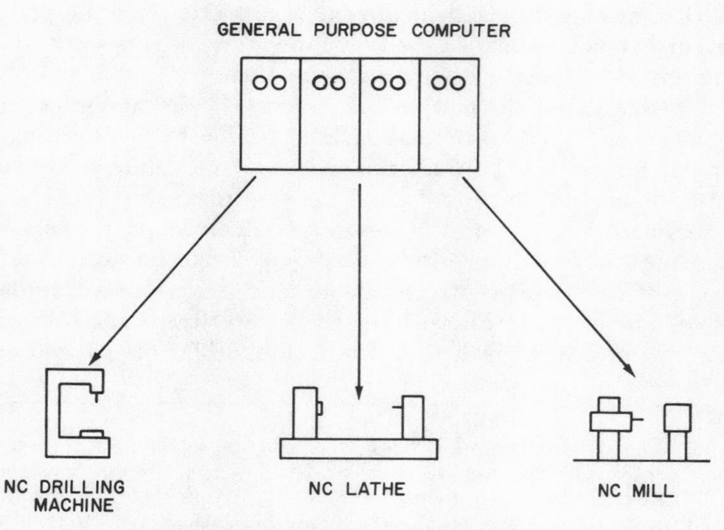

Fig. 7–16. DNC, which refers to Direct Numerical Control, is a numerical arrangement whereby a general purpose computer can operate a number of different types of machine tools at the same time. The machines are hooked directly to the computer.

on a time-share basis, it is possible to operate more than one machine tool from the same computer at the same time.

There are a number of advantages to this type of arrangement. One is that since the tape is eliminated, there is no storage problem for the tape or possibility of tape punching errors. Another is that corrections and modifications to the program can be made more readily by entering the change directly into the computer. A third advantage is that a shop manager can exercise better control over his shop operation since the computer can also automatically maintain records on, for example, the exact number of parts produced; their machining time; and whether the machine is producing ahead, on, or behind schedule.

Figure 7–16 illustrates a DNC arrangement. It will be noted that the machine tools need not be the same, or even of the same type. As shown, the single computer is operating a lathe, a mill, and a drilling machine.

A DNC operated lathe is shown in Fig. 7–17. The control unit, in this instance, is of the standard tape-controlled variety. However, lines from the computer are tied directly behind the tape reader, thus bypassing the requirement for the tape.

Courtesy of James J. Childs Associates

Fig. 7–17. A DNC operated lathe. In this instance lines from the computer are connected directly behind the tape reader of the conventional type control unit. Other types of DNC systems eliminate the tape reader entirely and a special control unit accepts the computer instructions.

Courtesy of Sundstrand Machine Tool

Fig. 7–18. Here the cathode ray tube displays the program stored in the computer, and the operator can make changes in the program while observing the actual cutting operation at the machine tool site.

One of the methods of communicating with the computer is shown in Fig. 7–18. In this case, the operator or part programmer can make changes in the part program instantaneously, while observing the actual cutting operation at the machine tool site. This is accomplished via the cathode ray tube, which displays the program stored in the computer, and the keyboard, which enables the operator to make changes in the program. The computer may be located in the same building or hundreds of miles distant. In the latter case communication is achieved via telephone lines.

Computer Numerical Control (CNC), differs from DNC in that the conventional hardware control system is replaced by a mini-computer. The mini-computer thus serves as the *electronic logic* for the control system. Although the mini-computer can handle more than one machine, the general practice has been to restrict its use to the operation of a single unit.

QUESTIONS CHAPTER 7

1. How does the computer help the part programmer?
2. What do the letters *APT* stand for?
3. What does the APT statement INTØL/.002 mean?
4. What function does the APT statement CUTTER/1 have?
5. How old is the development of the commercial electronic computer?
6. What does the term GIGØ mean and what is its significance?
7. What is the difference between an *analog* computer and a *digital* computer? What type of computer is used for numerical control programming?
8. What are the two major technical differences between a mini-computer and a so-called giant computer?
9. What is a teletypewriter terminal?
10. What is meant by the term *time-sharing*?
11. What does an assembly program do?
12. What does a compiler do?
13. How does the computer machine language compare with the character coding on a numerical control tape?
14. Using the straight binary code, how would the number 35 be expressed? (Use 0's and 1's.)
15. What are the four major elements of a computer?
16. What is the difference between a *part* program and a *computer* program?
17. Why would a part programmer want a computer listing?
18. What is the advantage of *magnetic core* storage?
19. Why was the ADAPT language developed?
20. What is the function of the keypunch operator?
21. What information is punched on the computer input cards?
22. What is the function of the *post processor*?
23. What is meant by the terms DNC and CNC?
24. Must all of the machines operated under a DNC system be exactly the same?

CHAPTER 8

Computer-Assisted Part Programming, the APT System

Chapter 7 described the computer's role as an assist to the part programmer for preparing numerical control tapes. In addition to reducing the number of calculations required, the computer also significantly reduces the errors that might be passed on to the tape if manual programming were used. Also, as noted earlier in the text, the computer does nothing that the part programmer could not calculate with the aid of a pencil and paper. The difference is that the computer can do it far faster, and with much greater accuracy.

The objective of this chapter and Chapter 9 is to acquaint the reader with some of the more important details of the APT part programming system. The information contained in these two chapters is intended to offer the student a foundation covering the concepts of the APT system, on which he may build a more detailed understanding. Although the student would need more advanced instruction to be an accomplished APT part programmer, from the information offered here he will be able to program simple parts involving point-to-point operations such as drilling, tapping, and boring as well as milling and turning operations involving straight lines and circular cuts.

The APT Programming System

The APT, or Automatically Programmed Tool system, was one of the first computer part programming systems developed. Although APT is not the simplest computer programming system, it has been chosen for description in this text because it is the system having the widest use and the greatest range of capabilities. It can be used with more different types of numerical control machines than any other programming system; and once the student has grasped the basic concepts, he will be able to use the same basic APT language to program a wide variety of numerical control machines—including milling machines, lathes, drills, punch presses, and even automatic drafting machines.

Fundamental Concepts of the APT Part Programming System

The APT system is a *three-dimensional* system, in which a cutter can be programmed so that it will move in all three axes, or dimensions (X, Y, and Z), at the same time. This is providing, of course, that the machine-tool cutting head is also capable of moving in all three axes at the same time. The part programmer must therefore think of the cutter as moving in three dimensions in space within an imaginary cube, rather than as moving over a flat surface. By visualizing three-dimensional motion, the part programmer is able to program the machine for cutting complex parts such as dies or molds. He can also program the cutter to avoid obstructions such as clamps or other holding devices. Even if the machine tool is capable of only two simultaneous axes of motion, that is, X and Y together, and Z by itself, the part programmer should still consider the part as fitting into a three-dimensional cube. Figure 8–1 shows a part as the part programmer should visualize it within the imaginary cube. The size of such a cube will be determined by the range of travel of the machine

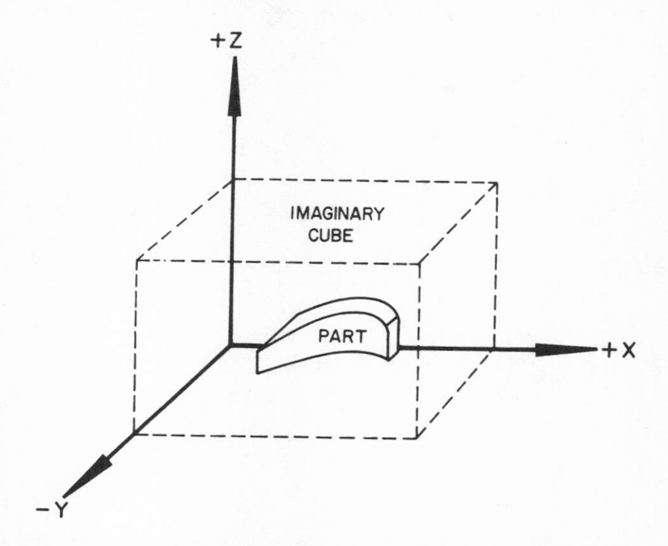

FIG. 8–1. The part programmer must imagine the part in its three-dimensional form.

along the X, Y, and Z axes. The cutting tool may be moved to any position within this imaginary cube, usually by a combination of tool head and table motions.

If the machine has a *free floating zero* capability, then the zero point, or origin, of the X, Y, and Z axes may be positioned anywhere within or without this imaginary cube. See Fig. 8–2.

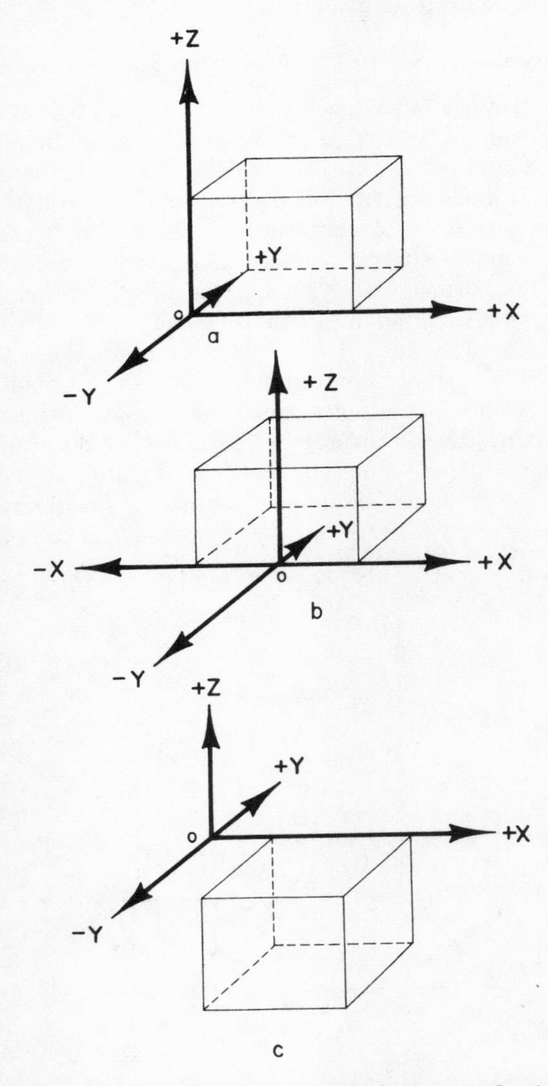

Fig. 8–2. Examples (a), (b), and (c) illustrate that with a free-floating zero system the origin of the X, Y, and Z axes may be located at any position with respect to the cube-like envelope.

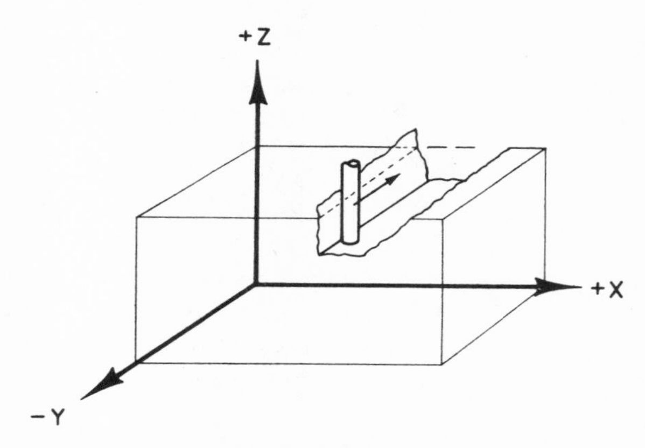

Fig. 8–3. The cutting tool may be guided along its path by two intersecting surfaces or planes.

Since the cutting tool is to be considered as moving in three-dimensional space, it must be continually guided along its path. In Fig. 8–3 it will be observed that if the cutting tool is to move along a straight-line path, it can be accurately guided by two intersecting surfaces, or planes. One plane guides the side of the cutting tool while the other plane guides the bottom. A close-up view is shown in Fig. 8–4. The surfaces, or planes, need not belong to the actual part. They may be considered as imaginary planes by the part programmer to guide the cutting tool along a specified path.

In order to change direction, or stop the cutter, a third imaginary surface is used, as seen in Fig. 8–5. Since each of these three surfaces has a specific function, they have been assigned names. Generally speaking, the surface that guides the side of the cutter is the *drive* surface, as this is the surface

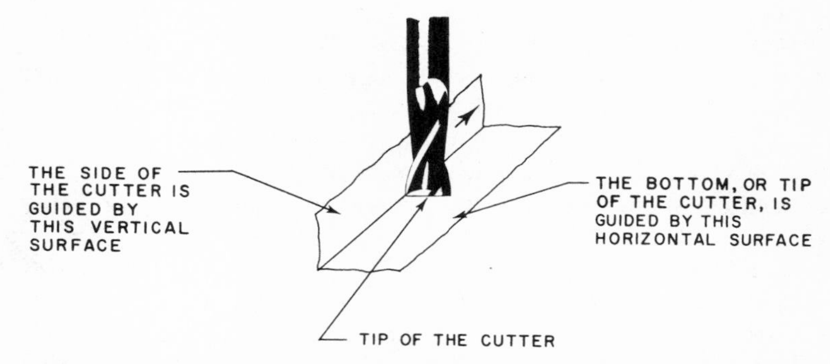

Fig. 8–4. The cutter is guided by two surfaces or planes. These may be imaginary surfaces, constructed by the part programmer to guide the cutting tool in space, or they may be the actual surfaces of the part.

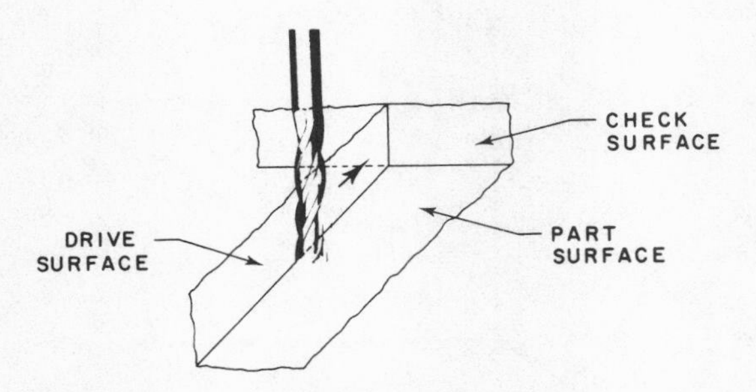

FIG. 8–5. The path of the cutter is guided by the *part* surface and the *drive* surface. To check the motion a third surface is used, and this is appropriately called the *check* surface.

that the cutter *drives* along. The surface that guides the bottom of the cutter is called the *part* surface. It is important to note that the part surface, along which the cutter is considered to move, need not necessarily be the surface of the part. It may even be an imaginary surface established by the part programmer to guide the cutter motion. The surface that stops, or checks, the motion is called the *check* surface. Like the part surface, the check surface and drive surface need not be actual surfaces of the part being machined. They are very real, however, as far as the computer is concerned and are used by the APT computer program to guide the cutter along a calculated path. Also, as far as the computer is concerned, these surfaces extend indefinitely and have no boundaries.

The cutter may move *to* the check surface, *on* the check surface, or *past* the check surface. If we looked directly down onto the part surface, the drive surface and the check surface would appear as lines, and the moves *to*, *on*, and *past* the check surface would appear as shown in Fig. 8–6.

After the cutter is moved *to*, *on*, or *past* the check surface, it may then be directed either to the *right* or to the *left* if it is to continue to move. It could also continue to move *forward* or even *back*. As mentioned earlier in the text, the APT language instructions that are fed to the computer are very much like the English language. This makes the APT language easier to learn. For example, *go to* is noted as GØ TØ. *Go on* is GØ ØN. *Go past* is GØ PAST. While a space *may* be shown between the two words, the common practice is to write them together. GØ TØ would then become GØTØ. GØ ØN would be GØØN. And GØ PAST would be GØPAST. It will be noted that all APT letters are capitalized and that there are slash (/) marks through the capital O's. The reason for the slash mark, as explained in Chapter 7, is to distinguish the letter O from a zero and avoid possible error. Thus, an O would have a slash mark through it; a zero would not.

Because of a strict rule of the APT programming language stating that there can be no more than *six* letters in an APT command or word, *go right* is shortened to GØRGT. *Go left,* while not exceeding six letters, nevertheless becomes GØLFT. *Go forward* is written GØFWD, and *go back* stays exactly as in English—GØBACK. These movements are illustrated in Fig. 8–7. The cutter may also be commanded to move up or down as shown in (G) and (H). The essential point for the part programmer to keep in mind is that he is always looking in the direction that the cutter is traveling. He might even imagine himself as riding along on the cutter and *steering* it.

Referring to Fig. 8–8(a), the APT command for the cutter to make a right turn from surface *A* to surface *B* would be GØRGT, which would direct the cutter to move along the new drive surface *B*, from the old drive surface *A*. For the cutter to make a right turn along the check surface *C*, the command would again be GØRGT. Then as soon as the cutter starts traveling along surface *C*, surface *C* becomes the new drive surface. Once the cutter moves just past surface *D*, as shown in Fig. 8–8(a), the command to move in the direction shown along surface *D* would be GØLFT. This last command might appear contradictory and perhaps a bit confusing, since an APT student might logically argue that the new direction is to the right and the command should, therefore, be GØRGT. The programmer must remember each time, however, that he is looking in the *direction of travel* of the cutter, and that the movement shown in Fig. 8–8(a), along surface *D* from surface *C* requires a left turn when looking in the direction of cutter travel along surface *C*. At first

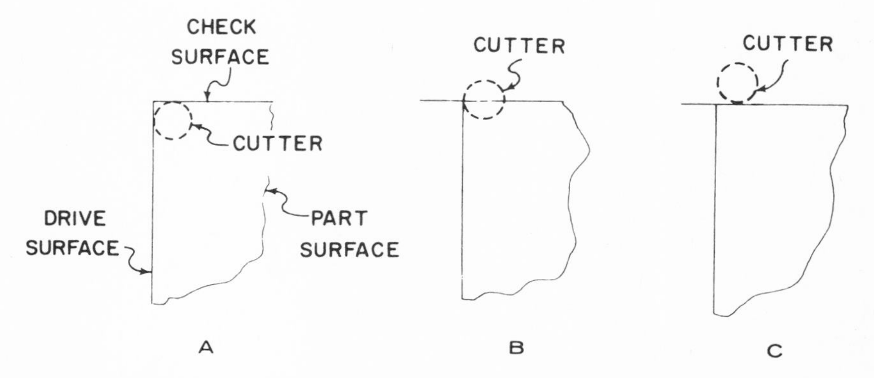

FIG. 8–6. Cutter moves in relation to the check surface. (A) A move *to* the check surface. The side of the cutter just touches the check surface. (B) A move *on* the check surface. The center of the cutter is on the check surface. (C) A move *past* the check surface. The side of the cutter just touches the check surface.

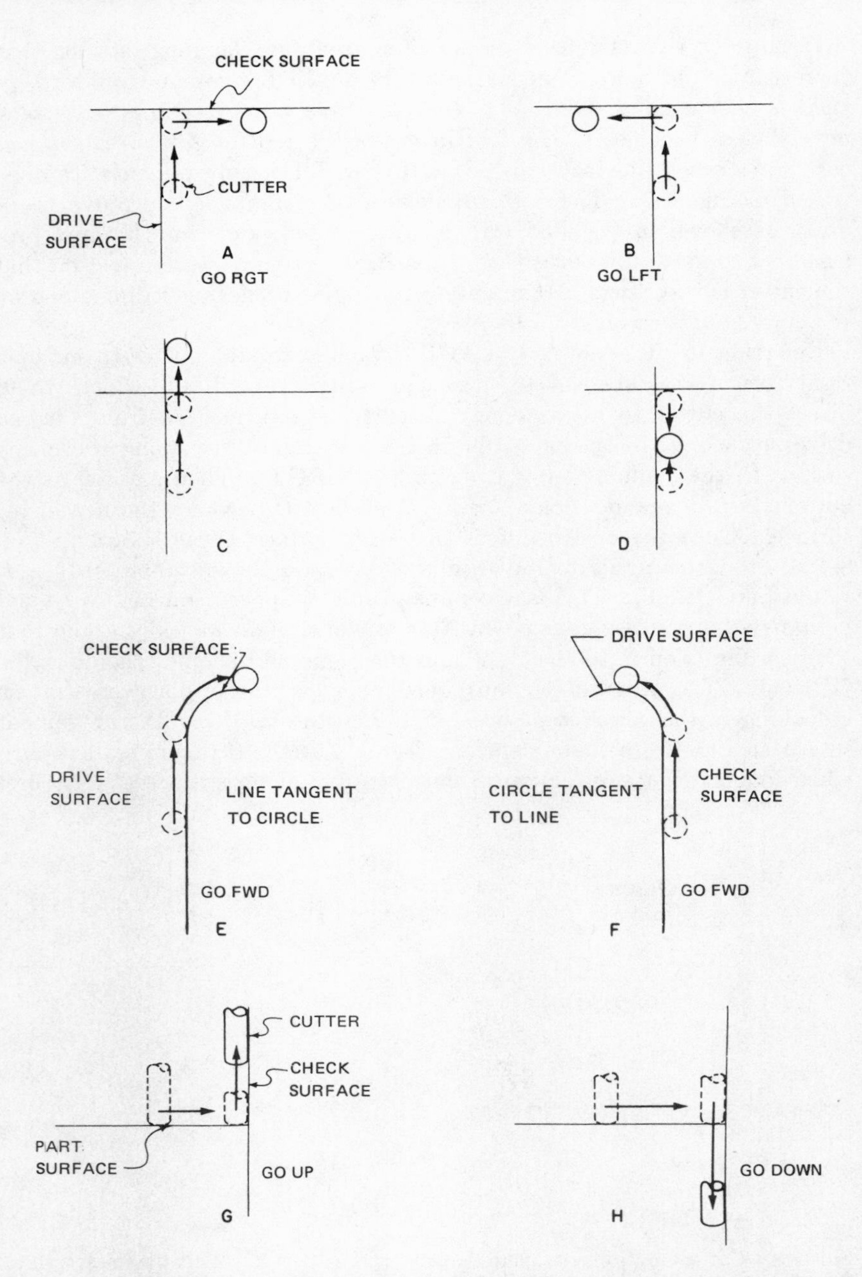

Fig. 8–7. In (A), (B), (C), (D), (E), and (F) the part programmer is looking directly down on the movements so that the check surface and drive surface appear as lines. In (G) and (H) the part programmer is looking from the side.

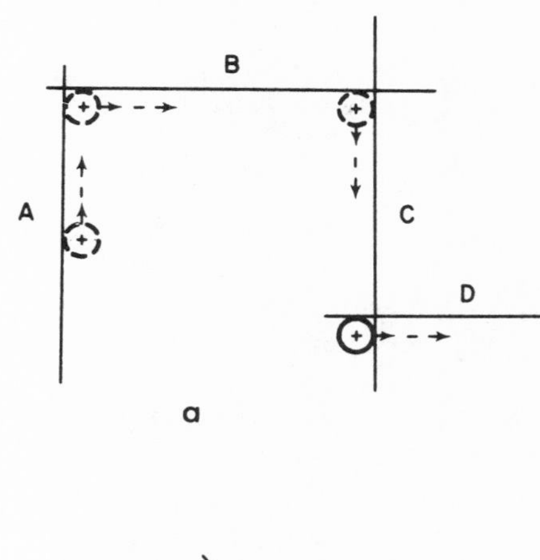

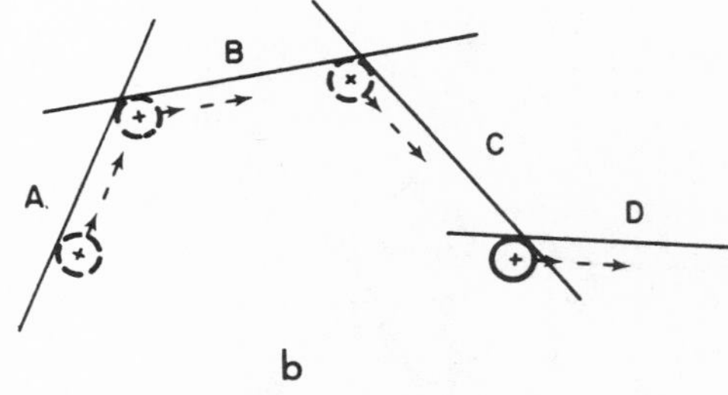

FIG. 8–8. The part programmer looks in the direction of travel of the cutter to determine the direction for the commands. In both diagrams from A to B would be GØRGT. From B to C would also be GØRGT. From C to D would be GØLFT.

the programmer may feel it necessary to "stand on his head" to get the proper viewpoint; however, a little imagination and practice will soon solve this problem and allow him to remain seated while writing a whole program. The lines or surfaces need not be at right angles to each other but can be oriented at any angle, as shown in Fig. 8–8(b). It is only necessary that these lines be properly defined before the cutter is directed to move about them.

Point-to-Point Motions

When a cutter is at rest and we wish to move it to a specified point, the motion command would be GØTØ. If there is no information in the computer regarding the coordinates of the point *from* which the cutter is to move, a FRØM command must be used. For example, if the cutter tip is to move from a point noted as P1 to another point noted as P2,

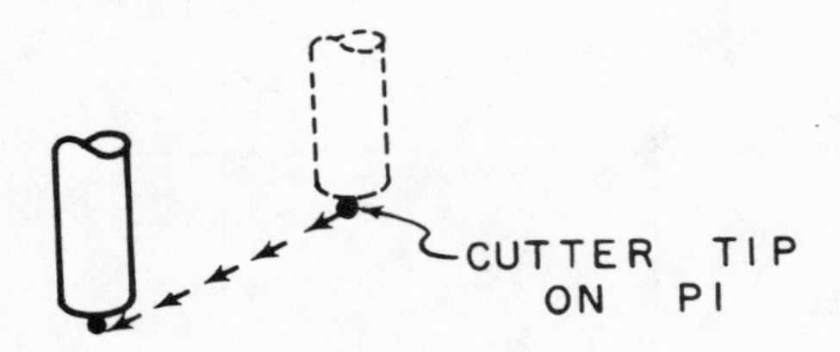

FIG. 8–9. The tip of the cutter is shown moving from P1 to P2.

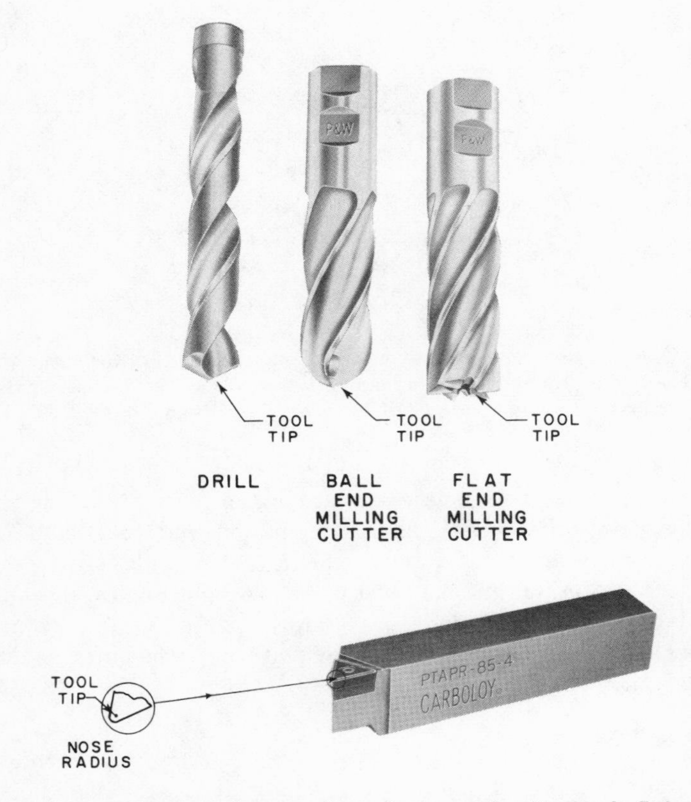

FIG. 8–10. Tool tip locations for several popular types of cutting tools. It is the movement of the tool tip that the computer calculates. With single-point lathe tools having a radius corner, it is the movement of the center of the radius that is calculated (as in Fig. 6–27). If the corner comes to a point or the radius is very small, then the corner of the cutter may be considered to be the "tool tip."

the *two* commands would be:

FRØM/P1

GØTØ/P2

This is illustrated[1] in Fig. 8–9. The word FRØM is an instruction and also asks the question *from where*. The answer, which appears after the slash (/) mark, notes that P1 is the location from which the cutter is moving.

The second statement:

GØTØ/P2

is also an instruction, or motion statement, and asks the questions *go to where*. The answer, shown after the slash (/) mark, is P2. In carrying out these instructions, it is the tip of the cutter that moves from P1 to P2. This applies to all machine tool motions, and the coordinate information on the numerical control tape would refer to the motions of the tool tip. Figure 8–10 illustrates tool-tip locations for several common types of cutting tools.

Moving a tool tip from one point to another can be continued indefinitely. This is seen in Fig. 8–11, where the cutter is moved from P1 to P2, then to P2, then to P4, then to P5, then to P6, and finally to P7. The APT programming statements would appear as follows:

FRØM/P1
GØTØ/P2
GØTØ/P3
GØTØ/P4
GØTØ/P5
GØTØ/P6
GØTØ/P7

For the computer to prepare the necessary tape commands to move the tool tip from one point to another, the coordinate locations of the points must be stored in the computer's memory. This means that the part programmer must feed the computer the coordinate locations of the points before referring to them. It should be pointed out that P1, P2, P3, etc., are *symbols* for the particular locations. Almost any other symbol could be used—even CAT1, or CAT2, or MOUSE3 if we care to. The symbols,

[1] In this case the tip of the cutter is shown as moving in a straight line from P1 to P2. This would happen when a numerical control machine has a three-axis contouring system. However, in point-to-point and two-axis contouring systems, the actual path that the cutter takes will most likely not be a straight line. Straight-line paths have been shown throughout the text in order to simplify the illustrations.

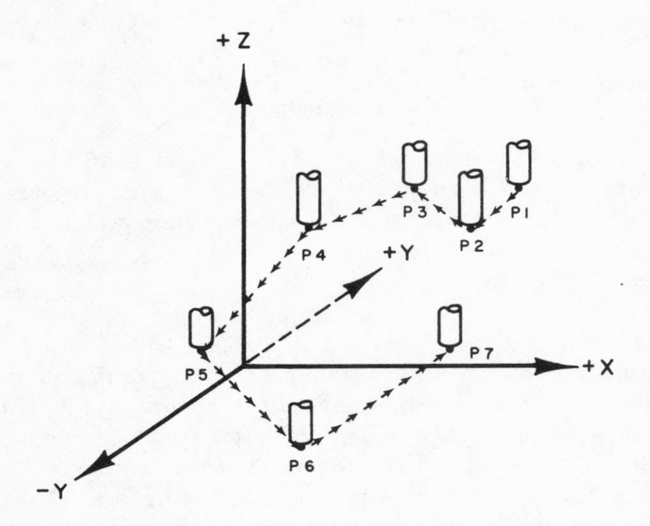

FIG. 8-11. The tool point or the tip of the cutter may be moved to any point within the range of the machine by using GØTØ statements. Here the tool tip moves from one point to another in a straight line. This would be true only for three-axes contouring control systems. For point-to-point or two-axis systems the path taken when moving from one point to another will usually *not* be direct.

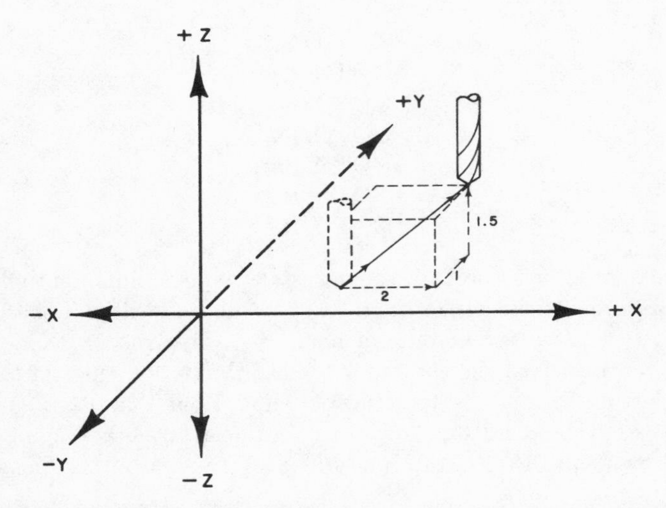

FIG. 8-12. The GØDLTA statement is used to move the cutter an *incremental* distance. In the sketch above the cutter has moved 2 inches in the $+X$ direction, 1 inch in the $+Y$ direction, and 1.5 inch in the $+Z$ direction. The APT statement would be: GØDLTA/2,1,1.5. It should be noted that, in order to move along the straight line shown, the NC machine would have to have three simultaneous axes capability.

however, have to be described *before* they are used in a statement. This is done by giving the X, Y, and Z coordinates. In this case, X, Y, and Z coordinates of P1 would have to be given *before* the symbol P1 is used in a statement such as FRØM/P1. Just how points and other geometry such as lines and circles are described in the APT language will be covered in the next chapter.

Another motion statement frequently used in point-to-point operations is called the GØDLTA statement, which instructs the cutter to move a given distance. For example, referring to Fig. 8–12, the tip of the drill is instructed to move a distance of 2 inches in the $+X$ direction; 1 inch in the $+Y$ direction; and 1.5 inch in the $+Z$ direction. The APT statement would be:

GØDLTA/2,1,1.5

This statement is particularly helpful when drilling, boring, tapping, or reaming holes, where the depth, or Z motion, has to be controlled. For example, if we were to program a three-axis machine (one having Z-axis control) to drill the holes shown in Fig. 8–13, the program would appear

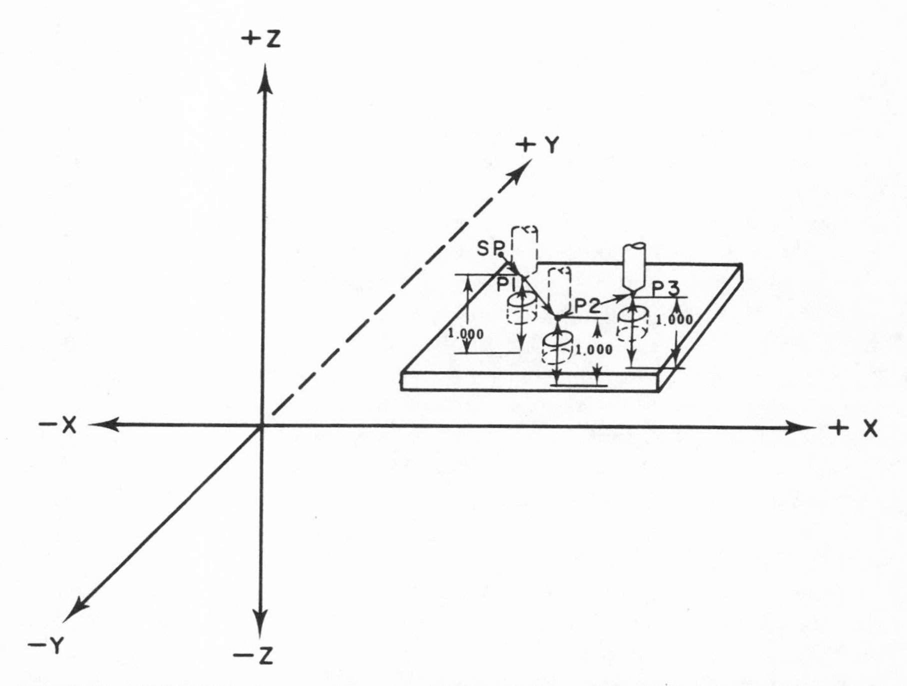

Fig. 8–13. GØDLTA statements are useful for point-to-point drilling, boring, tapping, and reaming operations.

as follows:

```
FRØM/SP
GØTØ/P1
GØDLTA/0,0,−1.000
GØDLTA/0,0,1.000
GØTØ/P2
GØDLTA/0,0,−1.000
GØDLTA/0,0,1.000
GØTØ/P3
GØDLTA/0,0,−1.000
GØDLTA/0,0,1.000
```

In using the GØDLTA statement, it is not necessary to determine the z coordinate point to which the cutter moves, but rather the incremental distance that it moves. With the first and following GØDLTA moves, for example, there is no motion in the X or Y direction, and therefore zeros (0) are shown.

Referring again to the GØTØ statements in which the cutting tool is directed to a specific point, the student might rightfully ask what is to be gained by using APT if the coordinates of each point have to be calculated and noted since this is precisely what has to be done with manual programming. The answer is that there is actually little to be gained if the cutter must move through an irregular pattern of points, except that the tape is punched automatically by the computer. However, APT is useful when the part programmer is concerned with *patterns* of points and when the computer can automatically *calculate* the coordinates of these points. An example of this would be a circular bolt hole pattern where the instructions on the tape must describe the coordinates of the center of each hole. In APT, essentially all the part programmer need note is the location of the center and the radius of the circular pattern on which the holes are located as well as the number of holes in the pattern. Having this information, the computer can calculate all of the coordinates and go on to prepare the necessary numerical control tape. Also, APT was designed initially for contouring applications such as those with milling machines and lathes, and it is in these machining applications that APT is most helpful to the part programmer.

Contour Motions

When considering a *contouring* application, such as milling, the part programmer must keep in mind the relationship of the *drive* surface, the *part* surface, and the *check* surface. Please refer to a description of this earlier in the chapter and to Fig. 8–5. As the cutter is directed over these surfaces, the computer calculates all of the coordinate points or incremental moves that must be punched on the tape. The computer also notes the associated proper code for turning the spindle *on* or *off*, for turning the coolant *on* or *off*, or for performing any other automatic function required of the machine.

Before *contour* motion commands, such as GØRGT, or GØLFT, can be given, several prior motion statements, called start-up statements, must be made to direct the cutting tool to the position from which it is to start operating. To bring the cutter to a precise location all three surfaces are

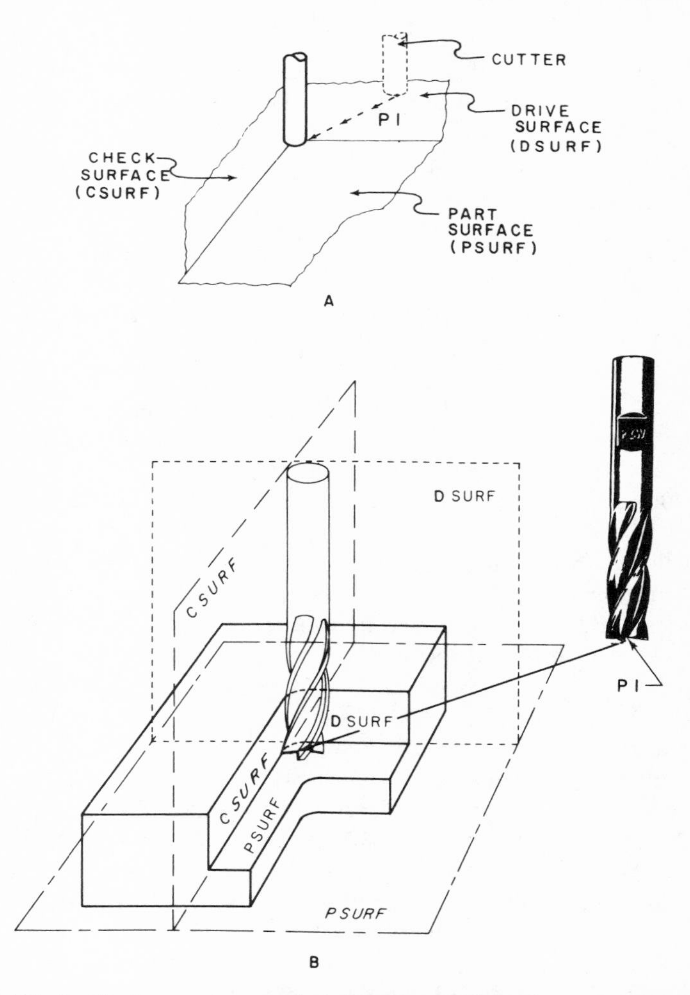

FIG. 8–14. (A) In starting contour motions the cutter is brought to the corner where the drive, part, and check surfaces intersect. (B) A three-dimensional illustration of the movement to the surfaces of a physical part.

used, namely, the *drive* surface, the *part* surface, and the *check* surface. See Fig. 8–14(A). The drive surface is noted by the symbol DSURF, the part surface by PSURF, and the check surface by CSURF. If we command the cutter to go from P1 *to* the corner of DSURF, PSURF, and CSURF, the cutter will move as shown in Fig. 8–14(B), with the tip of the cutter resting on PSURF and the *side* of the cutter touching DSURF and CSURF. The *start-up* statements for these motions are:

```
FRØM/P1
GØ/TØ,DSURF,TØ,PSURF,TØ,CSURF
```

The sequence in which the surfaces are noted in the GØ/TØ statement is most important, and must always be listed in the following order:

1. Drive surface
2. Part surface
3. Check surface.

Since the computer is programmed to receive the surfaces noted in the start-up statement in this sequence, the program will fail if they are not in this order. Repeating the example, the correct order is then:

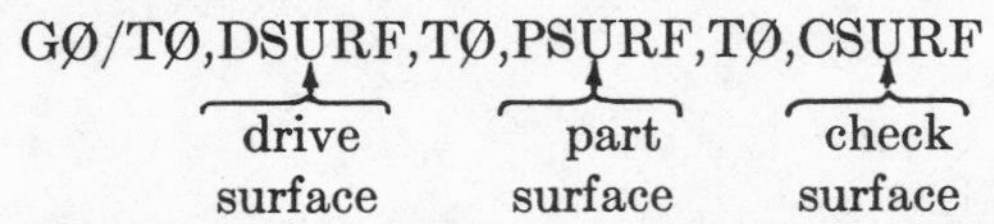

To review, the drive surface is the surface that the side of the cutter will move along after it has moved to the intersection of the three surfaces as a result of the GØ/TØ statement.

It will also be noted that the slash (/) mark lies between the GØ and the TØ rather than after the TØ as when the cutter is commanded to go to a point. Also, commas separate the words in the statement, no comma being necessary after the last word, CSURF.

DSURF, PSURF, and CSURF, like P1, P2, P3, and the symbols for the other points, are also symbols and refer to surfaces, or planes, defined earlier in the program. In fact, later in this chapter ASURF will be used as a symbol for the drive surface and BSURF for the check surface. Rules for describing symbols will be discussed in the next chapter.

It should also be pointed out that, in order for the side of the cutter to touch the drive surface and the check surface, the computer must know the diameter of the cutter. The reason is that the position of the tool tip lying on the center of the cutter, which is what the computer is calculating, will be *offset* from each of the two surfaces by a distance equal to one-half the diameter of the cutter—that is, the radius. This is illustrated in Fig. 8–15. Therefore, before giving the start-up instructions, the part programmer must note the actual *diameter* of the cutter with the following

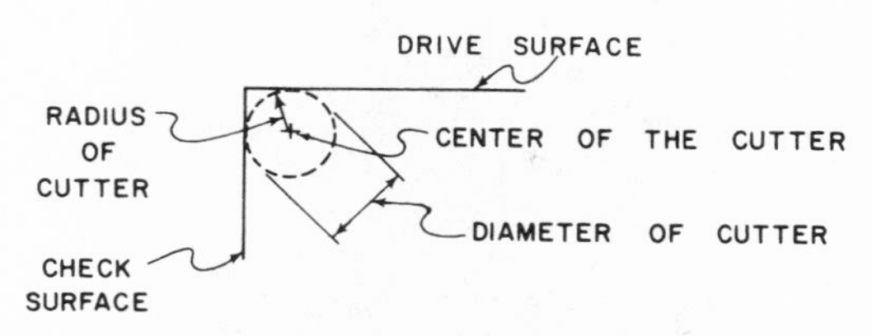

FIG. 8–15. The tool tip of the cutter is offset a distance equal to the cutter radius from the drive and check surfaces.

type of statement:

CUTTER/diameter

If, for example, the cutter is 1 inch in diameter the exact statement would be:

CUTTER/1

It will be observed that the slash (/) mark separates the *general* description, such as the word CUTTER, from the specific description, such as the number *1*, which states that the cutter diameter is *1* inch. As with the FRØM statement, already described, usually the *general* words are placed to the left of the slash (/) mark while the words specifying a particular operation or part being programmed follow the slash mark.

Following the GØ/TØ statement, the cutter must be directed along a path that will result in cutting out the necessary material to form the part. As explained earlier, the part programmer should imagine that he is looking in the direction of travel of the cutter and *steer* the cutter as though he were steering a car. The word GØRGT, for example, means that the cutter is to make a *right* turn. A left turn would be GØLFT. Other expressions, such as GØTØ, GØPAST, may also be used.

An entire cutting sequence involving start-up statements and follow-on statements is shown in Fig. 8–16. In Fig. 8–16(a) the cutter is moved about imaginary surfaces, or planes; and in Fig. 8–16(b) the cutter is moved about a block of material. In both cases the APT statements are *exactly the same* since it is immaterial to the computer whether the machine is cutting metal or not.

The first movement of the cutter is from the set-point noted by the symbol SP, to the intersection of the surfaces, or planes *A* and *B*. If the next movement is to be to the rear, then plane *B* is the *drive* surface, plane *A* is the *check* surface, and the bottom of the cutter will rest on the *part* surface as it moves around surfaces *B*, *C*, and *D*. As the cutter moves along surface *B* toward surface *C*, then surface *C* becomes the check surface. In this case the cutter should move just *past* the check surface *C*, after which the cutter is instructed to make a left turn with the command

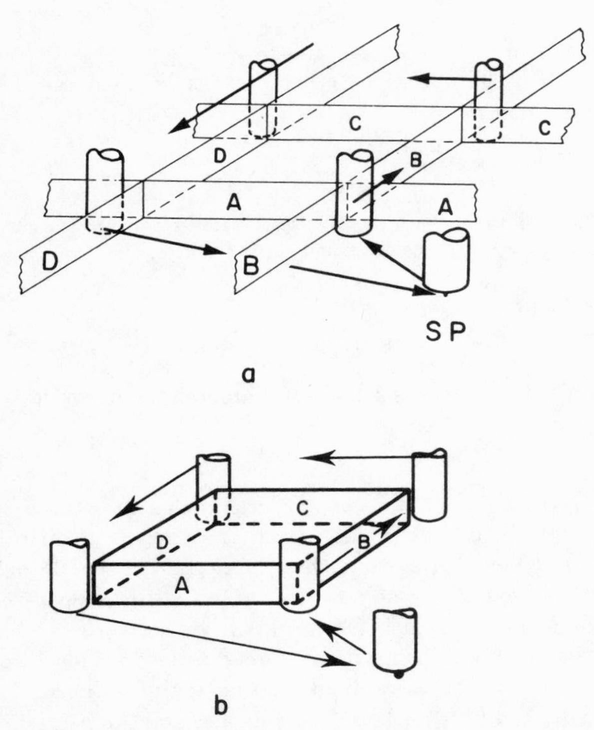

FIG. 8–16. (a) The side of the cutter is guided by surfaces, or planes, which are known as the *drive* surface, the *part* surface, and the *check* surface. In (b) the surfaces appear as a rectangular block of material. The part surface is not shown in either case.

GØLFT. The cutter would then move just *past* the new check surface D. Again, the cutter would make a left turn with the command to GØLFT and head past surface A. After just passing surface A, the cutter would be commanded to move back to the set-point, SP. Except when moving to and from SP, the cutter is being guided by a drive surface and a part surface. The part surface which controls the motion of the tool either *up* or *down* is constant. However, the drive surface changes each time the cutter makes a turn. It should also be pointed out that the surfaces extend indefinitely, as far as the computer is concerned, and are cut off to the desired dimension by the motions of the cutter. Thus, when the cutter makes a left turn onto surface C it cuts the length of surface B to the required dimension.

If the movements are viewed from above, the surfaces appear as straight lines and the cutter appears as a circle. See Fig. 8–17(a). The surfaces could therefore be reasonably considered as *lines*, which is more consistent with the blueprint description of a part. Also, since we do not need the extensions of the lines in a part program, these may be dropped, and the configuration would appear as a normal rectangle as shown in Fig. 8–17(b). It should be kept in mind, however, that the *computer* looks at the configuration as having surfaces (or lines) that extend indefinitely.

If the configuration shown in Figs. 8–16 and 8–17 were to appear as a part on a blueprint, or engineering drawing, it would look like that shown in Fig. 8–18. In this case the depth of the part is shown as well as the top surface. Figure 8–19 shows the top, side, and perspective views of the part that is shown in the blueprint of Fig. 8–18, as it might appear on the worktable of an NC machine. PSURF, the imaginary *part surface* in this case, is located below the bottom surface of the part. Thus, when the tip of the cutter is instructed to move to the part surface, the bottom of the cutter will lie below the part so that the side of the cutter will cover the full depth of the part.

Referring again to Fig. 8–16, if we designate surface *A* by the symbol ASURF; surface *B* as BSURF; surface *C* as CSURF; surface *D* as DSURF

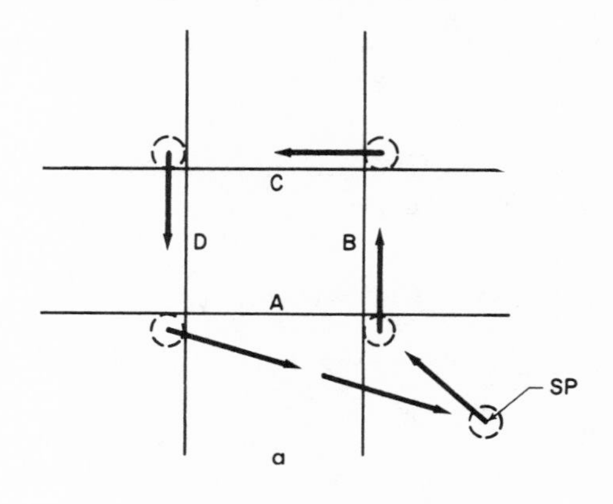

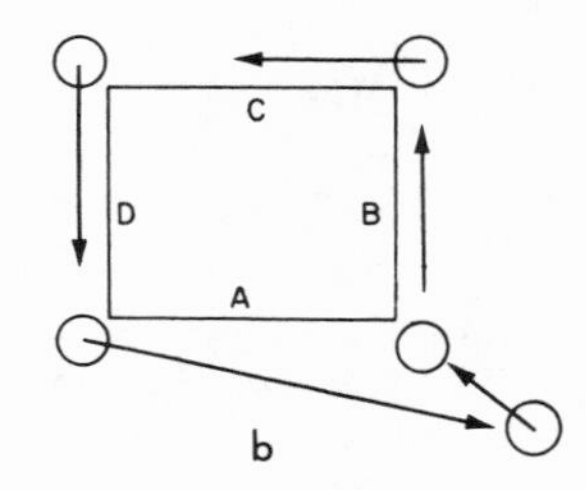

FIG. 8–17(a and b). The surfaces shown in Fig. 8–16 would appear as lines and the cutter as a circle if we were to look directly down on the figure.

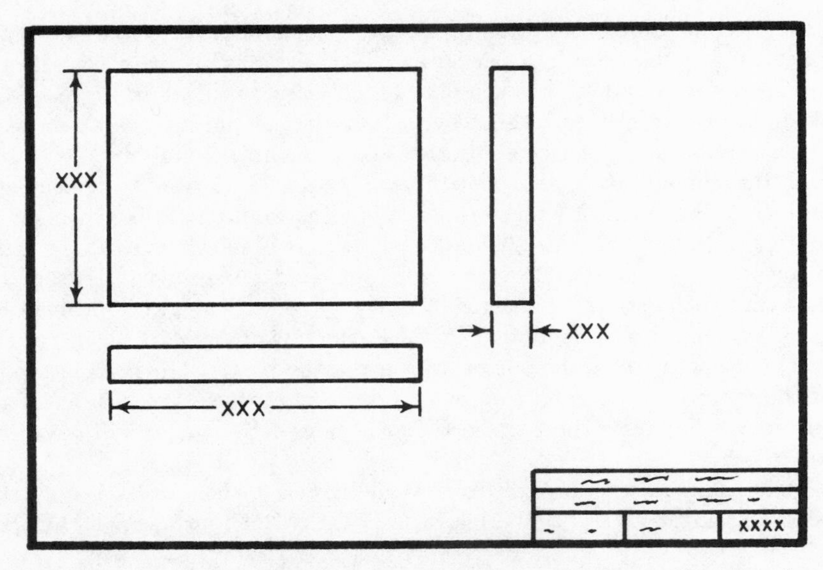

FIG. 8–18. A blueprint or engineering drawing of the rectangular configuration shown in Fig. 8–16 would appear as shown above.

and the part surface as PSURF, as shown in Fig. 8–19(A) and (B), then the start-up statements would be:

```
FRØM/SP
GØ/TØ,BSURF,TØ,PSURF,TØ,ASURF
```

In accordance with the rule for the order of designating start-up symbols, BSURF is the *drive* surface, PSURF is the *part* surface—which in this instance lies below the surface of the part as shown in Fig. 8–19(B)—and ASURF is the *check* surface, and the surfaces must be *shown in this order*. The next motion statement in the program would direct the cutter to make a right turn and go along BSURF, past CSURF, which is now the check surface. This statement would be

```
GØRGT/BSURF,PAST,CSURF
```

Next the cutter would be instructed to go left along CSURF, which is now the drive surface, and past DSURF, which is now the check surface. The next motion statement would instruct the cutter to make a left turn along DSURF and go past ASURF. These two statements would appear as follows:

```
GØLFT/CSURF,PAST,DSURF
GØLFT/DSURF,PAST,ASURF
```

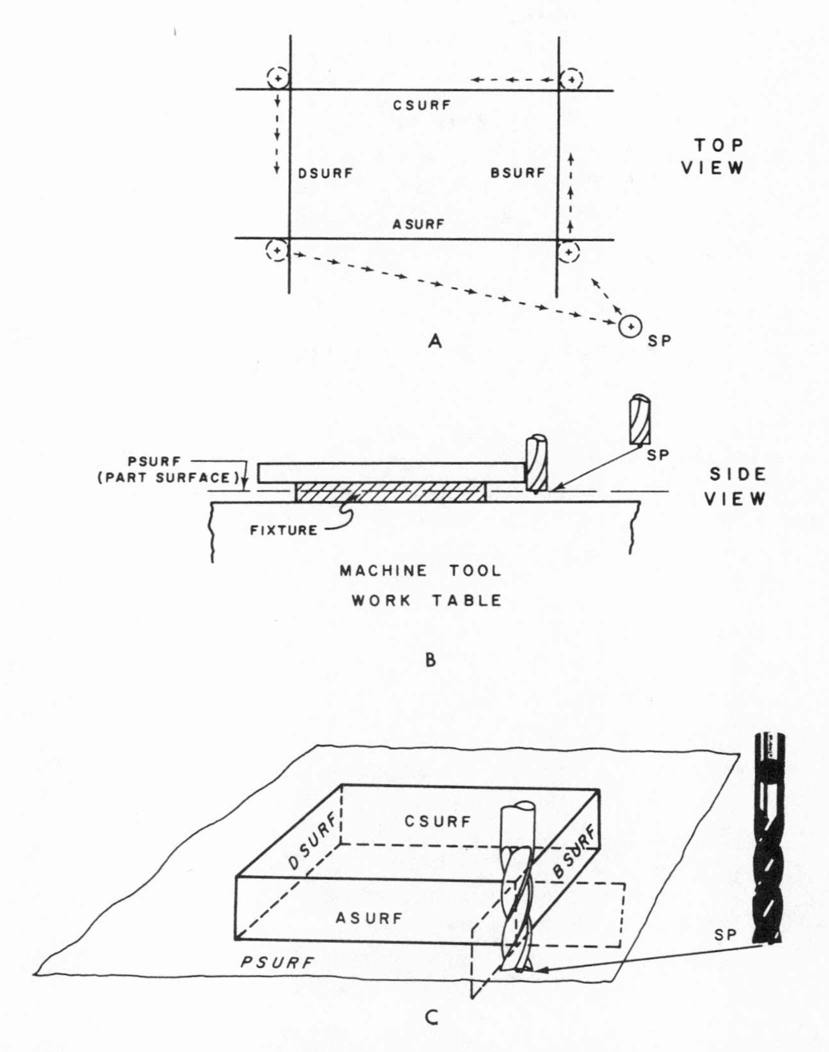

FIG. 8–19. (A) The cutter is instructed to move from the set-point, SP, to the corner of the part. (B) The bottom of the cutter is to move to a point below the bottom surface of the part so that the full edge of the part is covered by the side of the cutter. (C) The cutter is then to move around three surfaces of the rectangular part, and then back to SP.

The last statement would instruct the cutter to move directly to SP, which, in this case, is the starting point. This statement would be:

GØTØ/SP

All of the motion commands, as they would appear on the part programmer's manuscript, are as follows:

```
FRØM/SP
GØ/TØ,BSURF,TØ,PSURF,TØ,ASURF
GØRGT/BSURF,PAST,CSURF
GØLFT/CSURF,PAST,DSURF
GØLFT/DSURF,PAST,ASURF
GØTØ/SP
```

PRACTICE EXERCISE NO. 1 CHAPTER 8

1. Why is APT described as a three-dimensional system?
2. Is it necessary for the surfaces guiding the cutting tool to be actual surfaces on the part being machined?
3. Write the APT commands that correspond to the following illustrations. For example, (a) would be GØFWD.

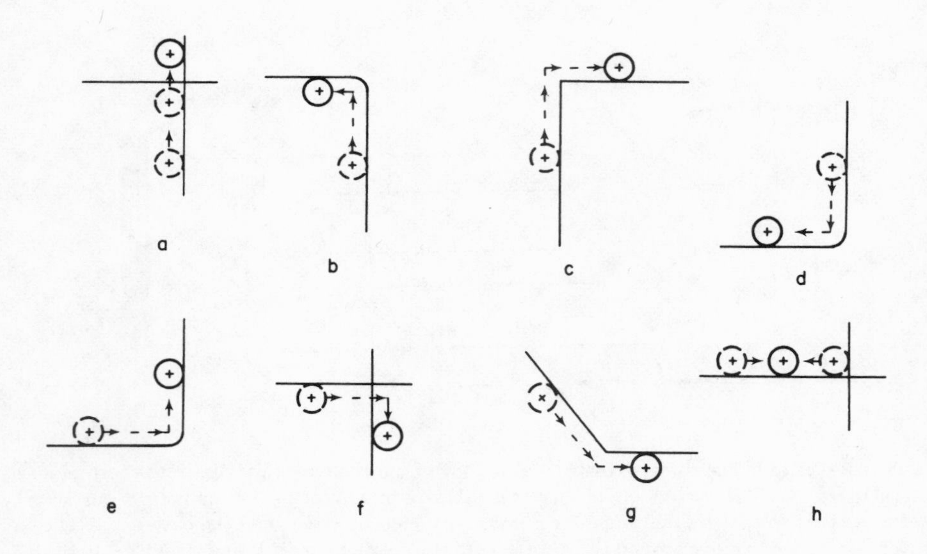

4. In the following illustration, note which is the *drive* surface and which is the *check* surface when the cutter is at the following positions: (a) position a, (b) position b, (c) position c, (d) position d, (e) position e.

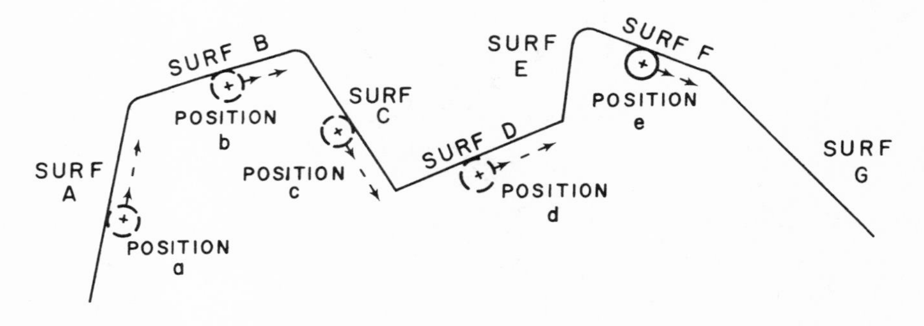

5. In the sketch below, the tip of the cutter is to move from the set-point (SP) to the three points shown and then back to the set-point. Write the APT *motion* statements for these moves.

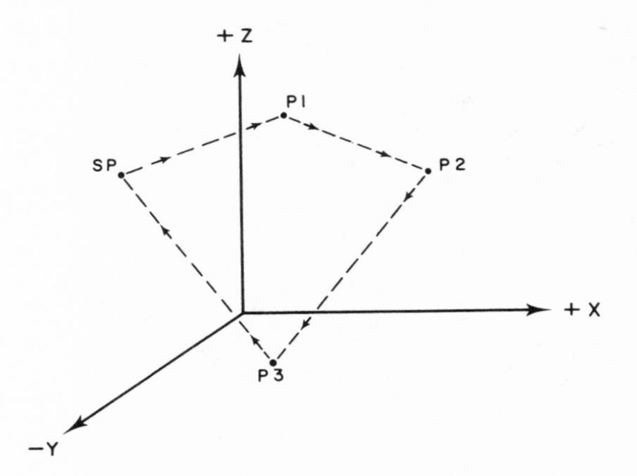

6. What are names for the three surfaces that the cutter moves to in a start-up statement?
7. Is it important that these surfaces be listed in any kind of order in the start-up statement? What will happen if they are not?
8. What is the cutter statement for a ½-inch-diameter cutter? For a ⅛-inch-diameter cutter? For a 2¼-inch-diameter cutter?
9. In the sketch below, list the motion statements for moving the cutter from a set-point (SP) to three intersecting planes, and then around the three sides of the triangle, passing ASURF, then back to SP, as shown by the direction arrows. The part surface, PSURF, lies below the bottom surface of the part.

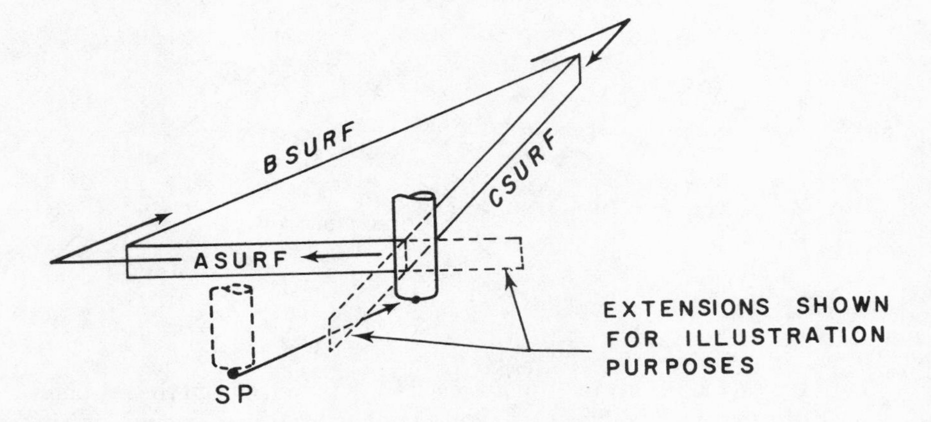

10. The rectangular block shown below requires both drilling and milling operations. Write the motion statements for drilling the holes and then for milling the perimeter of the part. The drill depth motion of 3 inches for all three of the holes will drill the holes through. The path of the drill and of the milling cutter is also shown. The drill is

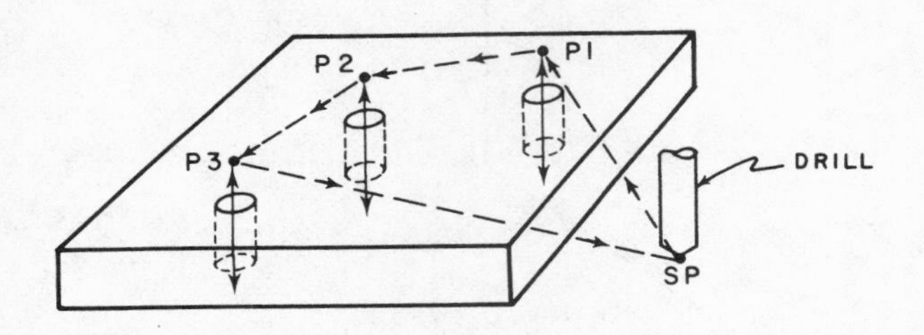

DRILLING OPERATION

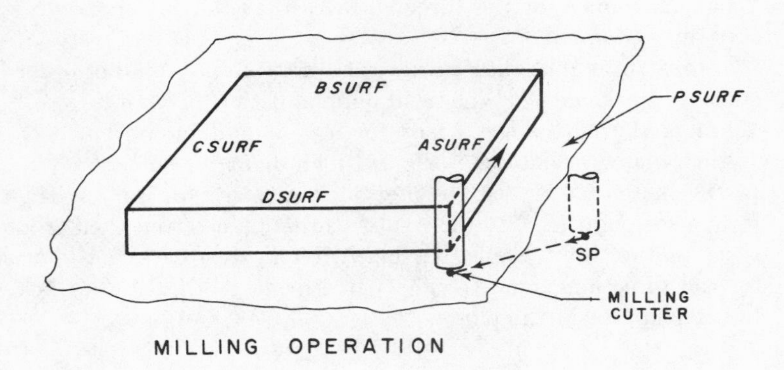

MILLING OPERATION

to move from SP; then to P1; to P2; to P3; and then back to SP where the motion is to stop so that the operator can change to the milling cutter. The APT word for stopping the machine is simply STØP. Machine operation may be started again by pushing the "start" button. Dimensions have not been shown since it may be assumed that the part has already been mathematically described in APT statements. Methods for describing the dimensions and geometry of a part are covered in Chapter 9.

Two-Surface and One-Surface Start-up Statements

Thus far, three-surface start-up statements have been mentioned wherein the drive surface, part surface, and check surface must be noted. It is also possible to have start-up statements involving only two surfaces—namely, the drive surface and the part surface. In this case the cutter would move from the set-point to the drive and part surfaces along the shortest path, which is often *perpendicular*, or normal, to the drive surface. This is illustrated in Fig. 8–20. The start-up statements would be:

FRØM/SP
GØ/TØ,DSURF,TØ,PSURF

Even if PSURF is not noted in the program, the cutter may be moved in a perpendicular direction to the drive surface. The part surface in this

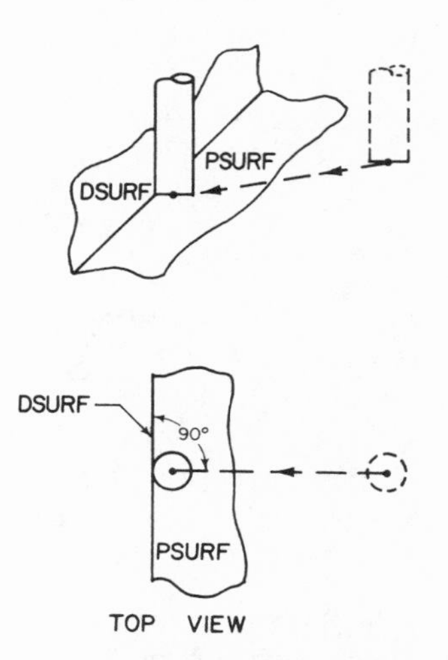

FIG. 8–20. When a two-surface, or one-surface start-up is used, the cutter will move along the shortest path to the drive surface. When a flat drive surface is considered, this path would be perpendicular, or normal, to the drive surface if the move is viewed from above.

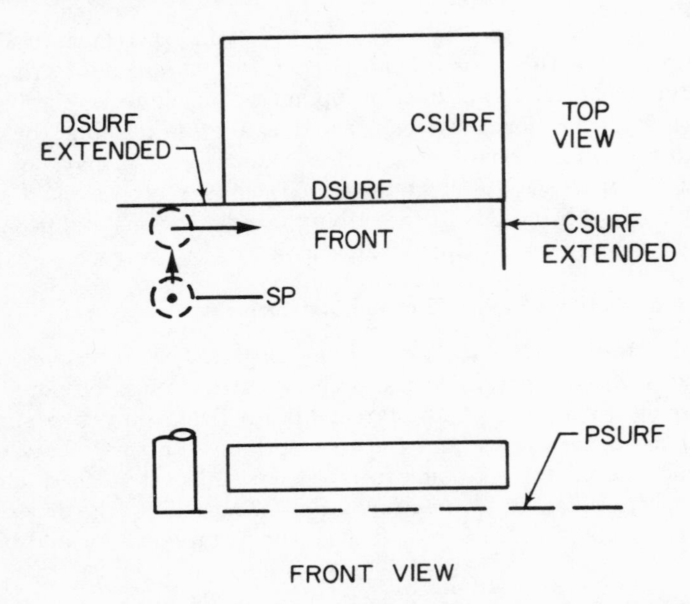

FIG. 8–21. An example of a two-surface start-up in which the cutter will move in a perpendicular direction toward the drive surface.

instance is assumed by the computer to be the plane that passes through the X and Y axes where z is equal to zero. Here the start-up statements would be:

```
FRØM/SP
GØ/TØ,DSURF
```

The tip of the cutter will now move to the X-Y plane which is now the part surface and where z equals zero.

While two- and even one-surface start-ups are permissible, beginners should use three-surface start-up statements in order to achieve a precise and known position from which to begin the cutter motions.

An example of a two-surface start-up is shown in Fig. 8–21, where it is desirable to start the cutting cycle *off* the part so that the cutter may move smoothly into the material. The start-up statements and the first motion statement are as follows:

```
FRØM/SP
GØ/TØ,DSURF,TØ,PSURF
GØRGT/DSURF,PAST,CSURF
```

In this case we could even use a one-surface start-up statement, and the first motion statement would be:

```
FRØM/SP
GØ/TØ,DSURF
GØRGT/DSURF,PAST,CSURF
```

PRACTICE EXERCISE NO. 2 CHAPTER 8

1. The start-up statements for a particular program are as follows:

```
FRØM/SP
GØ/TØ,DSURF,TØ,PSURF
```

   Which of the three sketches shown below describes the path of the cutter?

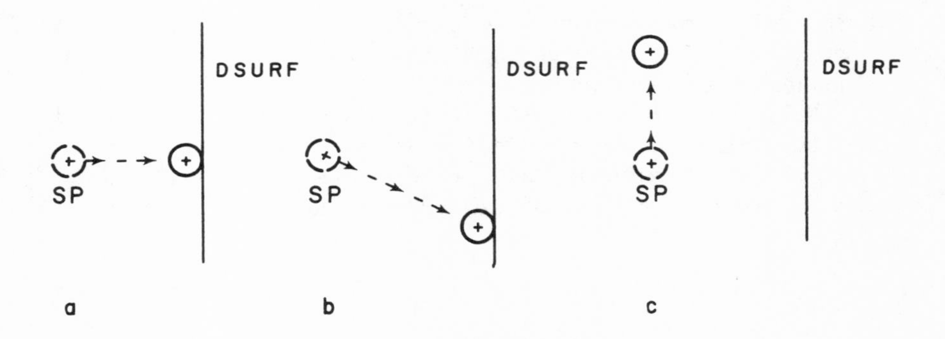

2. Using one-surface start-up statements, write the correct statements for the sketches shown below:

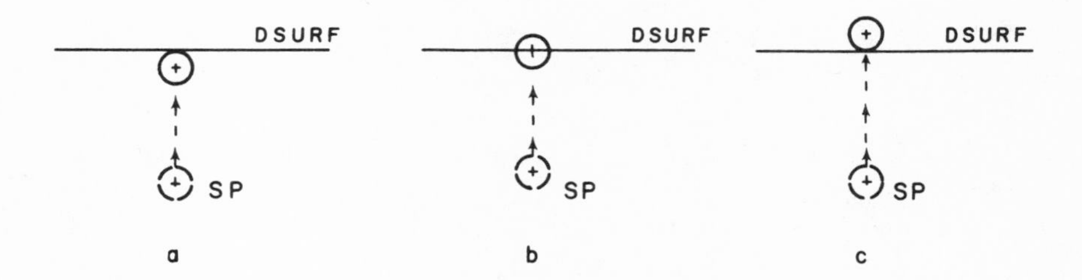

3. The part shown must have a rectangular groove cut in it. The cutter starts at the position shown and is at the proper depth which is the x-y plane. The position of the cutter at the start is also the set-point (SP). The outsides of the groove have also been labeled and are to be used in

the motion statements. Write the motion statements for cutting the groove.

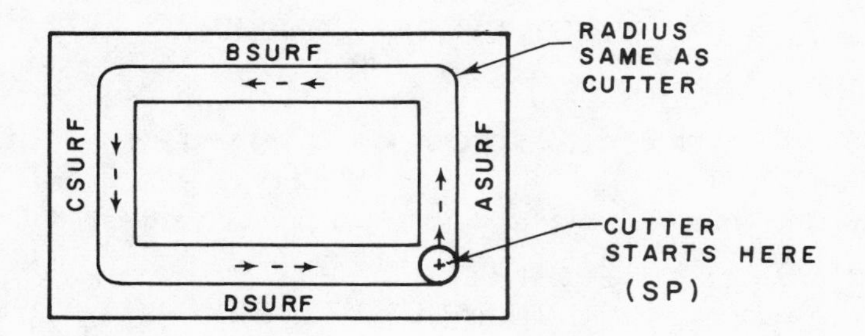

QUESTIONS CHAPTER 8

1. In addition to reducing the calculations required of the part programmer, what other chief advantage does the APT system have over manual part programming?
2. What are the two surfaces called that guide a cutting tool along a path?
3. When approaching a *check* surface what are the three alternatives regarding the location of the cutter with respect to the check surface?
4. What would be the appropriate APT commands in the illustrations below?

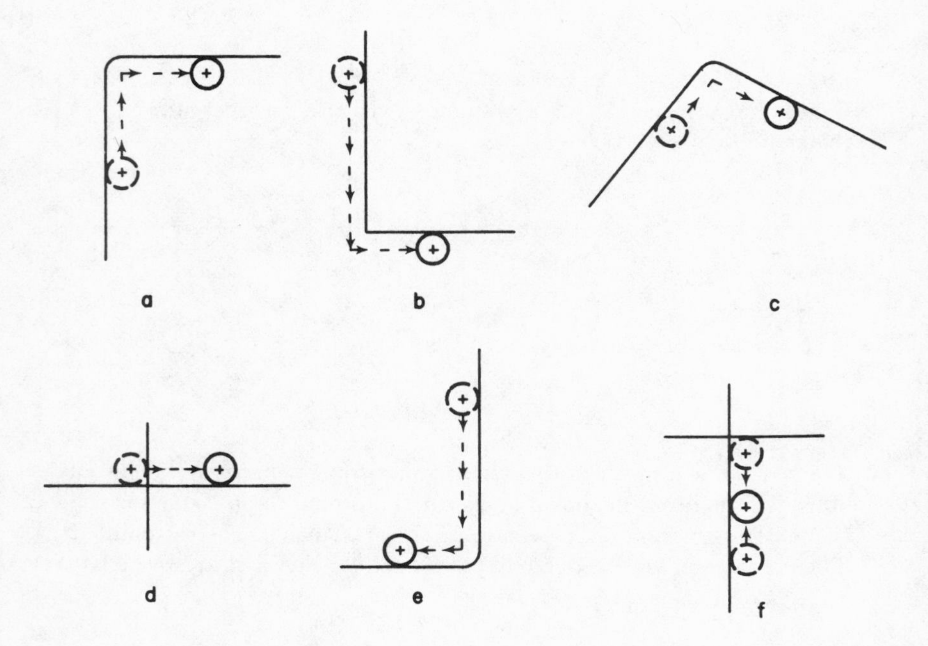

5. It is desired to drill a hole in the section of the part shown below. The tip of the cutting tool is to move from a set-point, SP, to a point directly over the hole, P1. The drilling cycle will bring the tip of the cutter to a point below the hole noted as P2. The drill tip will then retract to point P1. From P1 the drill tip will return to SP. Write the APT start-up and motion statements that describe this operation. (Note: DELTA statements could also have been used.)

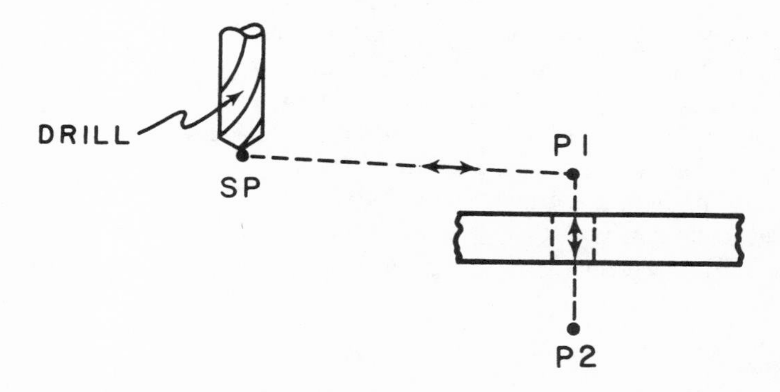

6. In the sketch below, the flat-end milling cutter is located at the corner of the intersection of the three surfaces shown. Which two of the three surfaces might the cutter be expected to move along from its

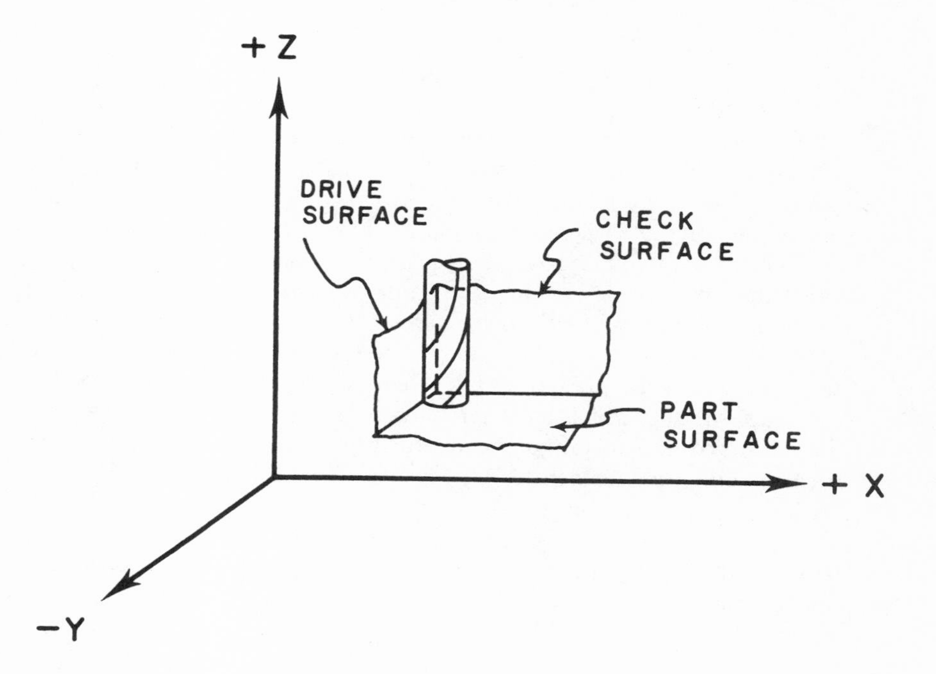

present location? If the coordinates of the corner point are $x = 3.000$, $y = 2.000$, and $z = 2.000$, what are the coordinates for the location of the tip of the cutter in the position shown? The diameter of the cutter is 1 inch.

7. To describe movement to three intersecting surfaces in a start-up statement, in what order must the surfaces be noted?
8. Note the *cutter* statement for each of the following cutters:
   (a) A flat end mill having a radius of .250 inch,
   (b) A flat end mill having a diameter of 2.000 inches,
   (c) A reamer having a constant radius of .500 inch,
   (d) A drill having a diameter of .500 inch.
9. Two imaginary surfaces have been established, as shown below, in order to move the tip of the cutter from the point shown as SP to the inside corner of DSURF and CSURF. What are the two start-up statements? The tip of the cutter will rest on the part surface (PSURF), although this imaginary surface has not been shown.

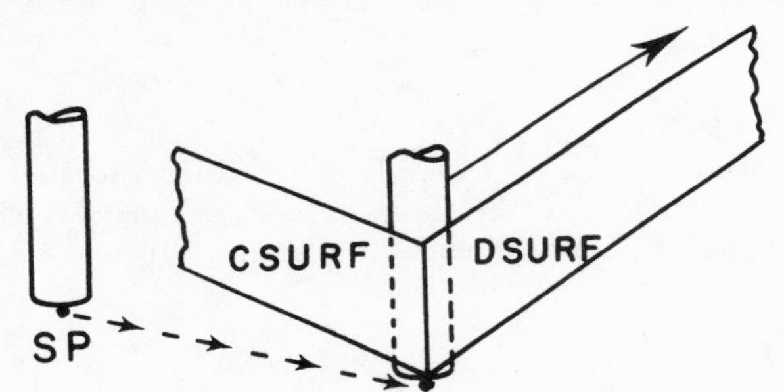

10. A pocket is to be cut in the part described in the drawings below. The radius in the corners of the pocket is .250 inch. Referring to the close-up view, also shown below, the tip of the cutter is to be targeted above one of the corners of the part. The cutter is then to move to any convenient point above the pocket so that the move to the three start-up surfaces can be made. This point is noted as P1. After going to P1 the tip of the cutter is to move to the bottom right-hand corner of the pocket. In this move the cutter will first be moving in air and then will cut material. When the tip of the cutter reaches the part surface, which has been chosen as the bottom of the pocket, it will move in a rectangular pattern in order to cut the sides of the pocket, and then to points P2, P3, P4, P5, P6, P7, P8, and P9, in order to clean out the pocket. From P9 the tip of the cutter is to move back to P1, and then to SP. Write the start-up and motion statements to perform this operation. Be sure to note the proper cutter statement.

10 (*cont.*)

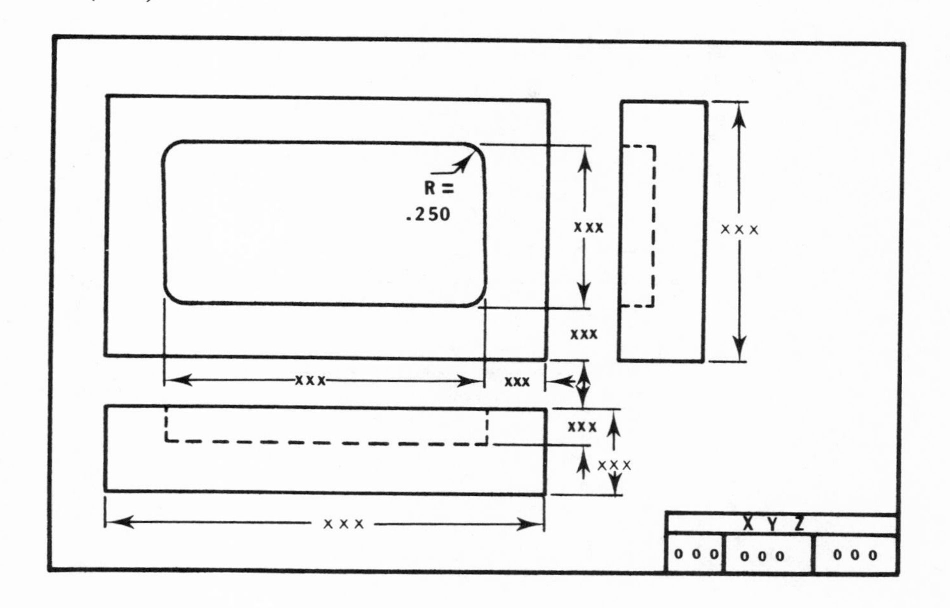

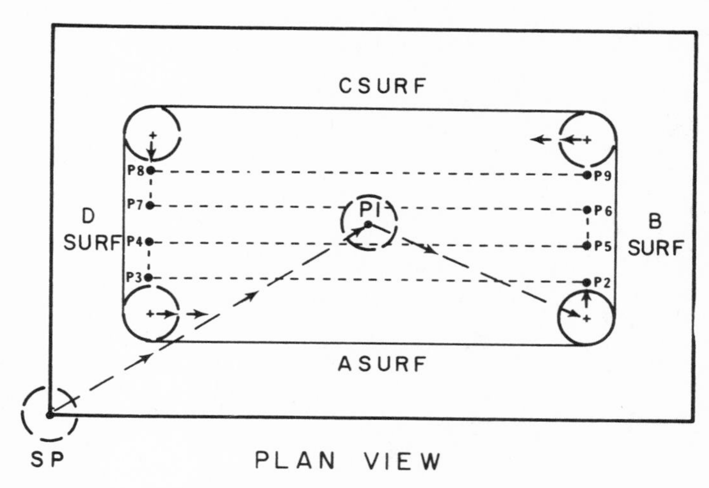

See next page for side view.

10 (*cont.*)

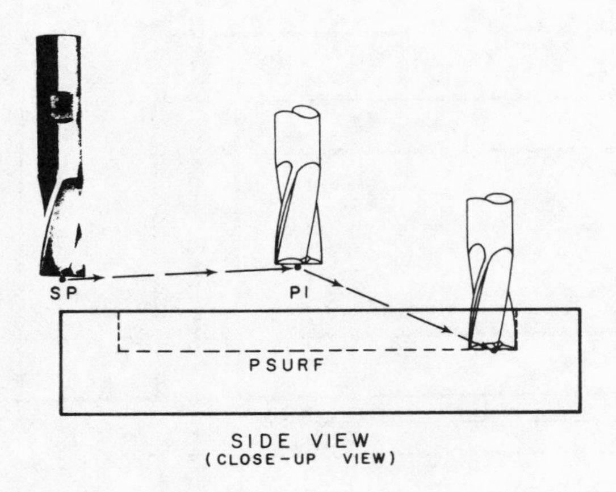

CHAPTER 9

Geometry, Auxiliary, and Post Processor Statements as used in APT

Chapter 8 dealt with the basic concepts of the APT system, particularly as applied to motion statements and the use of intersecting planes to guide the cutting tool. Actually there are *four* types of statements that make up a complete program. These are:

1. *Motion* statements, which describe the movements of the cutting tool.
2. *Geometry* statements, which describe the configuration of the part that the cutting tool is to move around.
3. *Post Processor* statements, which apply to the machine tool and control system.
4. *Auxiliary*-type statements, which cover just about anything that is not covered by the motion, geometry, and post processor statements.

An example of a *motion statement*, as we have seen in Chapter 8 is:

GØRGT/BSURF

Another example is:

GØTØ/P1

The first part of the statement, which lies before the slash (/) mark, tells the cutter *how* it is to move. The part of the statement that lies *after* the slash mark tells the cutter precisely *where* it is to move.

For the cutter to know precisely where to move, it is first necessary for the surface, or the point, or the line, or any other specific element, such as a circle, to be described. For example, if we were to instruct a cutter to move to a point noted by the *symbol* P1, we would have to describe P1 before using it in the motion statement. Otherwise the computer would not know where P1 was located and could not direct the cutter to it. Thus, if P1 appeared in a motion statement without being described beforehand in a geometry statement, the program would fail.

Geometry Statements

The purpose of the geometry statement is to describe the configuration of the portion of the workpiece to be machined. A typical geometry statement is given below:

```
SYMBOL                  GENERAL DESCRIPTION
          P1 = PØINT/5.000,-3.000,4.000
                                    SPECIFIC DESCRIPTION
```

The geometry statement has *three major parts:* First, a symbol (P1) must be given to the left of the equal-sign. Following this, a general description (PØINT) is given to the left of the slash (/). Finally, a specific description (5.000, –3.000,4.000) is written to the right of the slash.

The first part of the geometric description, or the symbol that appears ahead of the equal sign, is used to identify a specific geometry element of a part. This may be a point, a line, a circle, a plane, or other geometry configuration. Thus far, in this chapter and in Chapter 8, we have used P1, P2, P3, etc., as symbols to identify points. Actually, almost any symbol can be used, if one rule is followed: the symbol cannot have more than *six* (6) figures, and at least one of these figures must be a letter. For example, the symbol P12345 would be allowable. The symbol 56789 would *not* be allowable. ABCDEF would be allowable and ABCDEFG would not be. The symbol assigned to a particular geometric element, whether it be a point, line, circle, or plane, need have no similarity to the element's English language counterpart. For example, a point does not *have* to be assigned a symbol like P1, or PT1, or PNT1; the point *could* have the symbol C1, or CIR1, or L1. However, it makes sense to assign a symbol similar to the type of geometric element being represented. A point might logically have the symbol P1 or PNT2; a line, LN5 or LIN5; a circle, C6 or CIR6; and a surface or plane, PL6 or PLN6. The part programmer may assign whatever symbol he wishes as long as it complies with the rules.

The second major part of a geometry statement, which immediately follows the equal sign, gives the *general description*, usually one word, such as PØINT, LINE, CIRCLE, etc., and explains what the symbol stands for. The third part, which follows after a slash mark (/), provides a *specific description* of the geometry denoted in the general description. For example, this may take the form of the coordinates of a point or the radius of a circle.

An example of a complete geometry statement describing a given point is:

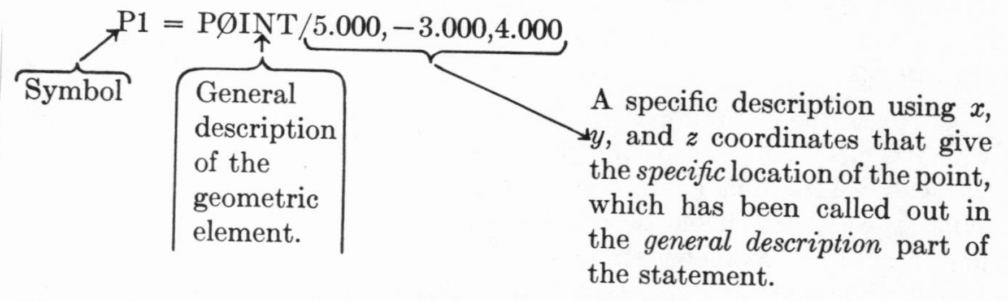

The word PØINT, which describes the general geometric element, must be spelled exactly as shown. If it is spelled incorrectly, the program will fail. The location of the point, with respect to the X, Y, and Z axes, in this case, is shown in Fig. 9–1. It will be noted that 5.000, which is the x coordinate, is expressed first. Next comes −3.000, the y coordinate, and then last, 4.000, the z coordinate. The coordinate values for the position of a point *must* be expressed in this order. It is not necessary to list the zeros after the decimal point. Thus, the statement could also read:

P1 = PØINT/5, −3,4

As in motion statements, commas are used to separate figures having different meanings. Therefore, the values of the three coordinates are separated by commas; and there is no punctuation at the end of the statement.

Many types of geometric elements or figures may be described using the APT language. Some of these geometric figures can be quite complex,

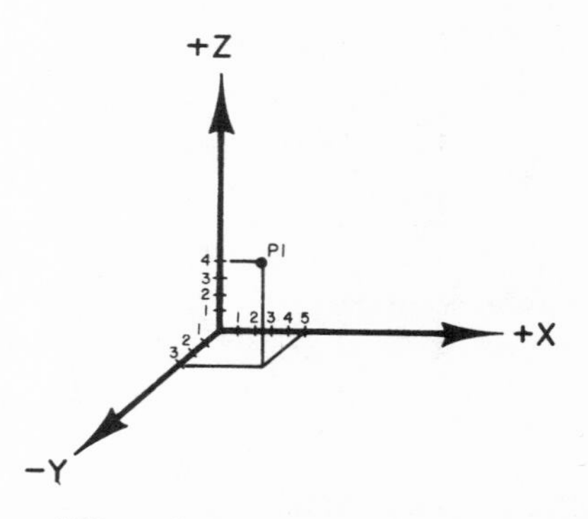

FIG. 9–1. The location of P1, with respect to the X, Y, and Z axes, is shown above.

such as an ellipsoid or an hyperboloid. However, since the great majority of parts are described with points, lines, and circles, and since this text offers essentially an introduction to the APT part programming language, only these three geometric elements—in addition to the plane—will be covered here.

The Straight Line

As far as the computer is concerned, a straight line is actually a plane which is perpendicular to the plane of the X-Y axes and which, when viewed from its edge, appears as a straight line. See Figs. 8–16 and 8–17. What appears on a blueprint as a straight line may therefore be noted as a line in an APT geometry statement. The most direct way of defining a straight line is to note *any* two points that the line passes through. The reason that *any* two points may be used is that one straight line and only one passes through two given points. As already noted, this line may be extended indefinitely in either direction.

In Fig. 9–2 the line, noted by the symbol L1, goes through points P1 and P2 and may be described by the APT statement:

L1 = LINE/P1,P2

Prior to being used here, however, P1 and P2 must each be defined in separate statements. The three statements defining the two points and the line would be as follows:

```
P1 = PØINT/5,6,0
P2 = PØINT/17,9,0
L1 = LINE/P1,P2
```

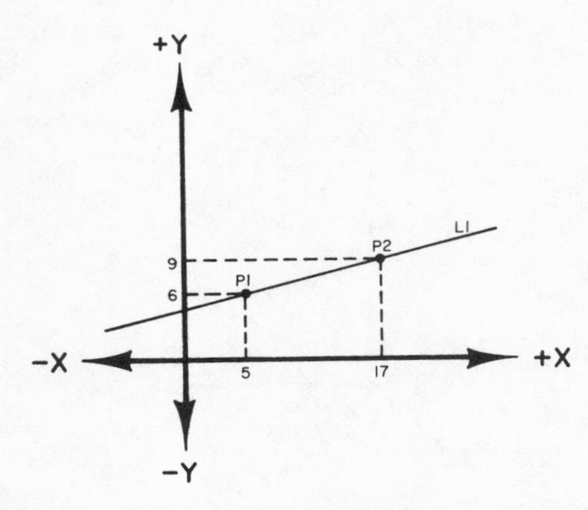

Fig. 9–2. The line L1 may be defined by the two points P1 and P2 that lie on it.

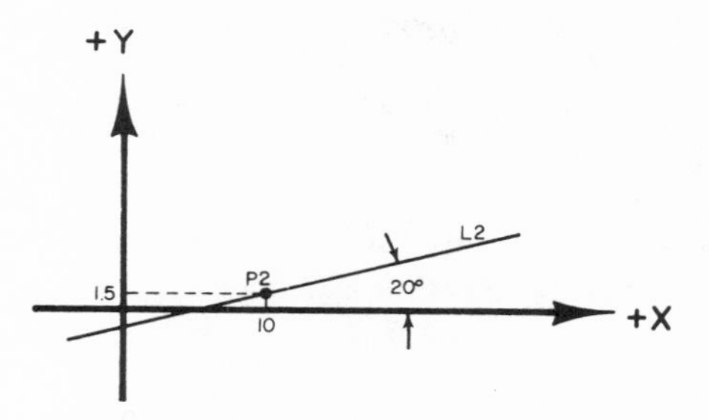

FIG. 9–3. A line may be defined by noting the symbol for a point on the line and the angle that the line makes with the X axis. The sign of the angle is plus (+) in the counterclockwise direction, and negative in the clockwise direction.

Unless otherwise noted, the z coordinate may be set to zero in the statement describing a point. This means that the line lies on the plane formed by the X and Y axes, and the z coordinate would therefore be zero.

There are sixteen different ways that a line may be defined in the APT system. And while sixteen may seem to be many, the different alternatives offered to the part programmer reduce the number of calculations required since he is able to select that definition which best fits the data shown on the blueprint. Several ways of defining a line are used more frequently than others, and these will be discussed later in the chapter. Also, additional line definitions can be found in the Appendix.

We have seen an illustration of the most direct way of defining a line—that is, by means of two points. Another way is to define one point on the line and also to note the angle that the line makes with the X axis. Refer to Fig. 9–3. Again, *any* point along the line would be satisfactory. In this case the coordinates of a selected point, P2, are $x = 10$ and $y = 1.5$. The statement for describing the line, denoted by the symbol L2, is:

L2 = LINE/P2,ATANGL,20

There is no degree (°) mark required after the 20, and minutes are noted as decimal parts of a degree. Thus, 20°30′ would be expressed as 20.5. It might also be noticed that, in order to follow the rule not to exceed six letters, the full term ATANGLE has been shortened by dropping the E. In addition, P2 must be defined in an APT statement before L2 is described, since P2 is used in the statement defining L2. The two statements would therefore be as follows:

P2 = PØINT/10,1.5,0
L2 = LINE/P2,ATANGL,20

PRACTICE EXERCISE NO. 1 CHAPTER 9

1. Note whether each of the following symbols is either correct or incorrect; if incorrect, explain why:

   a. 123 b. PLAN c. CAT34 d. MOUSE24 e. 683C

2. Select the most suitable symbol for the type of geometry listed:

Type of Geometry	*Possible Symbol*
Line	T3, P5, L9
Vertical line	HL7, VL7, BUG10
Circle	P1, CIR9, LIN
Plane	KIT, SIGMA5, PL5

3. Referring to the sketch below, write the statements that describe P1, P2, P3, P4, and P5:

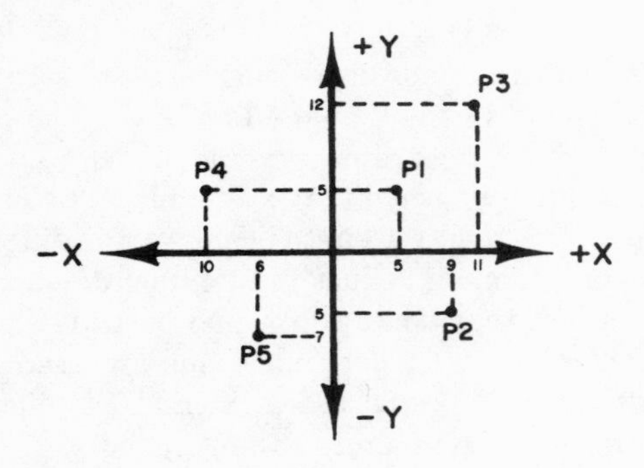

4. Write a statement for each of the five lines in the sketch below. Assume that the points have been defined except in cases where the distance along the axis is shown. In those cases, assign a symbol to a convenient point and also write the statement for the point:

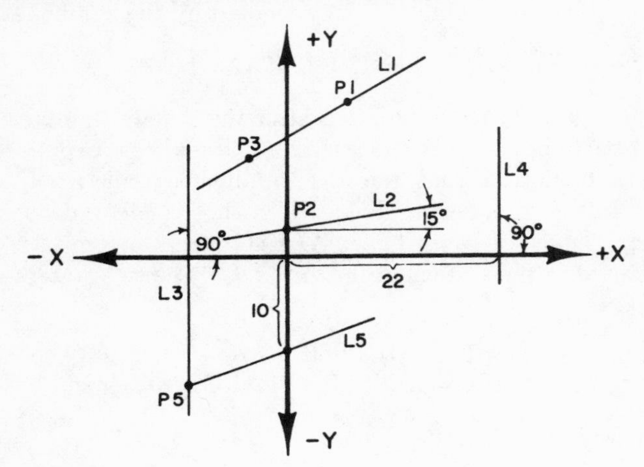

The Circle

The most common way of defining a circle is to note the coordinates of the center of the circle and to give the radius.[1] See Fig. 9–4, where the *z*

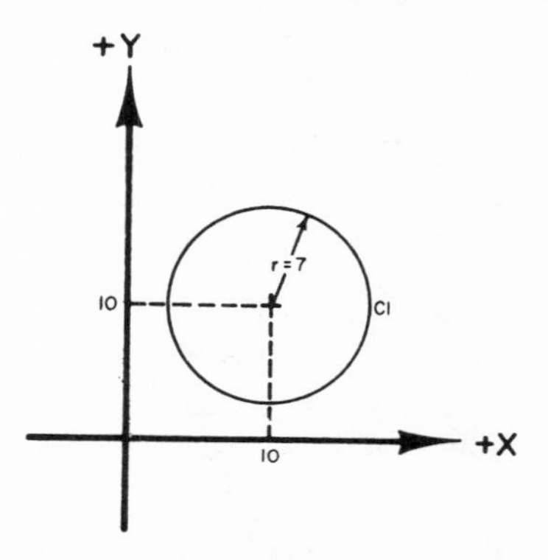

Fig. 9–4. A circle may be defined by noting the coordinates of its center and the radius.

coordinate, as in the statement for a point, is equal to zero. The statement for the circle, having a symbol C1, is:

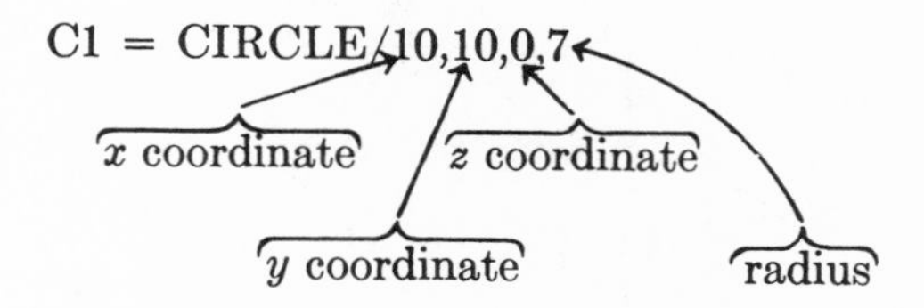

From the computer's standpoint, a circle is considered to be a full 360 degrees, just as a line is considered as infinite. It is the motion statements that determine that part of the circle or the actual number of degrees in the arc to be traversed by the cutter. Consider the following example:

TANGENT POINT
L2
C1
TANGENT POINT
L1
CUTTER

[1] Additional definitions are described in Appendix D.

If a cutter is to move along L1, around C1, and then along L2, the geometry statement[2] for C1 would note the coordinates of the center and the radius—without consideration for the tangent points of L1, C1, and L2, which would determine the degrees in the arc. The computer will calculate these points according to the described motions for the cutter. Therefore, although only a portion of the circle is used, the geometry statement describes the circle to the computer as though it were complete.

The Plane

We have seen how three surfaces—namely the drive, part, and check surfaces—are used to guide the motions of a cutter. However, before statements for surfaces can be used—such as in a start-up statement—they must be defined in geometry statements as PLANES: For example:

PLN1 = PLANE/P1,P2,P3

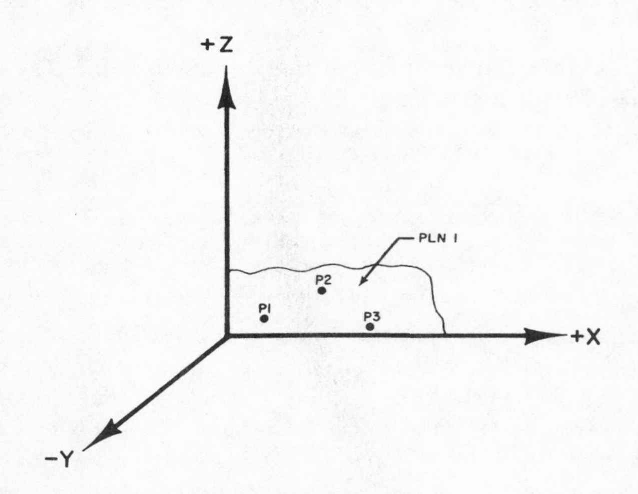

FIG. 9–5. PLN1 can be defined in an APT statement for a plane by noting any three points lying on its surface that are not in a straight line.

In this case, as shown in Fig. 9–5, the plane is described by three points that lie on it and that do not lie in a straight line—namely, P1, P2, and P3.

[2] The motion statements would be: GØFWD/L1

GØFWD/C1

GØFWD/L2

As for any APT statement which uses a symbol, the symbol must have been defined in a prior statement. For example, the three points could have been described as:

P1 = PØINT/4,0,2
P2 = PØINT/10,0,5
P3 = PØINT/15,0,1

We have seen how a cutter is guided by three planes called the drive, part, and check surfaces. Since the tip of the cutter usually moves over the part surface, it is general practice to describe the part surface as a PLANE. The part programmer may choose any plane, real or imaginary, as the *part* surface.

Another example is shown in Fig. 9–6. If we wanted to make PL5 a part surface, it would first be necessary to define PL5 in a statement, as follows:

PL5 = PLANE/P4,P5,P6

Assuming the points P4, P5, and P6 had been described, then PL5 could be used in the following start-up statement:

GØ/TØ,PL6,TØ,PL5,TØ,PL4
↑
Part Surface

Remember from Chapter 8 that the start-up statement is always given in the following order: 1. drive surface; 2. part surface; 3. check surface. In the preceding example the part surface, PL5, would be called out ac-

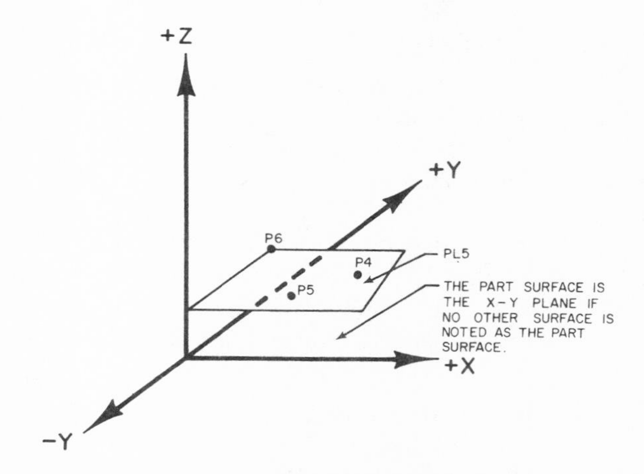

FIG. 9–6. The plane shown can be designated as a part surface by first defining it as a PLANE and then by using the symbol defining the plane in the start-up statement GØTØ, PL6, TØ, PL5, TØ, PL4.

cording to its position in the GØ/TØ statement. PL4 and PL6 could be the drive and check surfaces; or if already defined as lines, they could be noted in the start-up statement as:

GØ/TØ,L4,TØ,PL5,TØ,L6

Quite often it is necessary to *change* a part surface, or have more than one part surface, in order to machine a part. In Fig. 9–7(A) the first operation calls for cutting around the outside perimeter of the part. The coordinate axes have been positioned so that the bottom of the part rests on the X-Y plane. In this case we want the bottom of the cutter to be below the X-Y plane so that the side of the cutter completely cuts through the side of the part. It is, therefore, necessary to establish a part surface which is lower than the bottom surface of the part, so that the tip of the cutter is below the bottom surface and the cutter can be guided around the contour of the part on this surface. The plane noted by the symbol PLN2 in Fig. 9–7(B) must therefore be described as the part surface. This can be accomplished by the start-up statement:

GØ/TØ,PLN6,PLN2,PLN7

Of course, PLN2, as well as PLN6 and PLN7, would have to be described before being used in the statement above. This could be done by selecting any three points not in a straight line and lying on each plane. In the case of PLN2 the sequence of statements could be as follows:

```
P1 = PØINT/0,0,−2
P2 = PØINT/1,1,−2
P3 = PØINT/2,2.5,−2
PLN2 = PLANE/P1,P2,P3
GØ/TØ,PLN6,PLN2,PLN7
```

Any set of three points located on a plane that is parallel to and 2 inches below the X-Y plane could have been selected to define the part surface PLN2. As the points defined above and shown in Fig. 9–7(C) were selected arbitrarily, other points on this plane could be used if desired.

In order to machine around the rectangular section on the top of the part a new part surface must be established. This part surface may be noted as PLN1. As with the statements for describing the first plane, any three points on the plane and not lying in a straight line would be suitable for describing the plane. The statements might therefore be:

```
P4 = PØINT/0,0,3
P5 = PØINT/1,1,3
P6 = PØINT/2,2.5,3
PLN1 = PLANE/P4,P5,P6
GØ/TØ,PLN4,TØ,PLN1,TØ,PLN5
```

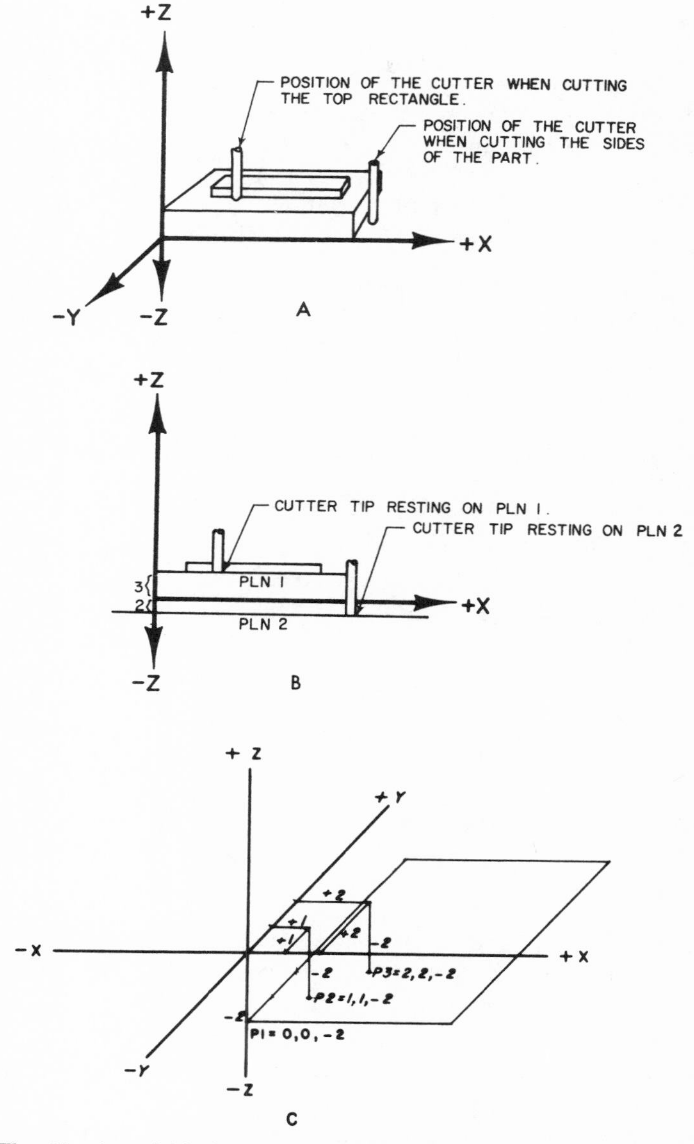

FIG. 9–7. The side view of (a) above is shown in (b). The bottom of the part rests on the plane of the X–Y axes. A part plane PLN2 is established 2 inches below the bottom of the part in order that the tip of the cutter may be below the bottom surface of the part and the side of the cutter may fully cover the side of the part. Another part surface PLN1 is established 3 inches above the bottom of the part so that the top rectangle can be cut. View (c) describes the points P1, P2, and P3, which have been arbitrarily selected.

Once the new part surface is defined in the GØ/TØ statement, the cutter is directed to move to it and then around the new contour.

While this is not a strict rule, PLANE statements are generally confined to describing part surfaces while the cutter is in a contouring mode, and may describe drive, part, and check surfaces when start-up statements are involved. Once the cutter is in motion, the drive and check surfaces normally appear as two-dimensional forms such as lines and circles and may be described as lines and circles in geometry statements.

Two important rules must be observed when writing APT programs:

1. A symbol can be used to describe one and only one element in the program. It cannot describe two different elements in the same program.
2. No more than one symbol can be used to describe the same element in the program.

For example, if a plane surface is described by PLN1 = PLANE/P2, P3,P4, the symbols PLN1, P2, P3, and P4 cannot be used to describe any other element in the program. Their meaning is fixed when they have been used to describe PLN1.

As an example of Rule 2, the following pairs of statements would be incorrect because more than one symbol is used to describe the same element in the program.

```
   C2 = CIRCLE/1,1,0,1
   C3 = CIRCLE/1,1,0,1
MOUSE = LINE/P1,P2
  CAT = LINE/P1,P2
```

Only *one* statement in each of the two pairs above would be allowable in a program.

More Ways of Defining a Line

There are a number of additional ways of defining a line besides the two already discussed. Frequently a line is defined as being tangent to a circle, as illustrated in Fig. 9–8. The line, noted by the symbol, L1, passes through the point, P1, and is tangent to the circle, C1. A statement, although not wholly correct, for defining the line might logically be:

```
L1 = LINE/P1,TANTØ,C1
```

This would mean that L1 is a line passing through P1 and tangent to the circle C1. The only problem here is that there are *two* lines passing through P1 and tangent to C1. L2, also shown in Fig. 9–8, is the other line. It is necessary, therefore, that the line to be described be distinguished from the line *not* to be described. This can be done by noting, in the

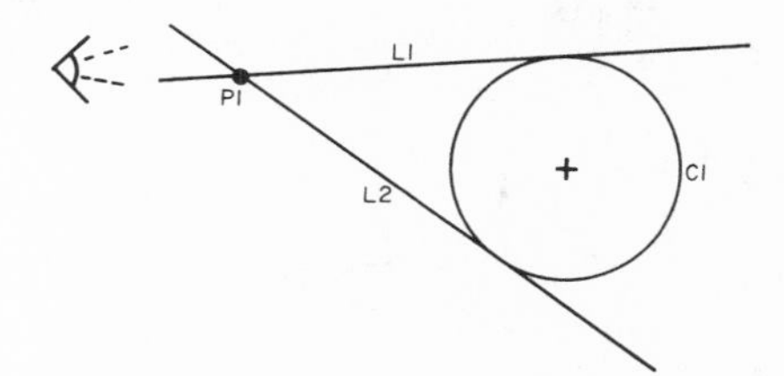

FIG. 9–8. A line may be described as going through a point and being tangent to a circle. The line, L1, goes through P1 and is to the left of the center of the circle, C1, when looking from the point to the circle. The line, L2, also goes through P1. However, L2 lies to the right of the center of the circle, when looking from P1 to the circle.

APT statement, that the line to be described lies either to the *right* or *left* of the center of the circle, when looking *from* the point *toward* the circle. Therefore, the correct statement for L1, shown in Fig. 9–8, would be:

L1 = LINE/P1,LEFT,TANTØ,C1

The word LEFT in the statement notes that the line L1 lies to the *left* of the center of the circle. If we wanted to describe the other line, L2,

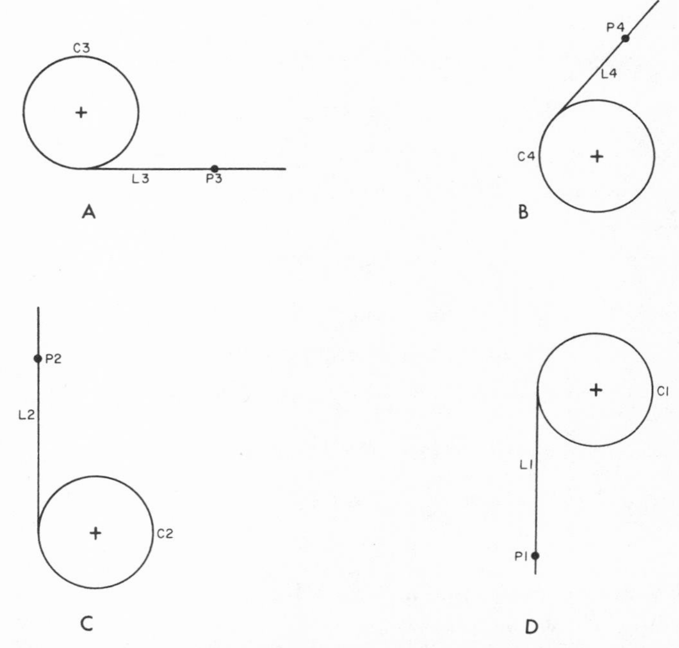

FIG. 9–9. In order to determine whether the line lies to the *right* or *left* of the circle, the part programmer must consider looking in the direction *from* the point *to* the circle.

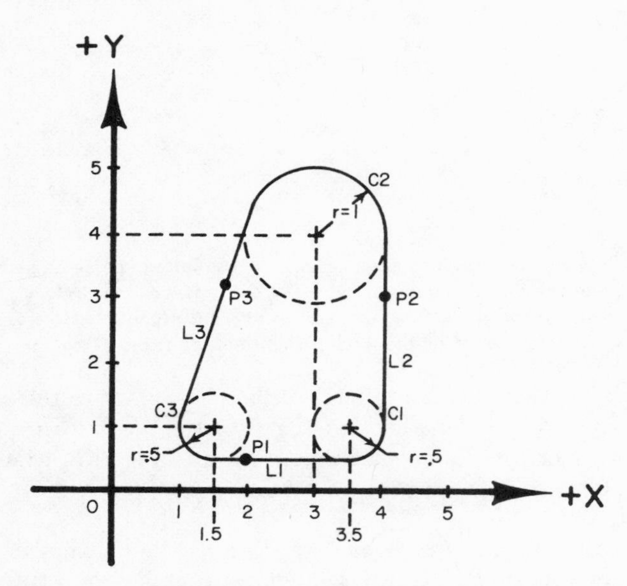

FIG. 9–10. The lines L1, L2, and L3 are defined as being tangent to the circles C1, C2, and C3.

which lies to the *right* of the center of the circle, the statement would be:

L2 = LINE/P1,RIGHT,TANTØ,C1

Consider the examples in Fig. 9–9. In Fig. 9–9(A) the line that passes through P3 lies to the *left* of the center of the circle, C3, so that the statement would be:

L3 = LINE/P3,LEFT,TANTØ,C3

In Fig. 9–9(B) the statement would be:

L4 = LINE/4,RIGHT,TANTØ,C4

In Fig. 9–9(C) the statement would be:

L2 = LINE/P2,RIGHT,TANTØ,C2

In Fig. 9–9(D) the statement would be:

L1 = LINE/P1,LEFT,TANTØ,C1

Another example is shown in Fig. 9–10, where the lines of the pattern L1, L2, and L3, are tangent to the circles C1, C2, and C3. In order to use the symbols for the circles in the statements for the lines, the circles must be described in statements. Therefore, C1, C2, and C3 would be:

C1 = CIRCLE/3.5,1,0,.5
C2 = CIRCLE/3,4,0,1
C3 = CIRCLE/1.5,1,0,.5

Next, the lines may be defined as:

L1 = LINE/P1,RIGHT,TANTØ,C1
L2 = LINE/P2,RIGHT,TANTØ,C2
L3 = LINE/P3,RIGHT,TANTØ,C3

The lines could also have been defined in the following statements:

L1 = LINE/P1,LEFT,TANTØ,C3
L2 = LINE/P2,LEFT,TANTØ,C1
L3 = LINE/P3,LEFT,TANTØ,C2

It is not necessary for the part programmer to calculate the tangent points of the lines and the circles, since the computer performs this function. It is only necessary that the part programmer note that the lines are tangent to the circles.

A line may also be described as going through a point and being perpendicular to another line, or as going through a point and being parallel to another line. In Fig. 9–11, the line L1 passes through the point P1 and is perpendicular to the line L2. The geometry statement that describes the line L1 would be:

L1 = LINE/P1,PERPTØ,L2

It should be noted that the points P1, P2, and P3 would also have had

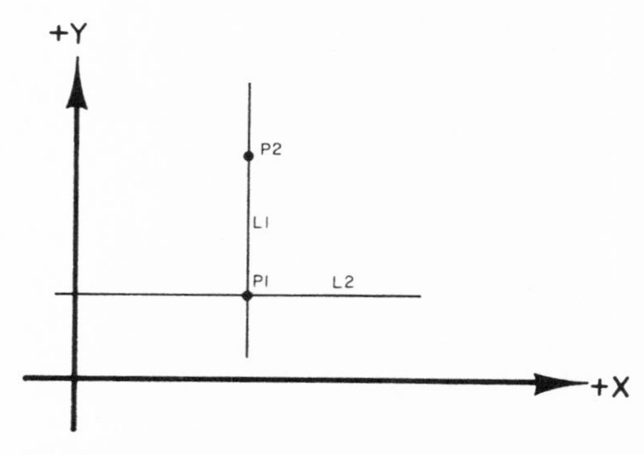

FIG. 9–11. L1 may be described as going through any point on L1 and being perpendicular to L2.

to be defined previously. It might also be pointed out that *any* point along the line L1 could have been used, as long as it had been described prior to use in the statement describing line L1. For example, the following statement, using the point P2 shown in Fig. 9–11, would have been allowable:

L1 = LINE/P2,PERPTØ,L2

By the same token, a line could also be described as passing through a point and being *parallel* to another line. In Fig. 9–12 the line L3 passes through the point P6 and is parallel to line L7. The statement in this case would be:

L3 = LINE/P6,PARLEL,L7

Combining Geometry and Motion Statements

Although geometry and motion statements can be combined in any desired order, as long as any symbols used have been previously defined, it is the common practice to list all of the geometry statements first and then list the motion statements. This is done in order to reduce the chance of error.

Consider the example shown in Fig. 9–13. The rectangular block in (A) would appear as a rectangular pattern when laid out with respect to the X-Y axes, as seen in (B). It is desired to move a milling cutter around the perimeter of the rectangular part. Symbols for selected lines and points are shown on the pattern in (B). To move a cutter around the perimeter, as shown in (B), it is necessary to define the four sides. This, in turn, requires that several points be described prior to defining the

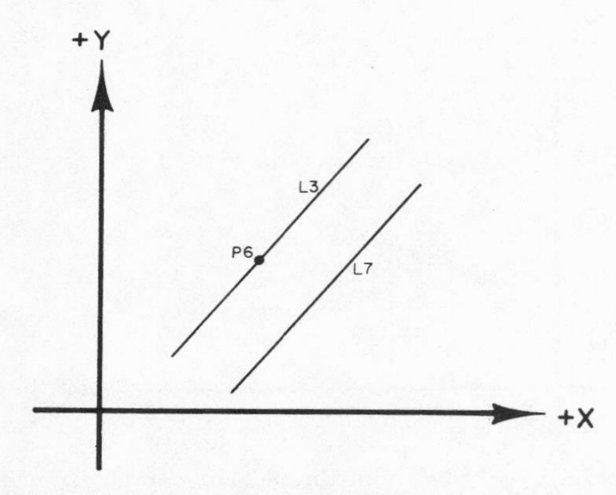

Fig. 9–12. Since L3 is parallel to L7, it may be defined in a statement as going through P6 and being parallel to L7.

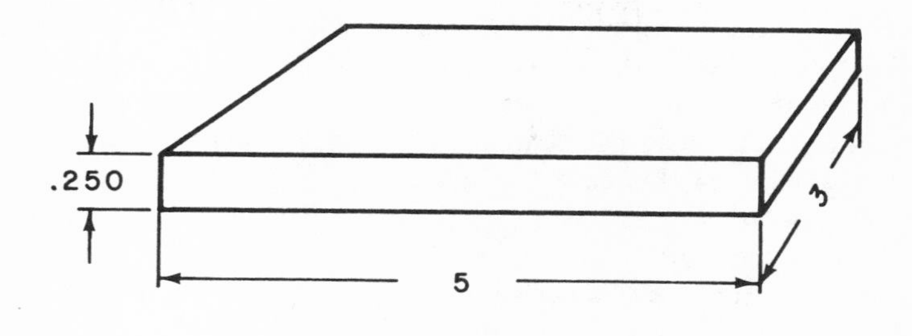

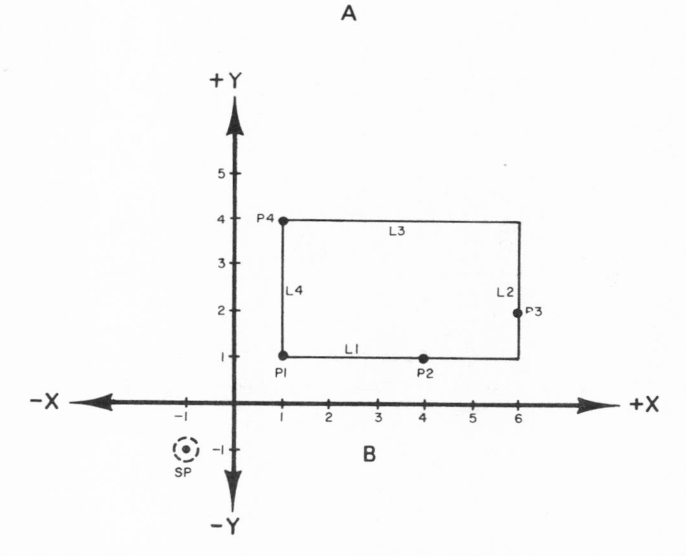

FIG. 9–13. The points and lines for the geometry are shown with symbols.

lines or sides. The geometry statements are as follows:

```
SP = PØINT/−1,−1,0
P1 = PØINT/1,1,0
P2 = PØINT/4,1,0
P3 = PØINT/6,2,0
P4 = PØINT/1,4,0
L1 = LINE/P1,P2
L2 = LINE/P3,ATANGL,90
L3 = LINE/P4,ATANGL,0
L4 = LINE/P1,P4
```

Since L2 is perpendicular to the X-axis, it is possible to use the statement shown for L2 in which the line passes through a point on the line and is at 90 degrees to the *X*-axis. L3 may also be conveniently defined as going through P4 and at zero (0) degrees to the *X*-axis. The statement for the set-point has also been noted. It is not necessary that the set-point be at the origin point where the X and Y axes cross. Any convenient point may be selected. In this case a location in the third quadrant has been chosen, and the symbol SP assigned to the set-point.

In Fig. 9–14 the cutter is directed first from SP to the intersection of the drive, part, and check surfaces. Next, the cutter makes a right turn along L1 until it passes L2. After passing L2, it makes a left turn and goes along L2 until it passes L3. Again, the cutter makes another left turn along L3 until it passes L4. Then it moves along L4 until it passes L1. After passing L1, the cutter is directed to move to the set-point, SP. The APT motion statements are as follows:

```
FRØM/SP
GØ/TØ,L1,TØ,PLN1,TØ,L4 (The part surface PLN1 is not shown)
GØRGT/L1,PAST,L2
GØLFT/L2,PAST,L3
GØLFT/L3,PAST,L4
GØLFT/L4,PAST,L1
GØTØ/SP
```

Since the instructions on the tape must describe the movements of the center of the cutter, an auxiliary statement describing the actual diameter

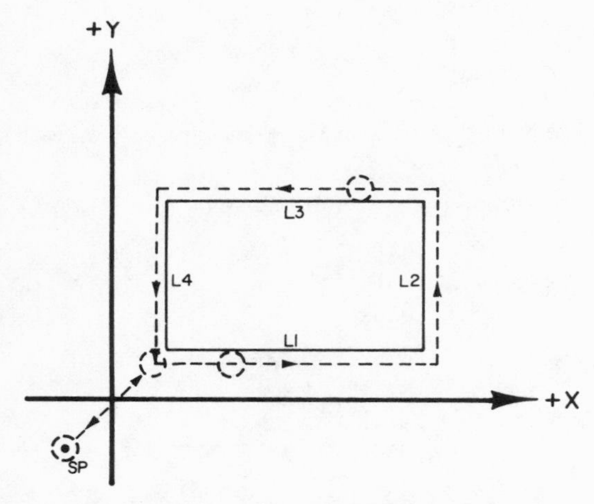

FIG. 9–14. After being set by the operator at SP, the cutter moves to the intersection of the drive surface L1, the part surface PLN1 (not shown), and the check surface L4, then makes a right turn along L1 and goes past L2, where it makes a left turn. Next it passes L3, makes a left turn on L3, passes L4, moves past L1, and then goes to SP.

of the cutter must be included in the program. If a ½-inch-diameter cutter were to be used, the statement would be:

CUTTER/.5

As already noted, zeros following the last significant number to the right of the decimal point need not be shown. For example, .500 inch may be noted as .5, and .250 inch may be noted as .25. This is true of *all* numbers in an APT program.

When a number of different cutters are to be used in the same program, the last cutter described in the CUTTER statement is the one that the computer recognizes. The CUTTER statement may also be used as a convenient means of making several cutting passes such as required with roughing and finishing cuts. This is accomplished by "lying" to the cutter. For example, referring to Fig. 9–14, we could have noted that the diameter of the cutter was .750 inch instead of .500 inch (cutter statement CUTTER/.750) yet use the .500-inch cutter. Then the center of the cutter would have moved to within .375 inch of the finish cut line instead of .250 inch, thus leaving an excess of material of .125 inch.

Auxiliary and Post Processor Statements

Two other types of statements besides geometry and motion statements are *auxiliary* and *post processor* statements. Auxiliary statements are those which are required for the program, but which do not apply to the geometry of the part being programmed or to the motions being described. For example, PARTNØ is an auxiliary word that must be used at the beginning of every program. A complete statement would be

PARTNØ FLANGE BRACKET NO. 5432 W

It should be pointed out that, in this rare case, there can be no spaces in the word PARTNØ, although any name, or combination of letters or numbers, may follow—including combinations of more than six numbers or letters. Another auxiliary statement is FINI, which must come at the *end* of every part program. Other auxiliary words include:

INTØL/.005

This means that the allowable calculated tolerance for the difference between a curved line and the approximated straight line on the *inside* of the curve should not exceed .005 inch. This statement is given prior to the motion statements in the program. See Fig. 9–15(A). The computer will calculate the *length* of the line so that the described tolerance is not exceeded. The computer will also calculate the x and y distances, which is information to be punched in the tape (in an incremental control system).

The auxiliary word meaning that the allowable calculated tolerance for

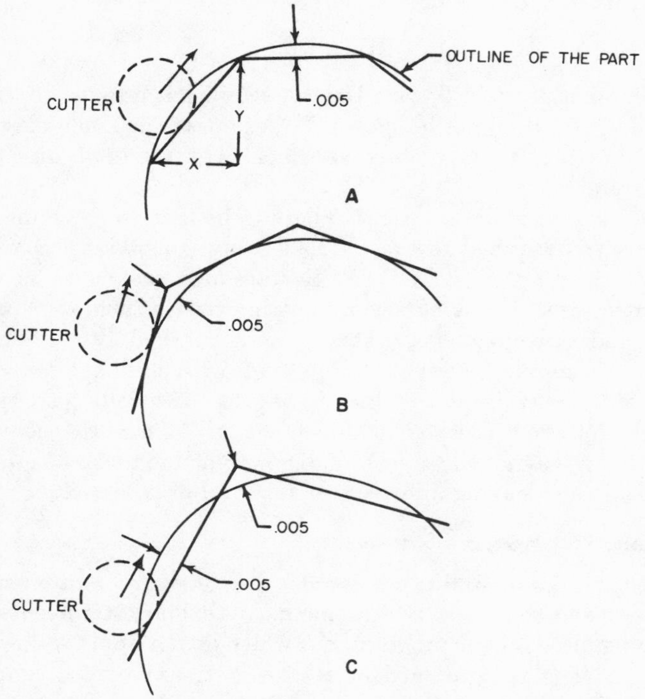

FIG. 9–15. In (A), shown above, the cutter would cut a little inside the boundary of the part. In (B) the cutter would leave a little excess material. In (C) the cutter removes a little material inside the boundary and leaves a little outside the boundary. These sketches are exaggerated for illustrative purposes, and the difference between the straight lines and the curves is very small and well within the tolerance limits. The programmer may use a tolerance value down to millionths of an inch, if he wishes. The tighter the tolerance, though, the greater will be the number of lines. This is also illustrated in Fig. 7–1.

the difference between a curved line and the approximated straight line on the *outside* of the curve must not exceed .005 inch is:

ØUTTØL/.005

See Fig. 9–15(B). Straight lines may also approximate the curve on the inside *and* the outside and, in this case, both an INTØL and an ØUTTØL statement are noted. See Fig. 9–15(C). The part programmer may select the tolerance required.

For the computer to print out the coordinates describing the movements of the center of the cutter, the word is:

CLPRNT

An auxiliary word required so that the computer may calculate the movements of the center of the cutter—since the geometry statements prepared by the part programmer describe the *part*—is given below. In this case, the diameter of the cutter is .250 inch, and the zero may be dropped.

CUTTER/.25

Post processor statements are those that apply directly to the operation of the machine tool. One example would be CØØLNT/ØN, which is an instruction to turn the coolant on. The coolant may be turned *off* by the statement CØØLNT/ØFF. Other post processor statements include:

FEDRAT/5,IPM

This means that the feed rate for the cutter to travel is to be 5 inches per minute (ipm). To instruct the machine to travel at a rapid traverse, or the maximum rate that the machine will travel, the statement is:

RAPID

To stop the machine's motions so that the operator may perform such operations as changing the cutting tool, the word is:

STØP

It should be noted that, in order for a particular post processor word to be effective, the machine tool must have the capability to perform the post processor, or auxiliary type, function. For example, if the part programmer instructed an NC lathe to turn to turret position 3 via the APT statement, TURRET/3, and the machine did not have a turret, nothing would happen. And, if the lathe did have a turret and the control system were not *wired* to handle this instruction, nothing would happen either. One must also realize that not all machine tools adhere to the exact same post processor words, and variations may exist between machines of different manufacturers. Usually, the manufacturer of the NC machine tool will furnish a listing of the post processor words that apply to the particular machine. The same situation would be true for auxiliary statements, though the variations would not be as numerous and would apply to the *computer* being used. A listing of these auxiliary statements can usually be obtained from the computer manufacturer or the computer service company being used. NOTE: *A list of auxiliary and post processor statements is shown in the Appendix D.*

PRACTICE EXERCISE NO. 2 CHAPTER 9

1. Referring to the circular patterns shown below, write the statements for the circles C2, C3, C4, and C5.

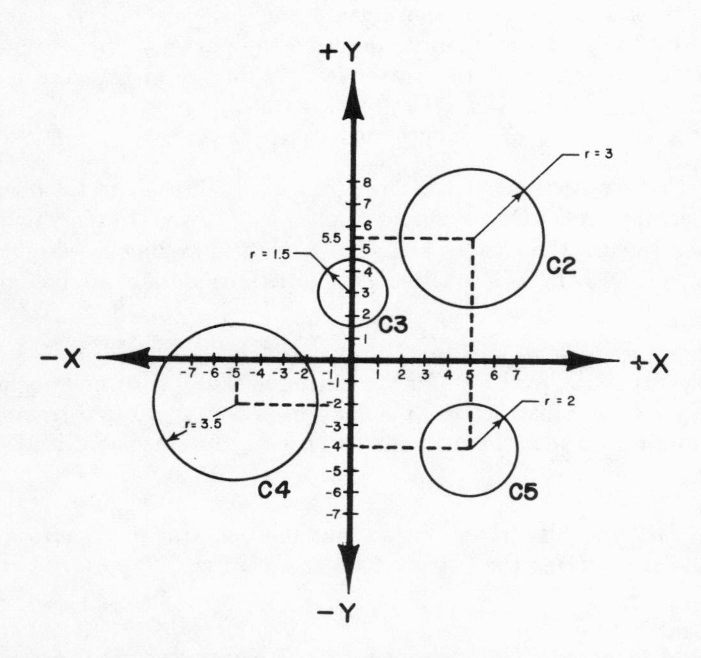

2. Write the statements for the circular arcs in the pattern below. The arcs have been assigned symbols C1, C2, C3, and C4. The lines are parallel to the X and Y axes. The radius of each of the corner arcs is $\frac{1}{2}$ inch:

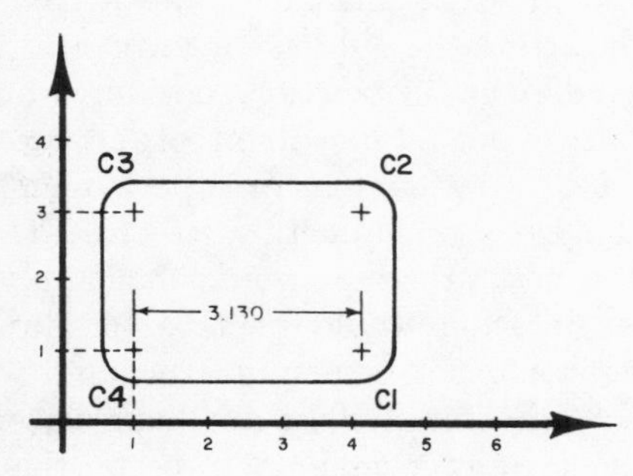

3. Referring to the sketch shown at the top of the opposite page, write the statement that describes the plane PLN7. Points P1, P2, and P3

lie on the plane and have been defined:

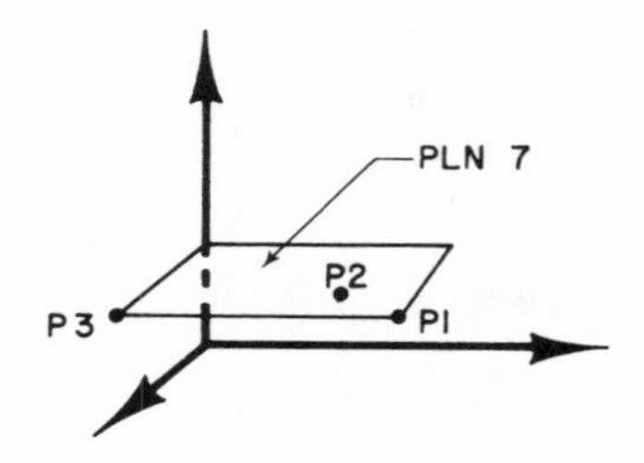

4. In the sketch below the plane, PLN8, is parallel to the plane of the X-Y axes and lies 4.5 inches below it. Write the statement that would describe PLN8. Also write the statements describing some points on the surface which would be necessary to define the plane PLN8 and put all statements in their proper order:

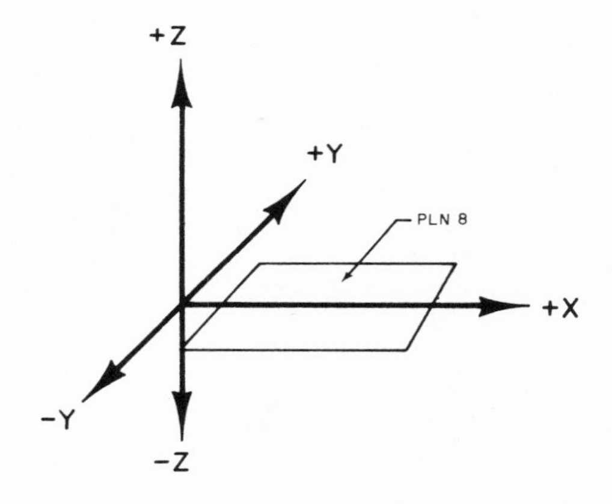

5. Why will the following set of statements in a part program not work? Indicate those which are incorrect and state why.

```
 SP = PØINT/0,0,0
 P1 = PØINT/1,1,0
PT2 = PØINT/5,4,2
 P3 = PNT/6,5,7
 L1 = LIN/P1,P3
 P1 = PØINT/2,2,3
```

6. What is the *first* statement in any APT program?
7. What is the *last* statement in any APT part program?
8. What is meant by the statement INTØL/.0005?

9. Write an APT statement defining L1 for each of the lines shown below:

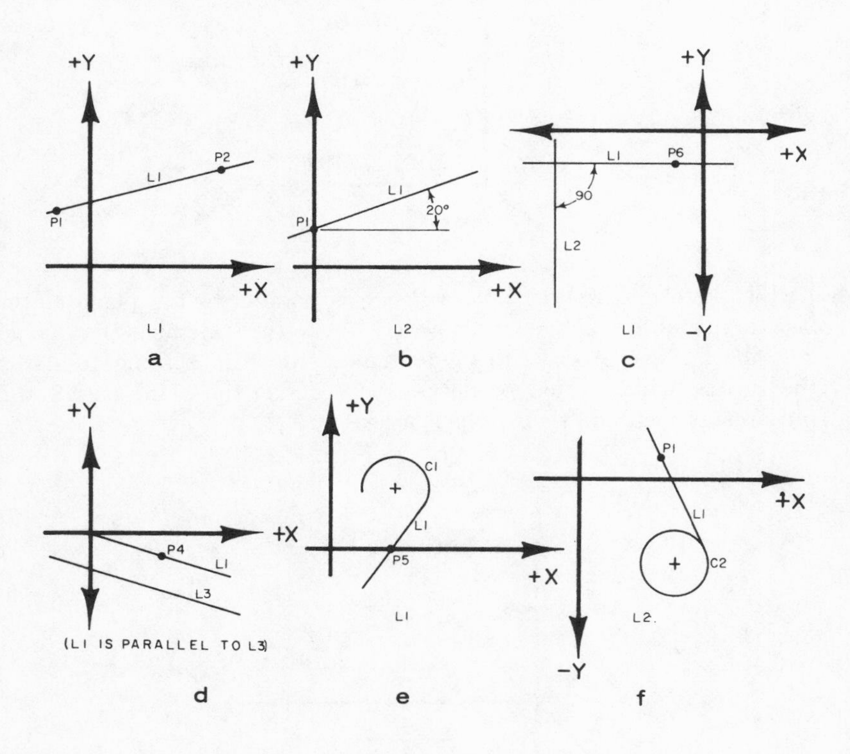

QUESTIONS CHAPTER 9

1. Which two of the following five symbols are incorrect? Why are they incorrect?

   DOG 1
   DOGGIE 2
   PLAN44
   654321
   651P42

2. The top view of a part is shown below. Position this part in the first quadrant of the *X*-*Y* axes with the sides touching the *X* and *Y* axes, and then write the statements for the points shown. Next, consider the part to be positioned in the third quadrant with the sides touching the *X* and *Y* axes, and then write the statements for the points shown: Consider the *z* dimension to be 0 in all cases.

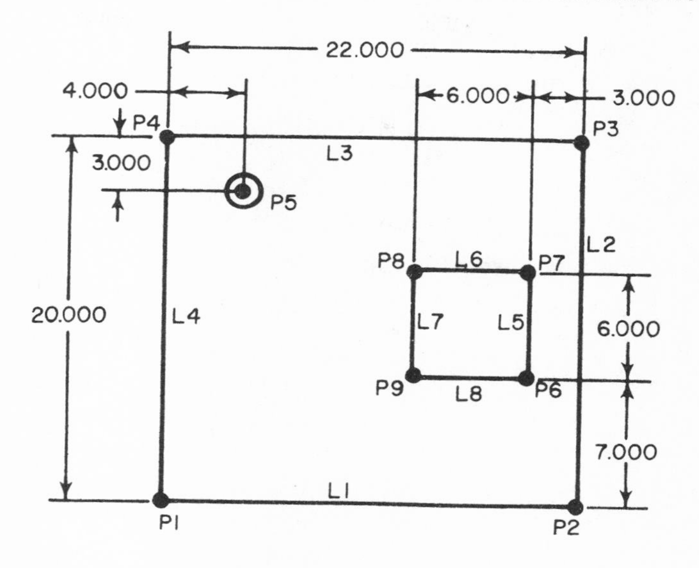

3. Refer to the sketch shown in Question 2 and write the statements describing the lines L1, L2, L3, and L4. Write statements for each of the symbols in as many ways as you can. Consider that the points P1, P2, P3, and P4 have already been described in the program. Also, consider that you are writing a program and do not use a symbol for a line unless it has been described in a statement previously. Begin by describing L1 and then move counterclockwise, and consecutively.

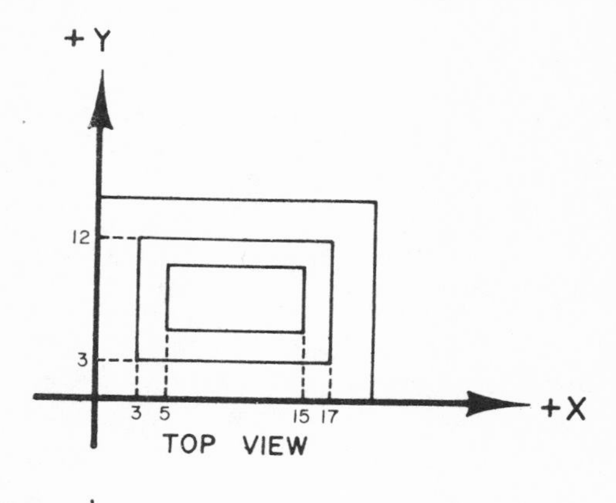

4. The part shown to the right is located in the first quadrant and touches all three axes. Write the statements for describing the symbols for the points used in the plane statements. Select any points that you feel are suitable:

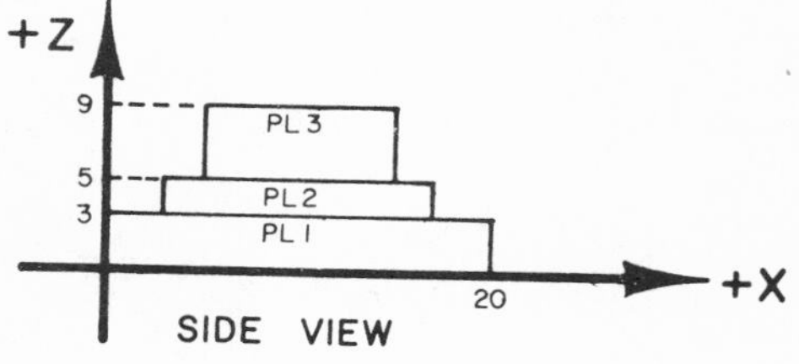

5. Referring to the sketch in Question 4, what kind of a statement would be used for changing the

part surface from PL1 to PL2?

6. Which statements in the following portion of a program are incorrect?

```
PART NØ  SK1234567
MACHINE/ABC
CUTTER/.5 DIA
FEEDRAT/7,IPM
```

7. What should the last statement be for any program?

CHAPTER 10

Writing a Complete APT Program

A complete part program, which would be suitable for preparing a tape via the computer, consists of four types of statements, as has been noted in the previous chapter. These are *auxiliary* statements, *post processor* statements, *geometry* statements, and *motion* statements. Usually the first statements of a part program are auxiliary and post processor statements, then follow geometry statements, and last motion statements. There is no fixed rule covering this, and auxiliary and post processor statements *may* also be scattered throughout the program, where necessary, and geometry statements *may be* mixed in with motion statements, providing any symbol used in a geometry or motion statement has been described *prior* to its use.

Referring to the outline of the part shown in Fig. 10–1 (A and B) which is a duplicate of that shown in Fig. 9–14, the complete part program would be as follows:

Auxiliary and Post Processor Statements:

```
PARTNØ  FLAT PLATE NØ 1
MACHIN/ABC
INTØL/.005
ØUTTØL/.005
CUTTER/.75
CØØLNT/ØN
CLPRNT
FEDRAT/5,IPM
```

Geometry Statements

```
SP = PØINT/-1,-1,0
P1 = PØINT/1,1,0
P2 = PØINT/4,1,0
P3 = PØINT/6,2,0
P4 = PØINT/1,4,0
L1 = LINE/P1,P2
L2 = LINE/P3,ATANGL,90
L3 = LINE/P4,ATANGL,0
L4 = LINE/P1,P4
PLN = PLANE/—,—,—
```

Motion and Post Processor Statements

```
FRØM/SP
GØ/TØ,L1,TØ,PLN1,TØ,L4
GØRGT/L1,PAST,L2
GØLFT/L2,PAST,L3
GØLFT/L3,PAST,L4
GØLFT/L4,PAST,L1
RAPID
GØTØ/SP
```

Auxiliary and Post Processor Statements

```
CØØLNT/ØFF
FINI
```

It will be noted that the post processor statement RAPID has been included along with the motion statements. This means that only the motion called out in the *following* motion statement is to be made under a rapid traverse mode. Where a motion statement is not preceeded by RAPID, the motion is to be made at the feed rate specified by the FEDRAT/ statement at the beginning of the part program. The reason that the word RAPID must appear before every movement to be made in a rapid mode is to help insure that a rapid movement is not made when the cutter is cutting metal, thus avoiding a potential accident.

Programming for the Machining of a Part on a Combination NC Milling and Drilling Machine

Most NC drilling machines can also perform milling operations. Whether or not the machine can perform contour cuts, such as moving around circles and other types of curved lines, depends on the type of control system. While it is possible to approximate an arc with a point-to-point system, as described in Chapter 6, a contouring system, which can guide the cutter along any angle, is far more suitable.

As an example of a contour machining operation, consider the part shown on the engineering drawing in Fig. 10–2. It is required to machine

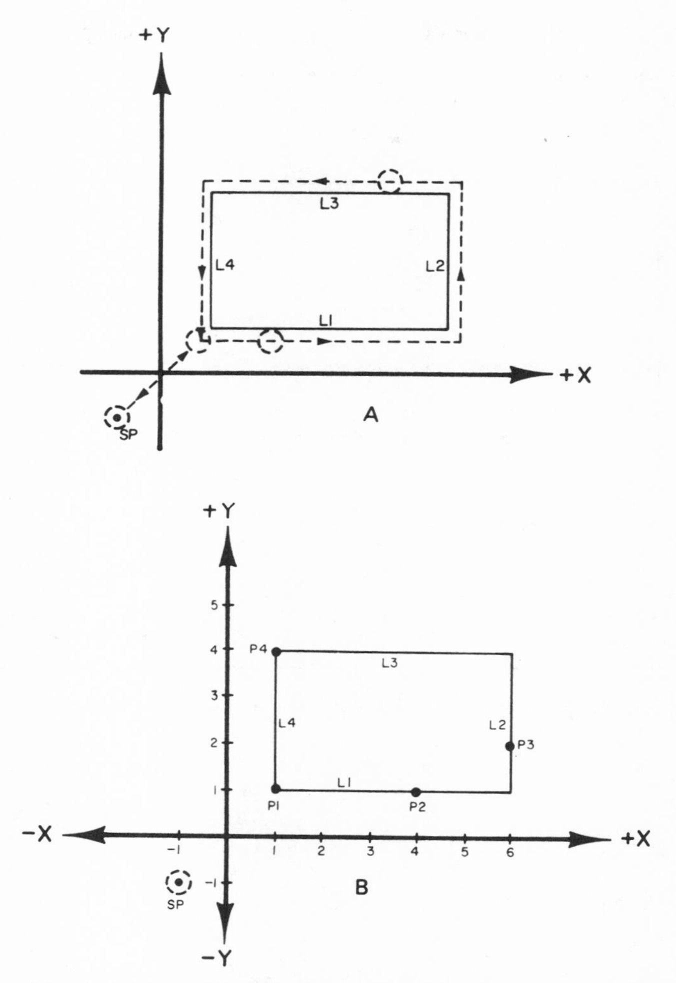

FIG. 10–1. After being manually set at SP, the cutter is directed to move around the perimeter of the part, as described by the two-dimensional pattern shown in (A). The coordinate values are shown in (B).

the perimeter of the part after it has been rough cut on a band saw. Only one pass will be necessary.

Fig. 10–3 shows the part as it would appear in the coordinate system. While the programmer would have the option of locating the part anywhere with respect to the axes, it has been found convenient in this case to locate the corner of the part at the origin so that the bottom of the part lies on the plane of the X-Y axes, where the z distance is equal to zero.

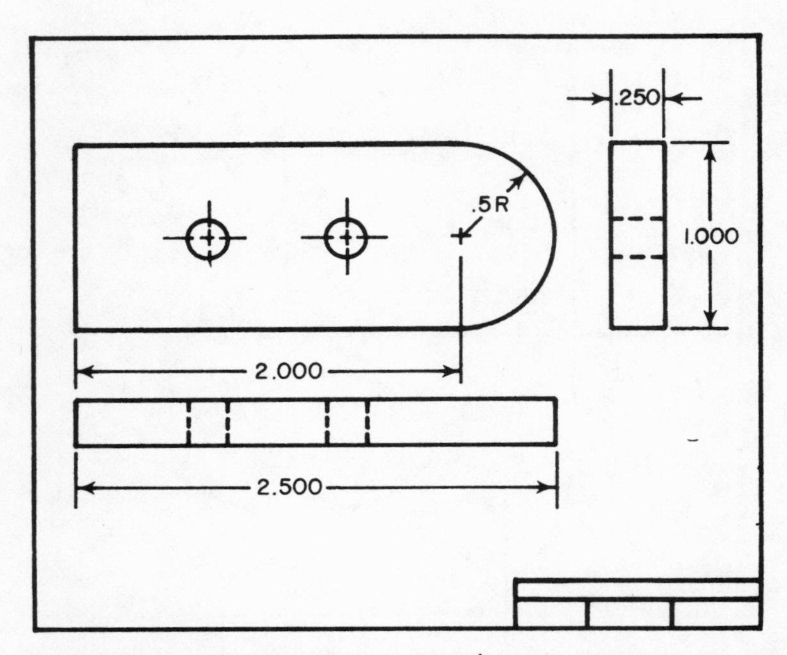

Fig. 10–2. Engineering drawing describing part to be machined.

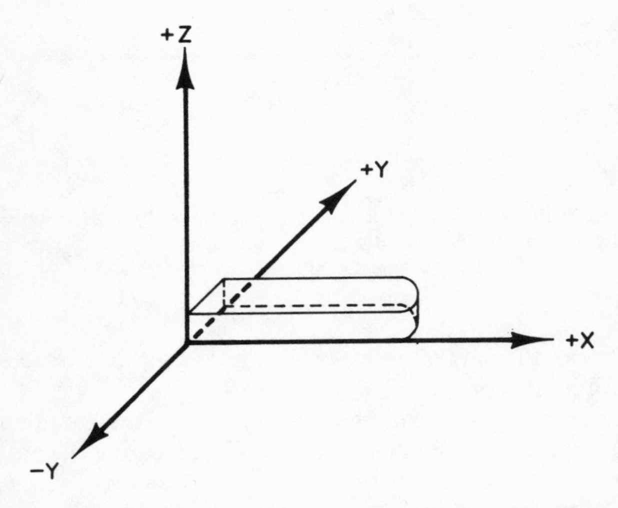

Fig. 10–3. The part, shown in Fig. 10–2, as it would appear with respect to the coordinate axes.

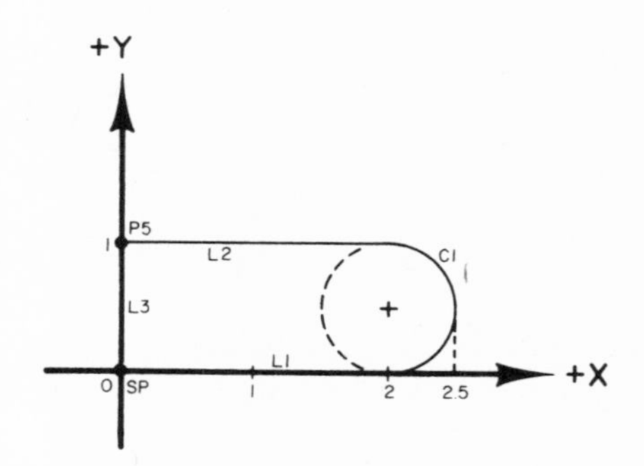

FIG. 10–4. The part as it would appear in the coordinate system if seen directly from above.

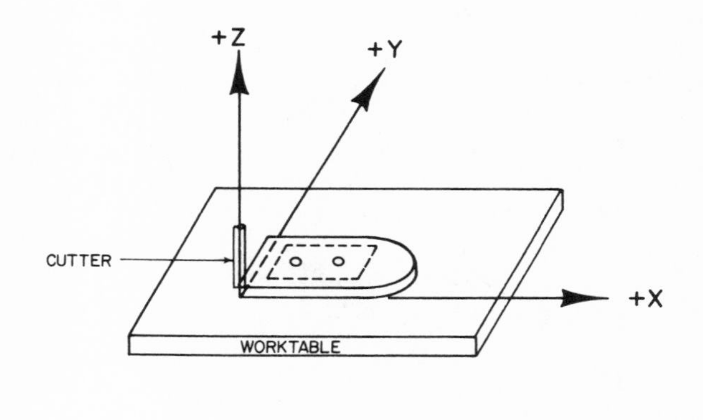

A

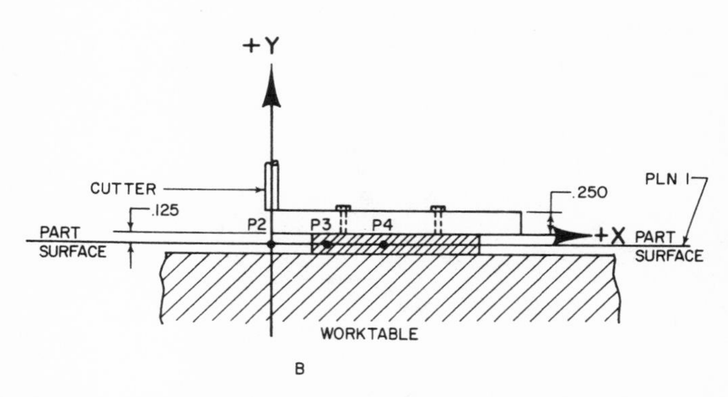

B

FIG. 10–5. (A) The part as it would appear on the table of an NC machine. (B) The part viewed from the front.

Fig. 10–4 shows a view of the part, located in the coordinate system, when seen directly from above. The sides of the part appear as lines and the circular end portion appears as a half-circle. We can therefore identify the lines with line statements and the half-circle with a circle statement. Here the lines have been labeled L1, L2, and L3. The circle has been labeled C1.

The part, as it would appear on the table of the N/C machine, is shown in Fig. 10–5(A). Viewed directly from the front, the part would appear as shown in Fig. 10–5(B). To machine the entire perimeter in one pass, two bolt holes have been drilled through the part and the part has been attached to a block-type fixture that allows the part to be raised above the worktable.

The first step requires that the operator set the cutter at the assigned set-point. A convenient point would be the top of the lower left-hand corner of the part. The cutter is then directed to a point away from the part so that it may then move to the intersection of the drive, part, and check surfaces. See Fig. 10–6(A). The view from above the part is shown in Fig. 10–6(B). It is necessary to move to the intersection of these three surfaces in order for the cutter to start contouring around the part. As illustrated in Fig. 10–5(B), the part surface has also been set below the

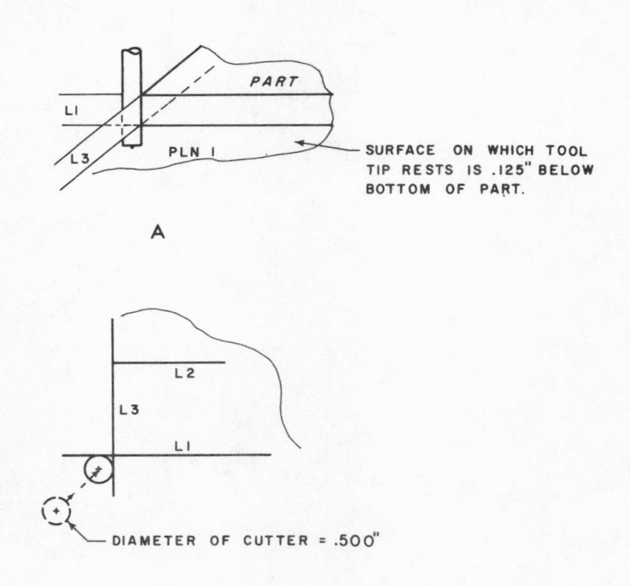

FIG. 10–6. (A) The cutter has moved to the intersection of the drive, part, and check surfaces. (B) The part as seen from above.

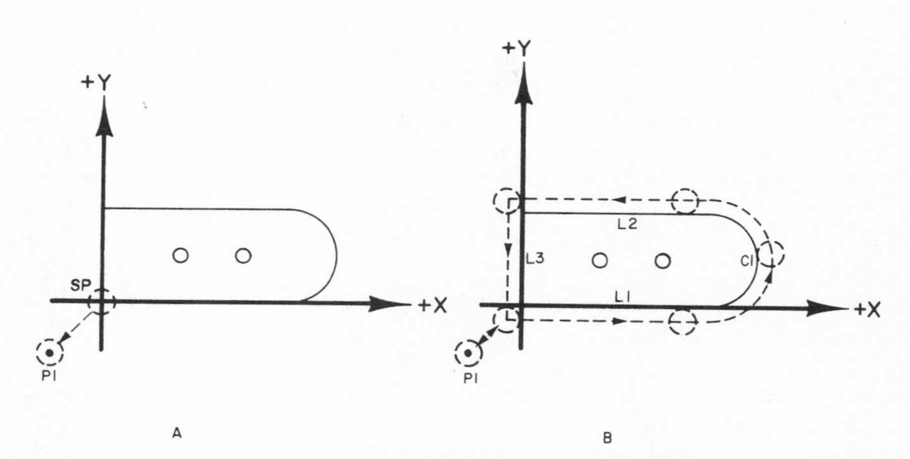

FIG. 10–7. (A) The cutter first moves from the set-point, SP, to P1. (B) The cutter next moves from P1 to the intersection of the drive, part, and check surfaces. Following this, the cutter moves on around the part and then back to P1, where it stops.

bottom of the part. The first point that the cutter is to move to has coordinates $x = -.5$, $y = -.5$, $z = 1$. This will put the cutter in a position to move to the drive, part, and check surfaces. The symbol for this point is P1 and the statement is:

P1 = PØINT/−.5,−.5,1

The statement for defining the plane, which lies .125 inch below the bottom surface of the part, is:

PLN1 = PLANE/P2,P3,P4

(Remember that the symbols P2, P3, and P4 must be defined prior to their use in the statement defining PLN1.) These points could have any coordinates as long as the points are not in a straight line and lie on the plane, which is parallel to the bottom surface of the part and .125 inch below it.

The statements for moving the cutter from the set-point, SP; to P1 (Fig. 10–7A); and then to the intersection of the drive surface, L1; the part surface, PLN1; and the check surface, L3 (Fig. 10–7B); are:

```
FRØM/SP
GØTØ/P1
GØ/TØ,L1,TØ,PLN1,TØ,L3
```

These statements would produce tape instructions so that the cutter would now be positioned as shown in Fig. 10–6(A). The sides of the cutter would be "touching" the imaginary extensions of surfaces L1 and L3, and the tip of the cutter would be resting on the plane, PLN1, which is the part surface and which lies below the bottom of the part.

Referring again to Fig. 10–7(B), the cutter is next directed to make a right turn along L1 to the point where L1 is tangent to the circle C1. The command for this GØRGT/L1,TANTØ,C1. The next command, which is GØFWD/C1,TANTØ,L2, will move the cutter around the circle to the point where C1 is tangent to L2. The cutter must then move along L2 and past L3. The command to perform this movement is GØFWD/L2,PAST,L3. The next command is GØLFT/L3,PAST,L1, and the last motion command is GØTØ/P1, where the cutter will stop. It should be pointed out that, although the cutter is instructed to make specific movements that direct it either TØ, PAST, or FWD, there is no dwell or stopping at the tangency points or at the intersections of the lines. The cutter moves along the full path, around the part, in a smooth continuous motion.

The complete APT program is as follows:

```
PARTNØ  FLAT PLATE WITH TWO TOOLING HOLES
MACHIN/ABC
INTØL/.005
ØUTTØL/.005
CUTTER/.5
CØØLNT/ØN
CLPRNT
FEDRAT/4,IPM

SP = PØINT/0,0,.25
P1 = PØINT/−.5,−.5,1
P2 = PØINT/0,0,−.125
P3 = PØINT/1,1,−.125
P4 = PØINT/2,3,−.125
P5 = PØINT/0,1,.25
C1 = CIRCLE/2,.5,.5
L1 = LINE/SP,RIGHT,TANTØ,C1
L2 = LINE/P5,LEFT,TANTØ,C1
L3 = LINE/SP,PERPTØ,L1
PLN1 = PLANE/P2,P3,P4

FRØM/SP
RAPID
GØTØ/P1
GØ/TØ,L1,TØ,PLN1,TØ,L3
GØRGT/L1,TANTØ,C1
```

```
GØFWD/C1,TANTØ,L2
GØFWD/L2,PAST,L3
GØLFT/L3,PAST,L1
RAPID
GØTØ/P1
CØØLNT/ØFF
FINI
```

Programming for a Lathe Part

From an APT part programming standpoint the motions of a lathe cutting tool are very similar to that of a milling cutter. In the case of a lathe, however, the center of the nose radius of the cutting tool is programmed rather than the center of the milling cutter. (Refer to Fig. 6–25.) There is a good deal of similarity, however, between the APT programming requirements for a milling machine and for a lathe. The points, lines, and circles that describe a part to be milled may also be used to describe a part that is to be turned. Consider, as an example, the part illustrated in Fig. 10–8.

Please note that the axes designations are similar to those used with a milling or drilling machine rather than what would be used when programming a lathe *manually*, as shown in Fig. 6–22. This arrangement is felt to be more convenient for the part programmer since the common *X*-*Y* coordinate arrangement will be the same for lathes as for other NC

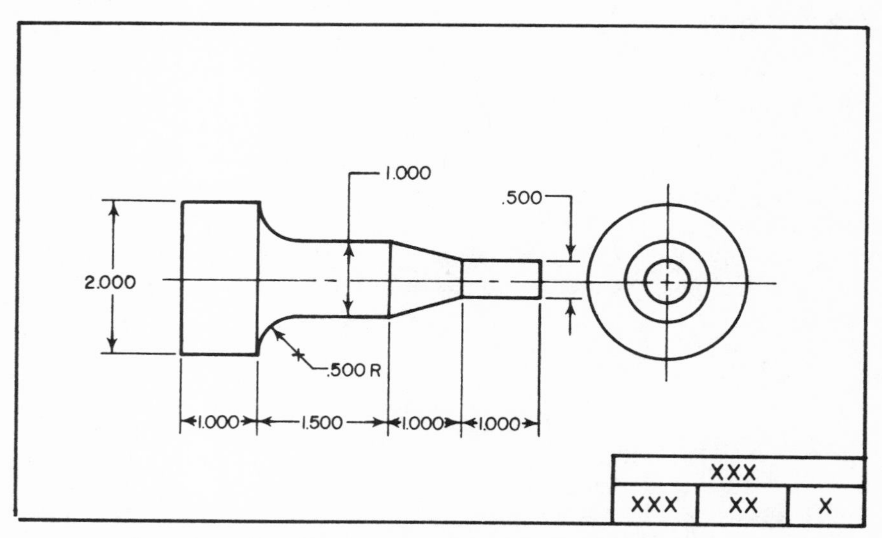

Fig. 10–8. The blueprint of a part to be turned on a lathe.

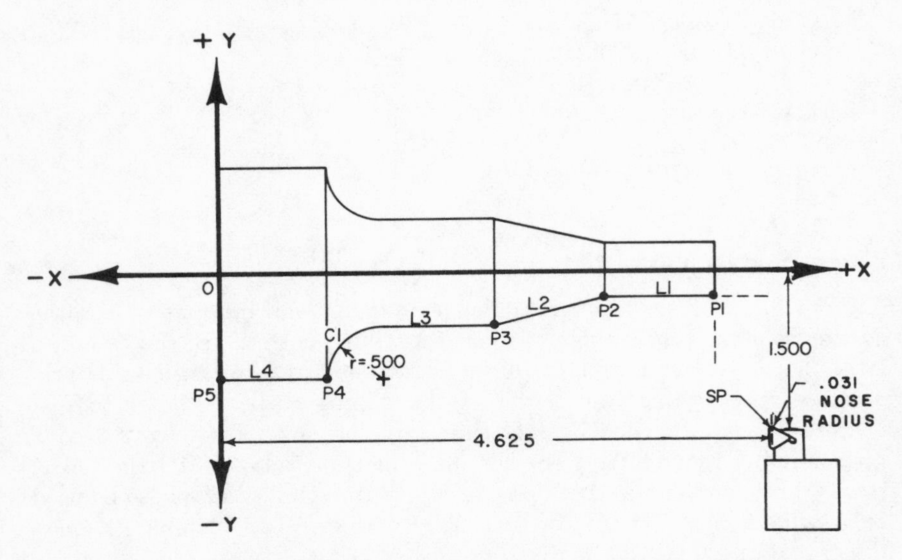

FIG. 10–9. After the part has been positioned with respect to the coordinate axes, symbols are assigned to the points, lines, circles, and other geometric segments. Since the part is to be turned, the portion of the part above the *Z* axis is identical to that below the *Z* axis; only the portion below the *Z* axis need be described.

machines—such as milling and drilling machines. The post processor in the computer program changes the *X-Y* coordinates to *Z-Y* coordinates, respectively, so that the tape may be read properly by the lathe control system.

If the center line of the part is noted as the *X* axis, then the part would appear, in the coordinate system, as shown in Fig. 10–9. The origin, or point where the *X* and *Y* axes cross, has been set at the left and center of the part. Also, since the part is rotating and will have the same shape above the *X* axis as it does below, only the contour for half the part need be considered.

As for any APT part program, the first step is to list the necessary auxiliary and post processor statements. These would be as follows:

```
PARTNØ  RØUND SHAFT NØ 1
MACHIN/DEF
INTØL/.0001
ØUTTØL/.00005
CUTTER/.062
CØØLNT/ON
CLPRNT
FEDRAT/2,IPM
SPINDL/200,SFM
```

Next, referring to Figs. 10–8 and 10–9, the geometry statements are listed:

```
SP = PØINT/4.625,-1.5
P1 = PØINT/4.5,-.25
P2 = PØINT/3.5,-.25
P3 = PØINT/2.5,-.5
P4 = PØINT/1,-1
P5 = PØINT/0,-1
L1 = LINE/P1,P2
L2 = LINE/P2,P3
C1 = CIRCLE/1.5,-1,.5
L3 = LINE/P3,RIGHT,TANTØ,C1
L4 = LINE/P4,P5
```

Please note that the Z coordinates, which remain at zero since the cutter moves in only the X and Y axes, need not be shown.

Next the motion statements are listed. The path describing the center of the tip of the cutting tool is shown in Fig. 10–10.

```
FRØM/SP
GØ/TØ,L1
```

NOTE: This is a one surface start-up, which was mentioned briefly in Chapter 9. Since the cutter cannot move up or down, it will automatically remain on the part surface. Thus, there is no need to note it in the motion statement. The cutter will also move perpendicular, or normal, to L1.

```
GØLFT/L1,TØ,L2
GØLFT/L2,PAST,L3
GØRGT/L3,TANTØ,C1
GØFWD/C1,PAST,L4
GØTØ/SP
```

Lastly, the closing auxiliary statements are added. These are:

```
CØØLNT/ØFF
FINI
```

Since the path that a lathe tool moves along is generally not as complex as that of a contour milling machine, the advantages of computer assisted programming for a lathe may not be as apparent at first as for a milling machine. The advantages, however, can be considerable. As an example, the machine tool operator or the part programmer need not be concerned about the proper adjustment of the spindle rpm with any change in radius of the part in order to maintain a near constant cutting speed in surface feet per minute (sfm) (also referred to as fpm) for, as the radius of the part changes, the proper rpm is automatically calculated by the computer

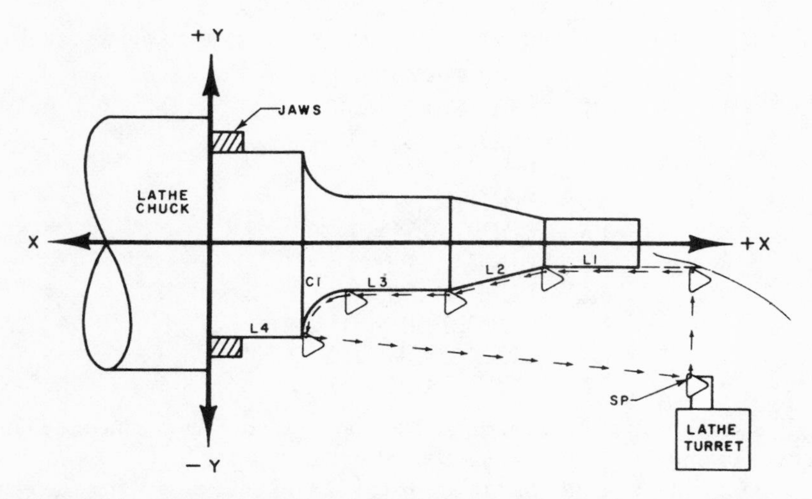

FIG. 10–10. The path of the center of the tip of the cutting tool is shown as it moves along in turning the part shown. For illustrative purposes, only the final pass has been shown. Actually, a series of cuts would be required to remove material in order to make the final cut. In the illustration the cutter is first commanded to move *to* L1; then left to L2. In order for the cutter to move along L3, as shown, the center of the tip of the cutter must move *past* the imaginary extension of L3. The cutter then moves along L3; along C, past the imaginary extension of L4, and then back to SP.

to maintain a near constant sfm. With a lathe having stepped spindle speeds the closest one would be selected. The APT statement for accomplishing this is:

SPINDL/100,SFM

which means that the rpm of the spindle is to be adjusted automatically in order to maintain a near constant sfm of 100.

Another computer calculation would be the conversion of the feed rate in inches per revolution (ipr) to inches per minute (ipm), which is the form required on the tape. The part programmer may note, for example:

FEDRAT/.02,IPR

The computer will then convert this to the corresponding IPM, which is dependent on the spindle speed and which is also noted in the program.

Threading may also be handled by noting the threads per inch and describing the threading statement. For example, if a thread having 8 threads per inch is to be cut, the statements would be:

PITCH/8
THREAD/TURN

The definition of *pitch* is the distance between threads which is the reciprocal of 8 (that is, one-eighth), and the statement therefore actually means: PITCH/8 threads per inch although only the 8 is given.

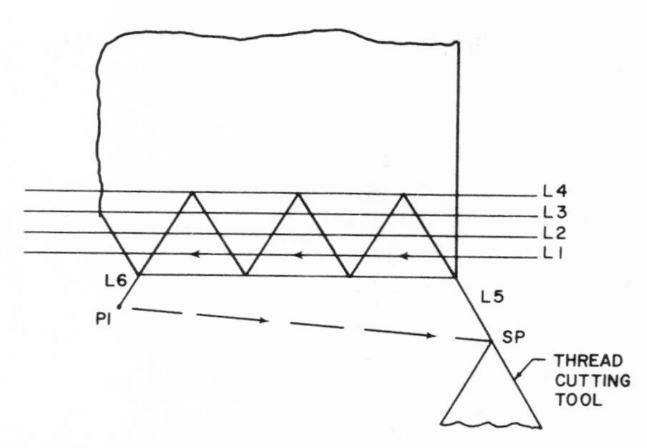

FIG. 10–11. Threading cuts may be made by programming a repetition of similar cutting cycles. The cutter shown is programmed to move to L1, then to L2, then to L3, and then to L4, where the full depth of the thread is reached.

As with threading on a conventional lathe, a number of passes are required to cut the thread properly. This may be accomplished on an NC lathe, using the APT program, by noting that the tip of the cutter is to move to a series of lines that have been defined previously in LINE statements. In Fig. 10–11 the cutting tool is directed to move to L1 and then left to L6; then left again to P1; then back to SP. On the next pass the cutter is to move to L2; then left to L6; then left to P1; and then again back to SP. The cycle is repeated with the cutter moving to L4, where the full depth of the thread is reached. There are APT programs that automatically handle repetitous cycles of this type. However, such coverage is beyond the scope of this text.

PRACTICE EXERCISE CHAPTER 10

Referring to the sketch shown in the blueprint below:

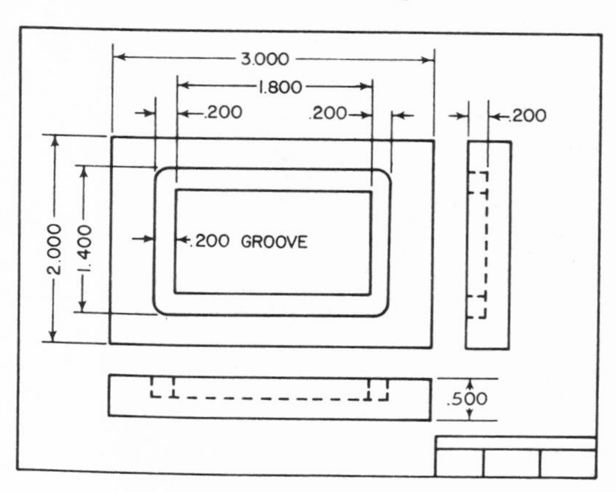

1.a. Draw a two-dimensional sketch of the part as it would appear in the first quadrant of the X-Y axes if two of the sides are to touch the axes.

1.b. Draw a two-dimensional sketch of the part as it would appear in the first quadrant of the X-Z axes if the bottom surface of the part is to be positioned .500 inch above the X-Y plane, with one side touching the Z axis.

1.c. Assign symbols to the required geometric elements, and then write the complete program to prepare a tape for machining the perimeter and the groove. Use the following information: First, the part shown below is to be machined around the perimeter using a .200-inch-diameter flat-end mill. Next, the groove is to be cut into the top surface of the part with the same tool. The part is to be held in a fixture that leaves the sides free to be machined in one pass. The groove is also to be machined in one pass.

QUESTIONS CHAPTER 10

1. The part shown is of .125-inch aluminum material and is to be machined on a contour milling machine. Show the part as it would appear in the second quadrant of the X-Y axes and as it would appear from the front and looking at the X-Z axes, showing the part surface and its relation to the plane of the X-Y axes. Also consider the following and write a complete part program.

   a. Feed rate = 5 ipm.
   b. Inside tolerance for the approximations of the circular arc = .0005 inch.

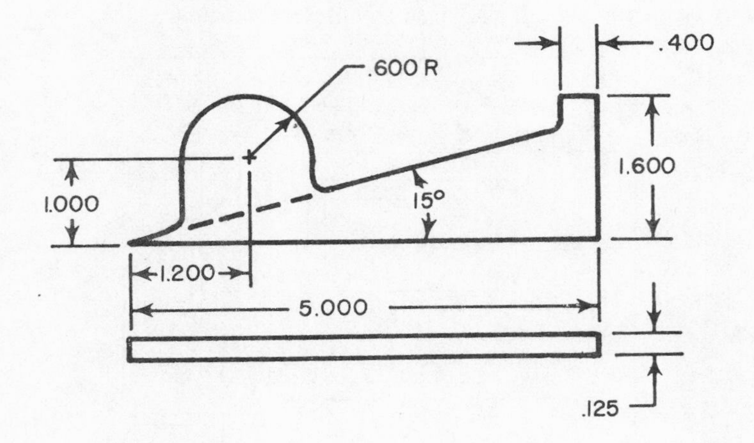

c. Outside tolerance for the approximation of the circular arc = .0001 inch.
d. The post processor statement is to be MACHIN/QRS.
e. The cutter is to be a ⅛-inch flat end mill.
f. The machine does not have a coolant cycle.

2. The part shown in the engineering drawing is to be rough cut on a band saw and then milled in one pass. The material is aluminum. Feed rates can be controlled from the tape; however, the speed is set by the operator. INTØL is to be .0005 and ØUTTØL is to be .001. The post processor call statement is to be MACHIN/ABC. The diameter of the cutter is ½-inch.

   NOTE: Check Appendix A for the correct feed rate if necessary. Also check Appendix C for the correct geometry statements.

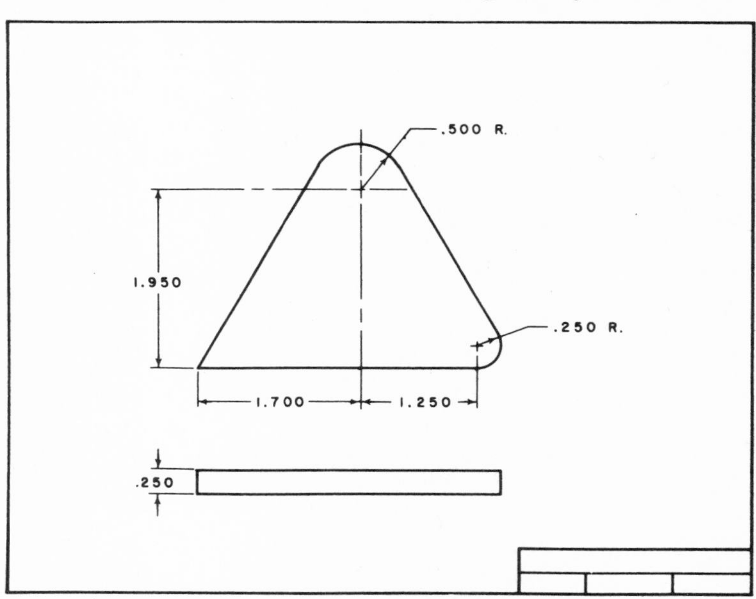

3. The plumb bob shown in the sketch is to be turned on an NC lathe Since the raw material is to be of 1-inch diameter bar stock, four rough cuts at the pointed end and two rough cuts and one finish cut at the rounded end must be made prior to the finished cut. The part is to be machined at the pointed end first; then removed from the chuck, turned and reinserted in the chuck; then the rounded end is machined. In this instance the material is brass. If harder material is used additional passes may be necessary. This will also depend, to a great extent, on the horsepower and rigidity of the lathe. The nose of the cutter in this example, for practical purposes, is a point having no radius.

3 *(cont.)*

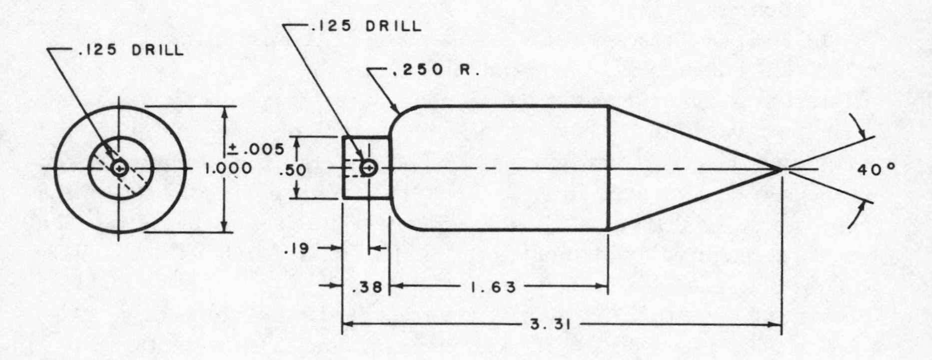

ALL DIMENSIONS ± .010 UNLESS OTHERWISE SPECIFIED

APPENDIX A

Feeds and Speeds for Cutting Operations

Drilling

Speed: The cutting *speed* of a drill is measured as the velocity of the circumference of the drill and is described in feet per minute. For example, the circumference of a 1-inch-diameter drill is approximately 3 inches.

The calculation for this is as follows:

$$\text{Circumference} = \pi \times \text{diameter}$$

$$= 3.1416 \times 1 = 3.1416$$

If the drill is to turn at a rate of 2 revolutions per minute (rpm), then the cutting speed, in *inches* per minute, would be:

$$\text{Circumference} \times \text{rpm} = \text{inches per minute}$$

or:

$$3.1416 \times 2 = 6.2832$$

To convert the inches per minute to *feet* per minute (fpm), we simply divide the 6.2832 figure by *12*. In this case the cutting speed, in feet per minute, would be approximately .52 fpm. Of course, 2 rpm is a very slow rate, and it is doubtful that any material would require this kind of cutting speed; the figure has been used only for illustrative purposes in this case. Approximate cutting speeds for a number of common materials, using high speed steel drills, are shown below. The exact cutting speed would depend primarily on the hardness of the particular material alloy. Care should be exercised so as not to overheat the cutting tool when machining plastics.

Material	Approximate Cutting Speeds (fpm) for Drilling
Hard wood	500
Aluminum	300
Brass	200
Plastics	150
Cast iron	90
Steel (low carbon type)	80

While most handbooks and speed tables specify cutting speeds in fpm, the speed for most numerical control machines is set in *revolutions per minute* (rpm). This means that the operator, or part programmer, must consider the diameter of the drill in determining the proper fpm for the material being considered.

The calculation of the spindle speed or drill speed from a specified cutting speed may be made by using the following formula:

$$N = \frac{12V}{\pi D}$$

where

V = speed of the circumference of the cutter, in feet per minute
D = diameter of the drill in inches
N = spindle speed of the machine or rotative speed of the drill in revolutions per minute.

Table A–1 lists approximate recommended speeds, in rpm, for the common materials described above, with respect to drill size. In the case of metals, selection of a more precise speed, in rpm, will depend primarily on the kind of metal and its hardness. Other factors, such as the surface condition of the metal, the rigidity of the machine tool, the type and condition of the cutter, and the coolant used, if any, will also have an effect. The figures shown in Table A–1 are proposed as a very general guide and the operator, or part programmer, may check more detailed tables in any one of a number of specialized handbooks.[1] Also, the speeds shown in Table A–1 refer to high speed steel (HSS) drills. If *carbon* steel drills are used, the rpm would be approximately one-half of the figures shown. When a *carbide* drill can be used, the rpm figures would be approximately twice those shown.

[1] For example, Erik Oberg and F. D. Jones, *Machinery's Handbook* (New York: Industrial Press Inc., 1971).

Table A–1. Approximate Drilling Speeds (rpm) for Use with High Speed Steel (HSS) Cutting Tools

Diameter of Drill (Inches)	Hard Wood	Aluminum	Brass	Plastic	Cast Iron	Steel
	Drilling Speed, rpm					
Less than 1/8″	16,000	10,000	6500	4500	2700	2600
1/8–1/4	10,000	6000	4000	3000	1800	1600
5/16–7/16	5000	3000	2000	1500	900	800
1/2–3/4	3000	1800	1200	900	540	480
7/8–1 1/4	2000	1100	720	525	315	290
1 1/2–2	1100	650	420	310	190	170

NOTES:

1. There are few machines that will go as high as the high ranges shown for the softer materials. In this case the maximum setting for the machine should be used.
2. Multiply the above figures by 2 for carbide drills and divide by 2 for carbon steel drills.

Table A–2. Approximate Drilling Feeds, in Inches per Revolution (ipr)

Diameter of Drill (Inches)	Drilling Feed (ipr)
Less than ⅛″	.001–.002
⅛–¼	.002–.004
¼–½	.004–.007
½–1	.007–.015
Larger than 1″	.015–.025

Feed: In addition to determining the speed, the part programmer, or operator, must also determine the feed, which is the rate that the cutter moves into the material. Feed is normally expressed in *inches per revolution* (ipr) to the nearest thousandth of an inch. The proper feed is dependent on the size of the drill and on the material being drilled. Table A–2 gives recommended feeds, in ipr, for ranges of different size drills. The lower end of the feed range should be used for the harder materials such as steel, while the upper end of the feed range applies to the softer materials, such as aluminum.

Most numerical control machine tools require that the feed be expressed in inches per minute (ipm) rather than in inches per revolution (ipr). Ipm is also a feed *rate* since it is associated with time. A relatively simple calculation will convert *ipr* to *ipm* when the spindle speed, *in rpm*, is known. For example, if the spindle speed is to be 1800 rpm and the feed called out is .010 ipr, then the feed rate in ipm can be arrived at by multiplying .010 by 1800 and the answer is 18 inches per minute (ipm).

Milling

Speed: The speed of a milling cutter is measured in the same way as for a drill—in the rate of travel of the circumference of the cutter in feet per minute (fpm). Again, as with drilling, the speed of a numerical control milling machine is usually set in revolutions per minute (rpm). Table A–3

Table A–3. Approximate Milling Speeds for HSS Cutters in fpm

Material	fpm
Aluminum	600
Brass	140
Plastic	300
Cast iron	100
Steel (mild)	90

Table A–4. Approximate Milling Speeds (rpm) for Use with High Speed Steel (HSS) Milling Cutters

Diameter of Mill (Inches)	Hard Wood	Aluminum	Brass	Plastic	Cast Iron	Steel
	Speed, rpm					
¼–⅜	10,000	7300	1700	3500	1200	1000
7/16–⅝	8300	4500	1000	2200	750	650
¾–1	3000	2600	600	1300	425	400
1¼–1¾	1600	1500	350	750	250	225
2–3	900	900	200	450	150	125
3½–5	600	500	100	250	80	75

NOTE:

1. There are few machines that will go as high as the high ranges shown for the softer materials. In this case the maximum setting for the machine should be used.
2. Increase the above speeds 100 percent for carbide cutters. Decrease to approximately 60 percent of the figures shown for carbon steel cutters.
3. Figures shown are for rough cuts. Increase by approximately 10 percent for finish cuts.

notes the approximate speeds for five common materials. Table A–4 lists approximate recommended speeds for various size cutters and different materials.

Feed: The feed for a milling cutter is normally specified in inches of material cut per tooth (ipt); however, most NC machines require that his factor be noted as a feed rate, or rate of travel of the cutter into the material, and is expressed in inches per minute (ipm). To convert, therefore, from *ipt* to *ipm*, it is necessary to know:

1. The number of teeth on the cutter;
2. The speed, in rpm (see Table A–4), that the cutter is to travel;
3. The specified ipt for the particular type of milling cutter and the material to be cut.

If, for example, an end mill cutter having two teeth is to cut aluminum and the recommended rpm is 1800, and the ipt called for is .011 inch per tooth, then the recommended feed rate, in ipm, would be calculated as follows:

$$\begin{aligned} f_m &= N f_t n_t \\ &= 1800 \times .011 \times 2 \\ &= 40 \end{aligned}$$

where

f_m = feed rate in inches per minute
f_t = feed rate in inches per tooth per revolution
n_t = number of teeth on cutter
N = revolutions per minute

While 40 inches per minute may seem high, it must be kept in mind that this is a theoretical cutting feed rate which would be possible only if the machine had both the power and rigidity to handle the cut. It may, therefore, be necessary to lower this figure while the part is being machined. This can be accomplished by a "feed rate override" knob usually located on the operator's panel of the control system.

Table A–5 describes chip-load ratings, in ipt, for different materials and for the two most common types of NC milling cutters. Table A–6 describes approximate feed rates, in ipm, for the two most common types of cutters used in NC, namely, end mills and face mills, with respect to their size and the material being machined. Fine finishes may be obtained by lowering the feed rate while maintaining the speed.

Reaming

The purpose of a reamer is to finish out a hole to a precise dimension and to provide a smooth surface finish on the walls of the hole. The amount

Table A–5. Approximate Milling Feeds in Inches per Tooth (ipt) for Use with High Speed Steel (HSS) Milling Cutters

Type of Cutter	Hard Wood	Aluminum	Brass	Plastic	Cast Iron	Steel
			Feed, ipt			
End mill (ball and flat)	.003–.012	.003–.010	.003–.008	.003–.010	.002–.008	.002–.008
Face mill	.012–.030	.012–.020	.008–.016	.008–.016	.006–.012	.006–.012

NOTE: The above figures should be increased by 25 percent when using carbide tipped cutters, or cutters having carbide inserts, and divided by 2 if carbon steel cutters are used.

Table A–6. Approximate Milling Feeds, in Inches per Minute (ipm), for Use with High Speed Steel (HSS) Milling Cutters

END MILLS							
Type	Diameter (Inches)	Hard Wood	Aluminum	Brass	Plastic	Cast Iron	Steel
		Feed, ipm					
2-Flute	¼–⅜	60	40	10	21	5	4
	7⁄16–⅝	80	35	8	17	4	4
	¾–1	40	30	7	15	4	4
	1¼–1¾	30	24	5	13	3	3
	2–2½	20	18	3	9	2.5	2
4-Flute	¼–⅜	120	80	20	42	10	8
	7⁄16–⅝	160	70	16	34	9	8
	¾–1	80	60	14	30	8	8
	1¼–1¾	60	48	10	26	6	6
	2–2½	40	36	6	18	5	4

FACE MILLS							
No. of Teeth	Diameter (Inches)	Hard Wood	Aluminum	Brass	Plastic	Cast Iron	Steel
		Feed, ipm					
6	2–3	65	65	10	22	5	5
8		85	85	13	29	7	6
10		100	100	16	36	9	8
6	3½–5	43	36	5	12	3	3
8		55	47	6	16	4	4
10		65	55	8	20	5	5

NOTE: The above figures should be increased by 100 percent when using carbide tipped cutters, or cutters having carbide inserts, and divided by 2 if carbon steel cutters are used.

of stock left in the hole for reaming should be approximately .008 to .012 inch in a 1/4-inch hole, .012 to .016 inch in a 1/2-inch hole, and .025 to .035 inch in holes up to 1 1/2 inches in diameter. The reamer will cut to the specified size that is required.

Speed: The speed for reaming should be approximately two-thirds (2/3) the drill speed used.

Feed: The feed should be approximately .0015 to .004 inch per flute per revolution of the reamer. Thus, the feed rate for a six-fluted reamer that is cutting at 100 rpm, using a .002 inch per tooth feed, would be 6 × .002 = .012 inch per revolution; or 100 × .012 = 1.2 inch per minute. When reaming, a cutting fluid of the correct type should always be used.

Tapping

The purpose of a tap is to cut threads on the inside of a hole. In order for the tap to cut threads, the hole size must be smaller than the tap size. The difference between the two diameters will depend on the number of threads per inch of the tap. One formula for calculating the size of a tap drill is:

$$\text{Tap drill diameter} = \text{Tap diameter} - \frac{.812}{\text{No. of threads per inch}}$$

For example, if we wanted to drill a hole for a tap that has an outside (major) diameter of 3/8 inch and the number of threads is 16 per inch, then the drill size would be:

$$\text{Tap drill diameter} = 3/8 - \frac{.812}{16}$$

$$= .3242 \text{ inch}$$

Speed: The exact speed at which a tap rotates cannot be given with certainty because it is based on factors that are unique to each job. However, care should be exercised so as not to use an excessive speed which may cause the tap to overrun its programmed depth movement. Most NC machines, therefore, operate at tapping speeds ranging from approximately 50 to 400 rpm.

Feed: The feed, which is specified in inches per revolution of travel into the material (ipr), will depend on the number of threads per inch on the tap and the rpm of the spindle. If, for example, there are 16 threads per inch, every time the spindle turns the tap will move 1/16 of an inch into the material. If the speed were set at 200 rpm, then the feed rate, in inches per minute (ipm), would be:

$$\text{Feed rate (ipm)} = 200 \times 1/16$$
$$= 12.5$$

On some numerical control machines the feed rate is not set but is left to be governed by the speed (rpm of the spindle) and the number of threads per inch on the tap. This is known as a free-floating tap arrangement. Still other NC machines offer a controlled feed rate for the tapping cycle in ipm. In most of these cases the tap holder has a built-in axial float allowance to compensate for a feed rate movement that might be greater than that programmed, because of a combination of rpm and threads per inch. In the instance described above, for example, the feed rate would be set at something lower than the 12.5 ipm, such as 12.0 ipm, if this is allowable with the particular machine.

Turning

Speed: As with all cutting tools the *speed* of a single-point lathe cutting tool is measured in feet per minute (fpm), and this figure will depend on a number of factors, most significant of which are the kind of material being cut, the type and material of the cutter, the rigidity of the lathe, and the tool life desired. Table A–7 describes approximate speeds for various types of material. Since the speed for most numerical control machines must be set in revolutions per minute (rpm), the part programmer or operator must calculate the conversion from fpm (*feet* per minute) to rpm. For example, if a steel bar, 3 inches in diameter, is to be machined, the speed should be approximately 110 fpm. The rpm to be set for the lathe would be:

$$N = \frac{12V}{\pi D}$$

Table A–7. Approximate Cutting Speeds for Lathes, in Feet per Minute (fpm), for Use with High Speed Steel (HSS) Cutting Tools

Material	fpm
Aluminum	600
Brass	180
Plastic	350
Cast iron	100
Steel	110

NOTE: Figures may be approximately doubled for carbide tools.

where

$$N = \text{Spindle speed, rpm}$$
$$V = \text{Cutting speed, fpm}$$
$$D = \text{Workpiece diameter before turning, rpm}$$

Thus

$$N = \frac{12 \times 110}{\pi 3}$$

$$N = 140 \text{ rpm}$$

The number 12 has been inserted into the formula so that the fpm would be changed to *inches* per minute in order to agree with the diameter of the bar which is measured in inches.

Figure A–1 is a graph which may be used to find the approximate rpm of the spindle when turning several materials at different diameters. As an example, if it were desired to find the approximate speed for turning a 2-inch steel bar, the part programmer or operator would check the hori-

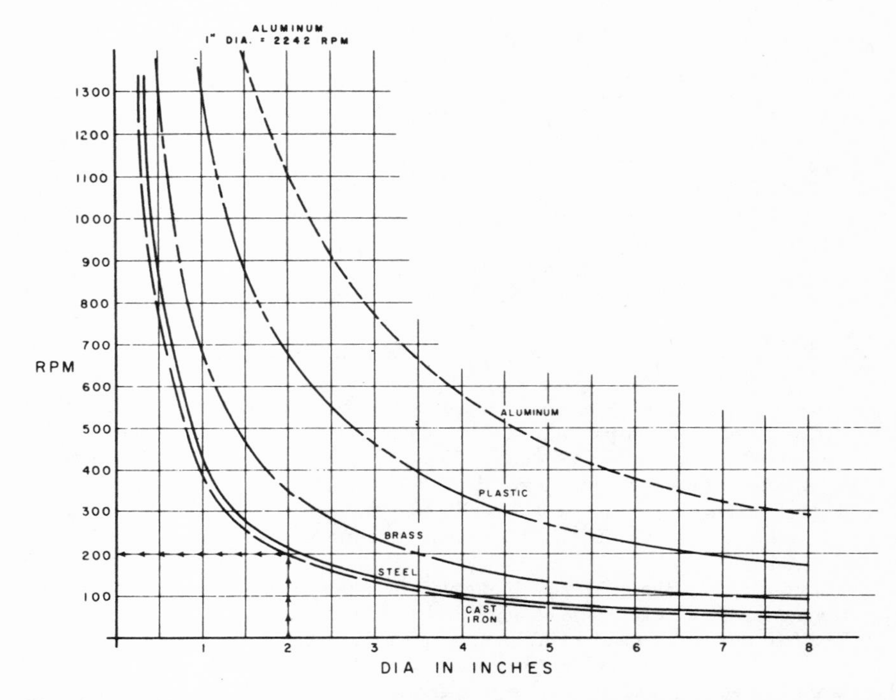

FIG. A-1. This graph shows turning speeds in rpm for various diameters of stock *before* the cut is made. The curves refer to High Speed Steel cutters. Carbide tools may be run at approximately twice the rpm's shown.

zontal scale for the 2-inch diameter figure; then, as shown by the dashed line, move vertically to the steel curve; then horizontally over to the left and read the speed on the vertical scale, which in this case is approximately 200 rpm.

The *depth* of cut to be taken when roughing *may* be ¼ inch or more; however, this would depend very much on the material being cut, on the tooling, on the rigidity of the lathe, and on the horsepower available on the lathe. The speeds shown in Table A–7 are for roughing cuts of ⅛-inch depth, or less. The depth for a finish cut is usually from .010 to .032 inch.

Feed: The *feed* of a lathe cutting tool is measured as the distance the cutting tool moves along the workpiece per one revolution of the workpiece, and is noted in inches per revolution (ipr). Table A–8 notes approximate cutting feeds for different materials.

As with NC milling and drilling machines, the feed for a lathe is generally programmed in inches per minute (ipm).[2] In order to determine the feed

Table A–8. Approximate Feeds for Turning, in Inches per Revolution (ipr), for Use with High Speed Steel (HSS) or Carbide Cutters

	MATERIAL				
	Cast Iron	Steel	Aluminum	Plastic	Brass
Type of Cut	Feed, ipr				
Roughing	.015–.060 (.020)	.015–.040 (.020)	.015–.040 (.020)	.010–.030 (.012)	.015–.030 (.018)
Finishing	.005–.012 (.008)	.005–.012 (.008)	.003–.010 (.006)	.003–.010 (.005)	.003–.012 (.005)

NOTES:

1. In the table the range of feeds normally used is given above and a suggested starting feed is given below in parentheses. For smaller lathes the suggested starting feed may have to be reduced by as much as 30 percent, if the lathe has insufficient power to cut using the suggested starting feed.
2. The feed in inches per minute is dependent on the feed in inches per revolution and the rpm of the spindle. The figures shown relate to speeds shown in Table A-7 and Fig. A-1.
3. The feed rate of the finish cut would also depend on the nose radius of the tool. Generally higher rates can be accomplished with larger radii.
4. The feed rate will depend upon the horsepower available on the lathe and upon the depth of cut used. However, feeds can be used when the horsepower available is larger and when the depth of cut is smaller.

[2] Some control system manufacturers offer an optional feature whereby feed can also be programmed directly in ipr.

rate, therefore, it would be necessary to multiply the speed (in rpm) by the feed (in ipr). As an example, the approximate speed in rpm (see Fig. A–1) for rough cutting an aluminum bar with a 3-inch diameter is 250 rpm, and the feed (see Table A–8) is .050 ipr. The feed rate, in ipm, would be:

$$
\begin{aligned}
\text{Feed rate (ipm)} &= \text{feed (ipr)} \times \text{speed (rpm)} \\
&= .050 \times 250 \\
&= 12.5
\end{aligned}
$$

Threading

The feed rate for single-point *threading,* as with a lathe, is determined by the speed of the spindle and the lead.[3] Threading on an NC machine is a relatively simple matter, with the programmer specifying the length of the thread and the lead. The feed is then automatically synchronized. However, the lathe must be equipped to feed the tool into the work at the end of each pass and to engage properly with the previously cut thread grooves.

[3] Lead, in this case, may be defined as the distance that the cutter will move along the Z axis for one revolution of the spindle.

APPENDIX B

Definitions for Numerical Control, Computer, and Machining Terms

TERM	DEFINITION	EXAMPLE
Absolute Dimensioning	A means of describing movement instructions as a distance from the axes. Word instructions on the tape are noted in absolute form.	In the example shown below, if we want to move from point A to point B, the absolute coordinates of B (distances from the X and Y axes) would be noted on the manuscript. Most PTP systems employ an absolute dimensioning system. COORDINATES PT X Y A 2.000 3.000 B 5.000 4.000

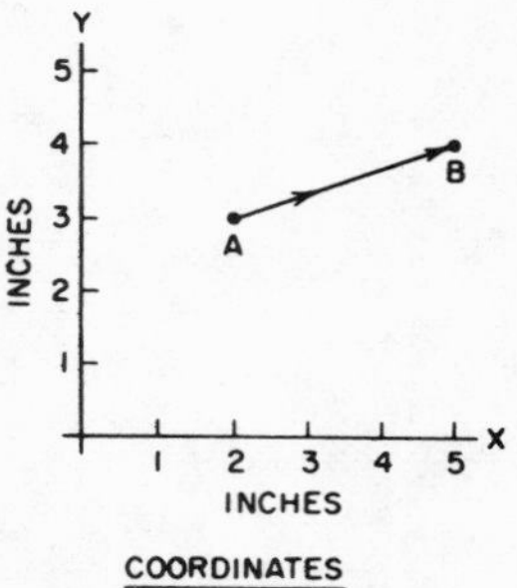

TERM	DEFINITION	EXAMPLE
Accuracy	Usually applies to the machine tool and describes its conformity to a position of indicated value. The latest accepted understanding applies to the accuracy of an individual axis movement. The National Machine Tool Builders' Association (7910 Westpark Drive, McLean, Virginia 22101) has developed a booklet describing a detailed and expanded definition covering accuracy and repeatability entitled "Definition and Evaluation of Accuracy and Repeatability for Numerical Control Machine Tools."	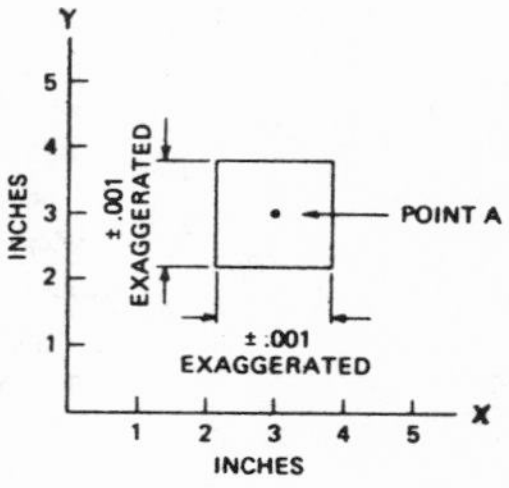 Y; 5; 4; 3; 2; 1; INCHES; ±.001 EXAGGERATED; POINT A; ±.001 EXAGGERATED; 1 2 3 4 5 X; INCHES Assume that the coordinate position of point A, as described on the tape, is X = 3.000 and Y = 3.000. If the machine accuracy is guaranteed ±.001" then the X-axis movement could fall between 2.999 and 3.001 and the Y-axis movement could fall between 2.999 and 3.001, or anywhere within the square shown above.
Adapter	A holding device for a cutting tool which is inserted into the spindle of the machine. The shank of the adapter is very often tapered.	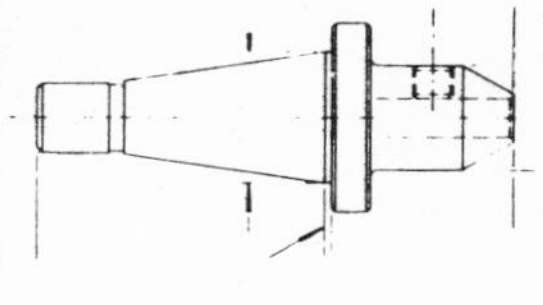
Address	Means of identifying information which is to be stored in a computer. With numerical control, the address usually consists of a letter noted on the tape which requires some action, the extent of which is described by numbers following the letter. In the example shown, x, y, f, and m are addresses which describe the x and y motions together with the feed-rate required (f), and the auxiliary function (m).	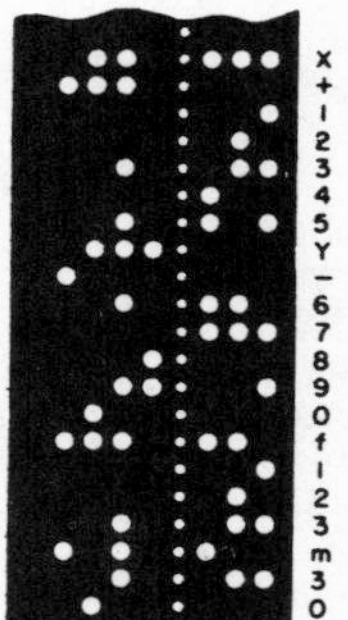 X + 1 2 3 4 5 Y – 6 7 8 9 0 f 1 2 3 m 3 0

TERM	DEFINITION	EXAMPLE
Analog	Analog data implies continuity of information such as would be developed directly from a rheostat or speedometer. An analog device is generally dependent for operation on a continuous signal which varies in magnitude with the variation in voltage or current as determined by the sensing unit. Although sufficiently precise for certain problems, it may not be accurate enough for others. The degree of accuracy obtainable is dependent upon the sensitivity of the measuring and indicating equipment employed (often contrasted with DIGITAL).	
APT	Stands for Automatically Programmed Tool and is a part programming language used with computer-assisted part programming. The words are very much like English words in order to make the language easier to learn. (See Chapters 8 through 10.)	Examples of APT words are: Line = LINE Point = PØINT Go right = GØRGT
Assembly Program	A computer program that translates a symbolic language into machine language. The symbol is generally more easily recognizable than a machine language instruction and is, therefore, referred to as a mnemonic form.	An assembly language instruction might be: ADD INCR The assembly computer program translates this into digits, which is the only form the computer understands, i.e., 63 4218 As can be seen, it is far easier for a programmer to recognize and handle "ADD INCR," meaning "add increments," than to recall that "63" means "add," and "4218" means "increments."
Auxiliary or Miscellaneous Function	An operation performed by or associated with the machine other than positioning or contouring.	1. Starting or stopping the spindle. 2. Turning the coolant on or off. 3. Positioning a turret. 4. Selecting a tool. 5. Initiating clamping devices.

TERM	DEFINITION	EXAMPLE
Backlash	Movement between interacting mechanical parts resulting from looseness.	Drive motor, if directly coupled, turns 1 degree with no motion of the table. Backlash would be equal to whatever linear motion should result from a 1-degree turn of the motor.
Ball End Mill	Also referred to as a ball nose mill. A type of milling cutter used for milling a curved surface. The "scallops" created by the radius of the milling cutter may be smoothed out by hand after the cutting operation.	
Binary	A system for describing numbers using only two digits.	Practically all computers operate on a form of binary system wherein a number can be expressed by a combination of "on" and "off" circuits. The concept is particularly well-suited for numerical control since a number may be expressed by either a "hole" or a combination of "holes" and "no holes" on a tape.

TERM	DEFINITION	EXAMPLE
Binary Coded Decimal Format (BCD Format)	A system of representing numbers comprised of a combination of four binary bits running across the tape. Letters are also expressed by a combination of binary bits.	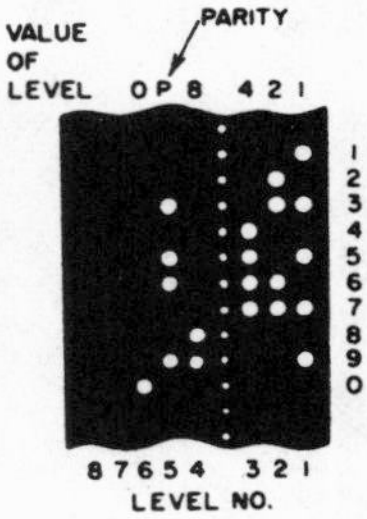 Considering only the four right-hand rows (levels, tracks) of the tape, numbers are expressed by combining holes across the tape in compliance with the numerical value of the row. Zero is expressed by a hole punched in one of the other four rows.
Bit	One of two possible states (i.e., either on or off). On the numerical control tape it can mean either the absence or presence of a hole.	
Block	A "word" or group of words" considered as a unit separated from other such units by an "end of block" character (EB). On a punched tape, it consists of one or more characters or rows across the tape that collectively provide sufficient information for a complete cutting operation.	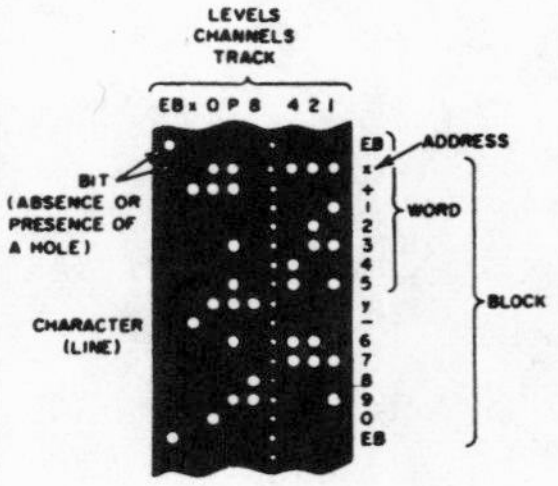
Block Count Readout	Display of the cumulative number of blocks that have been read from the tape. The count is triggered by the "end of block" character (EB). Has generally been replaced by a "sequence number readout."	

TERM	DEFINITION	EXAMPLE
Block Delete	This feature provides a means for skipping certain blocks by programming a slash (/) code immediately ahead of the block. The feature is useful when the operator desires to leave off certain cuts of a particular part configuration.	(A) (B) Assume it was desired to cut the perimeter of the above part and then drill holes (A) and (B) on one lot of parts and then on the next lot cut the perimeter and drill only hole (B). The same tape could be used for both lots by pushing the block delete button on the second lot and eliminating the drilling of hole (A) (assuming the (/) character had been incorporated before block (A).
Buffer Storage	A place for storing information in a control system or computer for anticipated utilization. Information from the buffer storage section of a control system can be transferred almost instantaneously to active storage which is that portion of the control system commanding the operation at the particular time. Buffer storage offers the ability of a control system to act immediately on stored information rather than wait for this information to be read into the machine via the tape reader which is relatively slow.	
Chad	The pieces of material that are removed when punching holes in tape or cards.	
Channel	See *Track*.	
Character	A number, letter, or symbol read on one line across the tape.	ONE LINE ACROSS TAPE = A CHARACTER

TERM	DEFINITION	EXAMPLE
Circular Interpolation	A simplified means of programming circular arcs in one plane which eliminates the necessity for segmenting the arc into calculated straight-line increments.	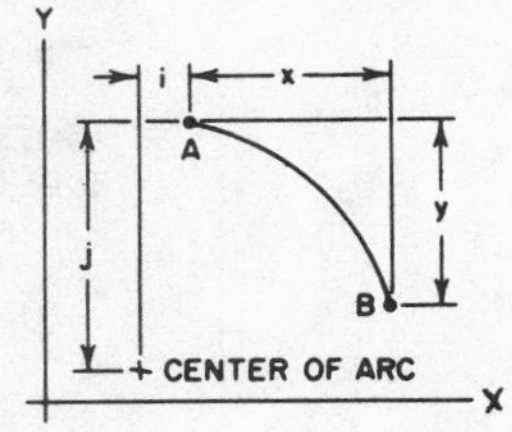 Y, i, x, A, y, j, B, + CENTER OF ARC, X The above circular arc (A–B) may be programmed by denoting essentially the x, y; i, and j dimensions. (For further details consult Chapter 6.)
CL Information	Stands for Cutter Location and describes the coordinates of the path of the center of the cutter resulting from a basic computer program. This information is common to all machine-tool-system combinations and is the input to the post processor.	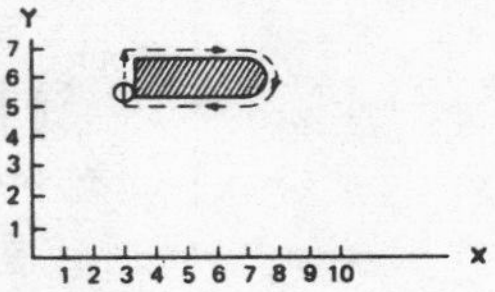 Y, 7, 6, 5, 4, 3, 2, 1, 1 2 3 4 5 6 7 8 9 10, X In accordance with the above, the CL computer print-out would include the following coordinate information: X Y 3.000 5.000 3.000 7.000 7.000 7.000 7.010 6.995 7.018 6.989 7.024 6.981 ↓ ↓ (7.010–7.024 rows:) Small straight-line increments rounding circular arc.

TERM	DEFINITION	EXAMPLE
Closed Loop System	A system in which the output, or some result of the output, is measured and fed back for comparison with the input. In a numerical control system, the output would be the position of the table or head; the input would be the tape information which ordinarily differs from the output. This difference would be measured and result in a machine movement to reduce and eliminate this variance.	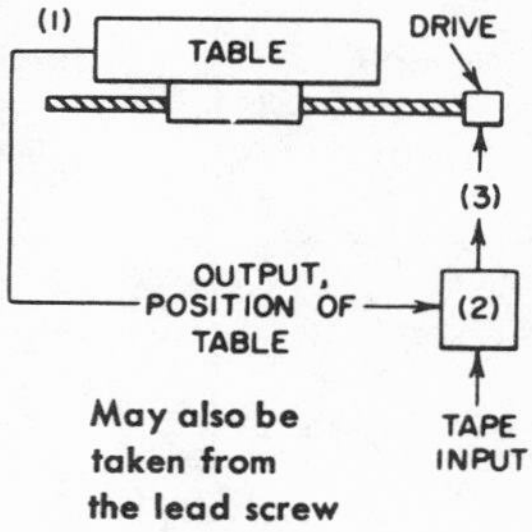 (1) TABLE DRIVE (3) OUTPUT, POSITION OF TABLE (2) May also be taken from the lead screw TAPE INPUT (1) Position or motion of the table fed back to (2). (2) Device for comparing tape input information with actual position. (3) Output of (2) actuates drive to move table until actual position agrees with tape input.
CNC	Stands for Computer Numerical Control, wherein one or more machines are operated directly by a computer—usually a mini-computer.	
Collet	A ring-type device that fits around a cutting tool. The assembly of these two pieces may then be inserted into an adapter, or directly into the spindle of the machine.	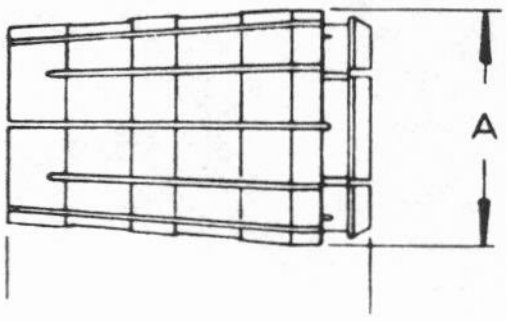 A
Command	A signal or group of signals or pulses initiating one step in the execution of a program.	

TERM	DEFINITION	EXAMPLE
Compact II	A proprietory part-programming system offered by Manufacturing Data Systems Inc., Ann Arbor, Michigan.	
Compatibility	A term used to describe the degree of interchangeability of tapes or numerical control language between numerical control machine tools and their respective control systems.	The ideal situation would exist wherein one tape could be utilized with various control systems of different manufacturers as well as different machine tools of the same type. Although a considerable degree of compatibility presently exists with regard to programming language, tape size, and, to a great extent, tape format, minor peculiarities between the systems as well as differences in the mechanical features of the machine tools generally prohibit the interchangeability of tapes between various systems and machine tools. The APT programming language offers an excellent example of compatibility in that once a manuscript is prepared for one system, the same write-up may be utilized for different systems by merely denoting the name of the control system at the beginning of the program.
Compiler	Similar in concept, to an assembler, in that the programmer employs a symbolic language that is more recognizable than a machine language. A compiler goes further than an assembler, however, in that it provides linkages to sub-routines, selects the required subroutines from a library of routines, and assembles these parts into an object program. The two most common compilers are FORTRAN, which is applicable to scientific type programming and COBOL, which is concerned with business-type problems. Numerical control	

TERM	DEFINITION	EXAMPLE
Compiler (Cont'd)	programming, because of its geometric and formular nature, utilizes the FORTRAN system. The significance of FORTRAN lies in its universal nature, as most computers will accept this language.	
Contour Control System	A system which continuously controls the path of the machine (e.g., cutting tool, pen or scribe, welding head or torching head) by a coordinated simultaneous motion of two or more axes.	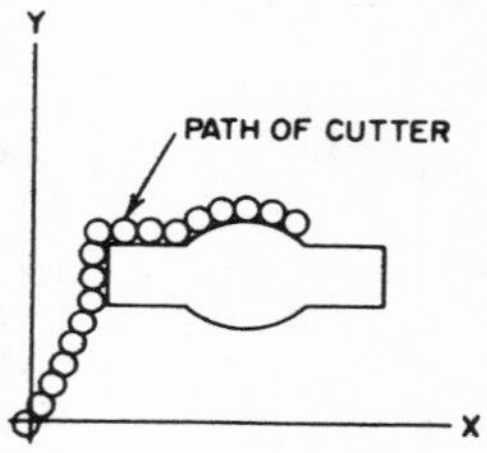 Y PATH OF CUTTER X
Coordinates	The positions of points in space with respect to X, Y, and Z axes.	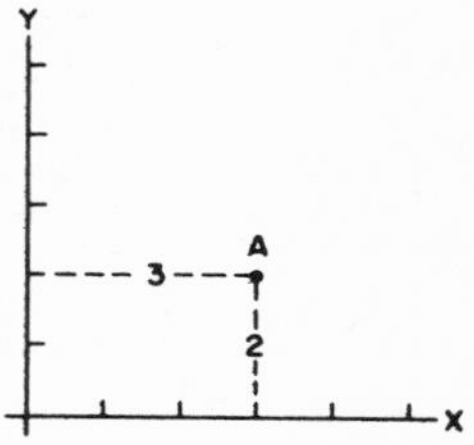 Y A 3 2 X The coordinates of point A are: x = 3 y = 2 Assuming the point lies on the plane formed by the x and y axes; the z coordinate would be zero.
CPU	Stands for Central Processing Unit and is the heart of the computing system. It contains the memory section arithmetic unit for computation; control circuits to direct operation of the system; and an operator console.	
Cutter Compensation	A means of manually adjusting the cutter center path on a contouring system so as to compensate for the variance in nominal cutter radius and the actual cutter	

TERM	DEFINITION	EXAMPLE
Cutter Compensation (Cont'd)	radius. The net effect is to move the path of the center of the cutter closer to or away from the edge of the workpiece. Special considerations must be noted on the tape if it is anticipated that cutter compensation will be used by the operator.	
Damping	A means of suppressing oscillation in a system which is caused by the system's response to correction signals.	
Data Link	Linkage of a teletypewriter or similar device with a remote computer, using public telephone lines. Generally on a time-shared basis wherein a number of Teletype units at different locations share the computer on an effective simultaneous basis.	
Dead Band	The range through which an input can be varied to the servo portion without initiating response at the machine tool. Generally the narrower the dead band the better the response of a machine-tool-system combination. Analogous to mechanical backlash.	
Decode	To translate from coded language into an easily recognizable language. (Opposite of encode.)	
Diagnostic Routine	A specific computer routine designed to locate an error or malfunction in the program. A "printout" can be made available which will describe the location of the error and prescribe its correction.	
Digital	Refers to discrete states of a signal (i.e., either on or off). Since a combination of on-off signals makes up a specific value, the magnitude of each signal is irrelevant. The theoretical extent of accuracy of a digital system is therefore unlimited and will depend on the amount and cost of electronics involved. Numerical control systems may be of the analog, digital, or combination analog-digital type of construction	

TERM	DEFINITION	EXAMPLE
Digital Computer	A computer that utilizes discrete numbers rather than related variable quantities in processing data.	
Digitize	The process of optical sensing of physical surfaces (lines, points, and curves) to produce Cartesian coordinate approximations of these surfaces.	The inverse of a numberical control drafting/plotting machine wherein an optical device follows a line and prints out the coordinates, or automatically prepares a tape which can then be run on a numerically controlled milling and/or drilling machine.
DNC	Stands for Direct Numerical Control and refers to the operation of a number of machines from a single relatively large computer. Unlike CNC, in which a computer, usually a mini-computer, replaces the more conventional electronic components, a DNC system normally *supplements* the conventional electronics and automatically regulates the operation of a number of machines. There can be as many as several hundred machines.	
Encode	To translate from an easily recognizable language into a coded language.	
End of Block Character (EB)	A character punched on the tape which denotes the end of a block of data (see definition for *Block*).	
External Memory	A storage device such as punched cards, punched tape, or magnetic tape which is external to the computer.	
Face Mill	A relatively large, flat type of cutter generally used for taking a flat cut over over the surface of a part. The cutter shown in the example has inserted teeth.	

TERM	DEFINITION	EXAMPLE
Feed	The rate of travel of the cutter through the material. In NC the feed is most often programmed in inches per minute (ipm). This very often requires a conversion from figures noted in inches per revolution (ipr).	The formula for the conversion is: $\text{ipm} = \text{ipr} \times \text{rpm}$
Feed Back	That part of a closed-loop system which feeds back information regarding the condition that is being controlled for comparison with the input values. (See *Closed-Loop System*.)	
Feed-Rate Override	A manual function, usually a rotary dial, which can override the programmed feed rate. The range is usually from approximately 10 to 100 percent and is infinitely variable. (See *Manual Feed-Rate Override*).	If the programmed feed rate as described on the tape is 20 ipm, the operator could lower it to 2 ipm. There is a justifiable tendency, therefore, for part programmers to specify feed rates "slightly on the high side."
Fixed Sequential Format	A means of identifying a word by its location in the block. Words must be presented in a specific order and all possible words preceding the last desired word must be present in the block. An address code is therefore not necessary; however, more characters are generally required per block than for the Word Address format. The fixed Sequential format system has steadily been losing favor to the preferred Word Address format.	
Flat End Mill	A popular milling cutter which can be used for cutting pockets, slots, and the perimeter of parts.	

TERM	DEFINITION	EXAMPLE
Floating Zero (also described as Free Floating Zero)	A characteristic of a numerical control system which allows the zero reference point (target point) to be established readily by manual adjustment at any position over the full travel of the machine tool. The control system retains no information on the location of any previously established zero. One advantage of a Floating Zero system is the ability to utilize negative as well as positive coordinate values, thus reducing the part programming time required on left- and right-hand parts and parts that are symmetrical about center lines.	TABLE The "0" or target point may be established at any position over the entire working surface.
FORTRAN	Stands for FORmula TRANslation and is probably the most popular computer program for scientifically oriented problems. Because of its universal nature, statements in this language will be acceptable to practically all scientific type computers. The program is machine-independent and the programmer, therefore, need not know the details of how the computer operates. FORTRAN is used in handling the APT system.	As will be seen from the example below, the FORTRAN statement is very similar to the conventional arithmetic statement. Algebraic $x = a^2 + bc - 1.5$ FORTRAN x = a**2 + b*c − 1.5 The FORTRAN statement must be translated into the specific machine language required by the computer. A special translation routine must, therefore, be prepared for each different computer.
Gain	The amount of increase in a signal as it passes through a control system or a control element. The gain in a control system would refer to its sensitivity and its ability to raise the power of a signal to a required output.	

TERM	DEFINITION	EXAMPLE
Incremental Dimension Word	A word defining a dimension or movement with respect to the preceding point in a sequence of points. Most contouring systems utilize incremental dimensioning. (Most Point-to-Point systems utilize Absolute Dimensioning.)	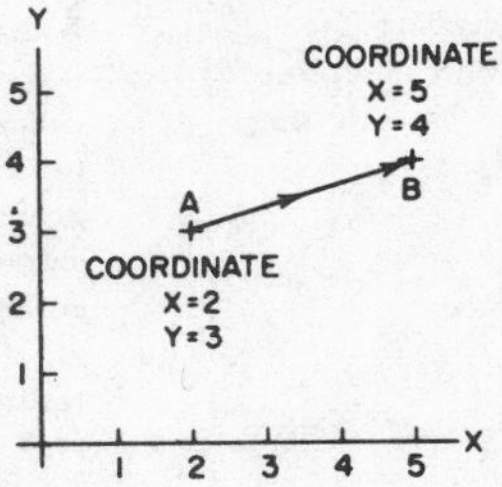 Y; 5; 4; 3; 2; 1; X; 1 2 3 4 5; A; B; COORDINATE X=5 Y=4; COORDINATE X=2 Y=3 Moving from A to B in the example shown would require programming the incremental X and Y distances between A and B. Incremental distance = x movement = 3. Incremental distance = y movement = 1.
Input	Transfer of external information into the control system.	
Integrated Circuit	A complete functional circuit consisting of transistors, diodes, capacitors, and resistors all constructed within or on the surface of a microlitic chip of silicon. A refinement of its predecessor, solid state, having greater reliability and requiring much less control space due, primarily, to a process resulting in condensed circuit modules.	
IO's	Stands for Input Output devices which are means of entering or extracting information from the CPU.	Magnetic tape units, printers, card readers, cathode ray displays, special typewriters, card punches.
Level	See *Track*.	
Machine Language	A computer term referring to exact digital instructions to a computer which it can execute directly, without modification or translation. Most programmers prefer not to program in machine language which requires keeping track of all memory locations as well as being difficult to identify since only numbers are generally used.	A machine language statement might be: 634218 where 63 is the code for addition, and 4218 is the address of a memory location.

TERM	DEFINITION	EXAMPLE
Magnetic Tape	A tape made of plastic and coated with magnetic material which can store information. The most common magnetic tape used is ½ inch wide and stores information in a digital or pulse format similar to the format on a punched paper tape.	
Manual Data Input	A means of inserting complete format data manually into the control system. This data is identical to information that could be inserted by means of a tape (including all auxiliary functions, feed-rate number, etc.).	
Manual Feed-Rate Override	Enables the operator to reduce the feed rate should the programmed rate noted on the tape be excessive for the particular piece of material being machined. This feature generally consists of a dial on the operator's console which will enable the operator to adjust or "override" the programmed feed rate. The percentage of override generally varies from approximately 5 to 100 percent of the programmed feed rate. The range is usually infinitely variable between these values. (Usually pertains to contouring systems.)	
Manuscrpit	A form used by the part programmer for listing the detailed instructions which can be transcribed directly by the tape preparation device or fed to a computer for further calculation.	
Memory	Synonymous with *Storage* and pertains to a computer device into which data can be entered, held, and/or retrieved. Usually refers to the *internal* capacity of the computer and is handled by a magnetic core arrangement in contrast to *external* storage which refers to such mediums as magnetic tape, discs, or drum.	A computer that can handle 32,000 words in internal storage is said to have a memory of 32K. The words may be of different bit lengths, two common ones being 32 and 48. The present APT system, as an example, requires a minimum of 64K. For some time it has been 32K; however, additional features have necessitated greater storage capacity.

TERM	DEFINITION	EXAMPLE
Mnemonic Code	Instructions for a computer written in a form which is easy for the programmmer to remember, such as FORTRAN or APT, but which must later be converted into machine language.	
Noise	Random electrical impulses similar to radio static, occurring within the control system, which can disturb its normal operation.	
Numerical Control System	A system in which actions are controlled by the direct insertion of numerical data at some point. The system must automatically interpret at least some portion of this data.	
Numerical Data	Data in which information is expressed by a set of numbers or symbols that can only assume discrete values or configurations.	
Off-Line Operation	Usually applies to peripheral equipment which operates independently of the central computer. This equipment is generally utilized where it is not necessary to operate the full capabilities of the major portion of the computer, thus saving time and expense.	Key punch device, card sorter, printer, tape preparation devices.
On-Line Operation	Applies to computer operations and calculations which are performed by the computer itself or the major portion of the computer.	
Open-Loop System	A control system that has no means for comparing the output with the input for control purposes (i.e., there is no feedback).	DRIVE MOTOR TABLE COMMAND INSTRUCTIONS ONLY LEADSCREW
Operation Sheet	Also called a "routing," this is a form on which the general operations for producing a part are noted. It differs from a manuscript in that a manuscript is much more detailed and describes every move of the cutter. The operation sheet might note the type of cutting tools to use and the holding fixture arrangement.	Sample operations noted would be: Drill 5 holes. Mill perimeter of part.

TERM	DEFINITION	EXAMPLE
Optional Stop	A miscellaneous function command similar to a program stop except that in this instance the control ignores the command unless the operator has previously pushed a button to recognize the command (generally denoted as an m01 code in accordance with standards).	
Overshoot	The amount of overtravel beyond the command position. The amount of overshoot is related to factors such as: system gain, servo response, mechanical clearances; and inertia factors relating to mass, feed rate, and strain.	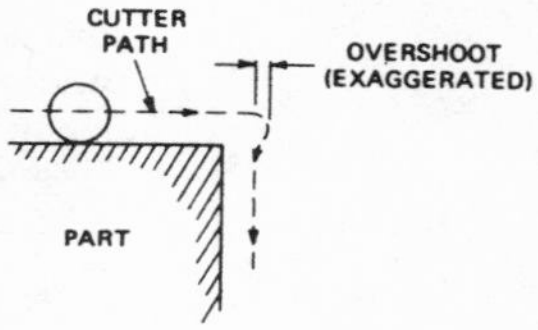
Parity Check	Considering the common numerical control standard RS-244-A, a hole punched in one of the track columns (channels) whenever the total number of holes representing a character would be even. The net result would be an odd number of holes in the character. This is utilized as a check when reading the tape since an even number of holes would indicate a source of error in the punching. The control system recognizes only an odd number of holes in the character and will automatically stop if an even number of holes in a character is read. The opposite applies with the USASCII code in that an even parity is sought and the control system will automatically stop when an odd number of holes is read across the tape.	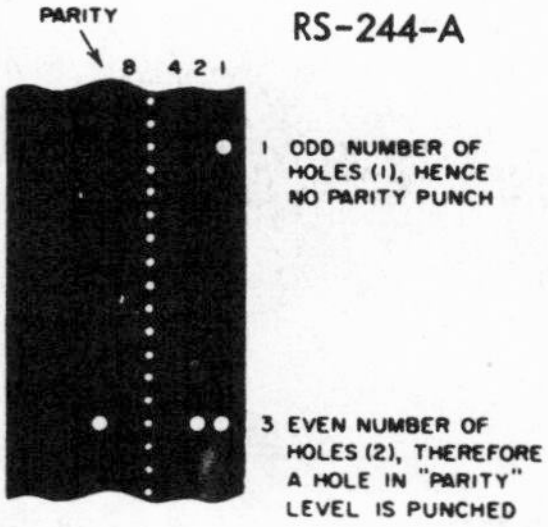 The number "3" is denoted as a hole in the first channel having a value of 1, and a hole in the second channel having a value of 2. A third hole is required and is punched in the parity column. All three holes are punched simultaneously in the character. The number "7," for example, would not require a parity punch since there are an odd number of holes (values 4, 2, 1) which comprise the character.
Point-to-Point Control System (PTP)	A system in which controlled motion is required only to reach a given end point with no path control during the transition from one end point to the next. The most common application is in a numerically controlled drill press.	

TERM	DEFINITION	EXAMPLE
Positioning System	See *Point-to-Point Control System*.	
Position Transducer (Resolver)	A device usually geared to a precision ball-nut lead screw for measuring position and converting this measurement to an electrical signal for transmission back to the control cabinet where a comparison is made with the input instruction. This cylindrical device measures approximately 1½ inches long by 1 inch in diameter.	
Post Processor	A set of computer instructions which transforms tool center-line data (CL) into machine commands using the proper tape code and format required by the specific machine/control system.	The APT instruction COOLNT/ON would be translated to the auxiliary word m07 by the post processor. Also, other codes peculiar to the machine tool/system combination such as the feed-rate number (fxxx), automatic tape rewind (m30), and tool change (txx). In instances where the control operates on incremental dimension words, the post processor makes the conversion from the CL coordinate form.
Precision	See *Repeatability*.	
Preparatory Function	A command on the tape changing the mode of operation of the control which is generally noted at the beginning of a block and consists of the letter character "g," plus a two-digit number.	Some preparatory functions would be: g01 Linear interpolation g02 Circular interpolation g05 Hold g08 Acceleration g09 Deceleration
Programmer (Part Programmer)	A person who prepares the planned sequence of events for the operation of a numerically controlled machine tool. The programmer's principal tool is the manuscript on which the instructions are recorded.	
Programmer (Computer Programmer)	A person who prepares computer programs which are to be used internally in the computer in order to solve the specific problems presented by the	

TERM	DEFINITION	EXAMPLE
Programmer (Computer Programmer) (Cont'd)	part programmer. A numerical control computer programmer generally requires a formal advanced mathematical background in addition to a thorough knowledge of computer operations.	
Program Stop (Sometimes referred to as Planned Stop)	A miscellaneous function to stop the coolant, spindle, and feed after completion of other commands in the block (denoted as an m00 code). The operator may restart the cutting cycle without loss of accuracy.	
Quadrant	Defines the fourth part of a circle or a fourth part of a graph in reference to the X and Y axes.	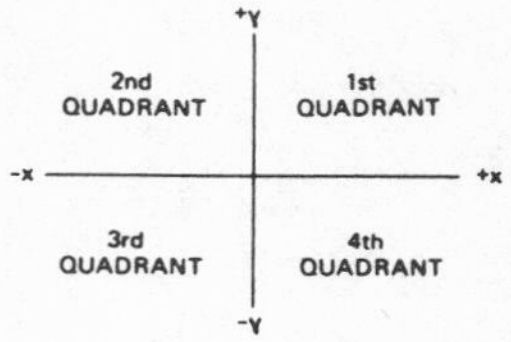
Read-Out, Command (Display)	Generally applies to contouring systems and to the display of absolute position as derived from the position commands within the control system. In effect, the display is an accumulative read-out of the incremental movements which have been fed into the control system. Assuming that the machine tool is operating in accordance with the control system, it can be reasonably assumed that the read-out represents the actual position of the table.	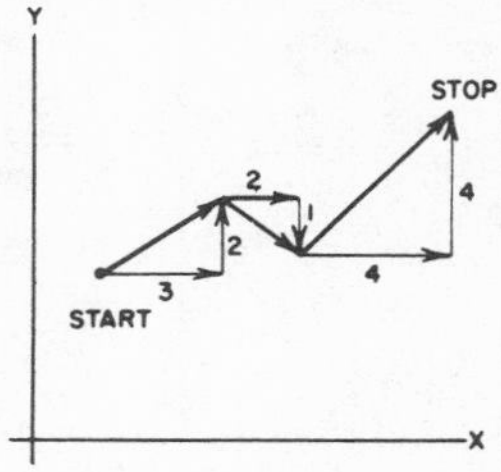 At the STOP point the read-out display would appear as follows: X 9.0000 Y 5.0000 Thus, the incremental x and y movements as noted on the tape have been added, algebraically.
Read-Out, Position (Sometimes referred to as Absolute Read-Out)	Display of absolute position as derived from a position feed-back device such as a transducer. The transducer is normally attached to the lead-screw. Although theoretically more reliable than the command read-out, it is generally more expensive since a separate feed-back circuit is required.	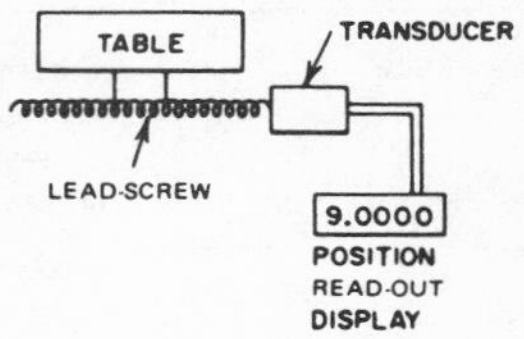

TERM	DEFINITION	EXAMPLE
Read-Out, Slide Location	Display of absolute position as derived from a sliding scale measurement device such as an inductosyn which is attached directly to the sliding moving part, i.e., the table. Theoretically it is the most reliable of the three read-out devices (command, position, slide location). Also it is the most expensive of the three.	SCALE ATTACHED TO SIDE OF TABLE SLIDING SCALE "PICK-UP" DEVICE
Reamer	Tool for rounding out a hole where a drill may not have left the required diameter tolerance.	
Repeatability	Closeness of agreement of repeated position movements to the same indicated location and under the same conditions.	Y REPEATABILITY (EXAGGERATED) ACCURACY (EXAGGERATED) A X Assuming A is the point programmed: The accuracy is the ability of the machine-system combination to position to A. The repeatability is the error between different repeated attempts to hit point A. It may also be defined as the consistency of inaccuracy.
Resolution	The smallest increment of distance that will be developed by the numerical control system in order to actuate a machine tool.	A system having a resolution of .0001″ should theoretically move a machine tool .0001″ for each .0001″ pulse.

TERM	DEFINITION	EXAMPLE
Roadmap	A drawing which describes the path of the cutting tool as it progresses around a part.	SP
Row	A path perpendicular to the edge of the tape along which information may be stored by presence or absence of holes or magnetized areas. A character would be represented by a combination of holes and no holes across the tape which would fall along a row.	
Sequence Number	A number identifying the relative location of blocks or groups of blocks on a tape. The sequence number is identified by the letter "n" as the address and usually falls at the beginning of a block. A control system incorporating a sequence number display would "read out" the number of the block being read by the tape reader which would correspond to the number following the address letter. This feature enables the operator to identify the tape location with respect to the position of the machine.	n 2 3 4 X + 1 2 3 4 5 Y – 6 7 8 9 0 EB 234 above, following "n," would identify the block and would be displayed.
Servo Mechanism	A type of closed-loop control system in which mechanical position is the controlled variable.	
Single Point Cutting Tool	A common type of cutting tool used with a lathe.	DoALL

TERM	DEFINITION	EXAMPLE
Software	Computer programs used to assist in part programming. The APT program is an example of an automatic programming software package.	Contrasted to hardware which refers to the machine tool, the control system, and other pieces of equipment.
Solid State System (also designated as a "Second Generation" System)	A control system comprised solely of solid state electronic and electrical components (i.e., having no moving, vacuum tube, or gaseous tube components. Mechanical tape readers or photo tape reading tubes are not considered).	
Speed	Term referring to the rotation of the spindle in revolutions per minute (rpm). Since the speed of an NC machine is programmed in rpm it is often necessary to convert from the conventional form of describing the speed in feet per minute (fpm).	The formula for the conversion is: $\text{rpm} = \frac{\text{fpm} \times 12}{\pi \times \text{Dia. of Cutter}}$
Straight Cut Control System	A system which has feed-rate control only along a single axis and, therefore, can control cutting action only along a parallel to the linear (or circular) machine ways. Cannot coordinate two or more axes to produce true contouring. Also designated as "Picture Frame" milling.	
Sub-Routine	A section of a computer program which is stored in the computer memory and can be used over and over again to accomplish a certain operation, e.g., square root, cube root, sine, log, etc.	
Tab Sequential Format	Means of identifying a word by the number of tab characters preceding the word in a block. The first character of each word is a tab character Words must be presented in a specific order but all characters in a word, except the tab character, may be omitted when the command represented by that word is not desired.	
Tachometer	A speed-measuring instrument generally used to determine revolutions per minute. Tachometers may be used in conjunction with contouring systems as a supplemental control for governing feed rates.	

TERM	DEFINITION	EXAMPLE
Tap	A tool used for cutting a thread on the inside of a hole.	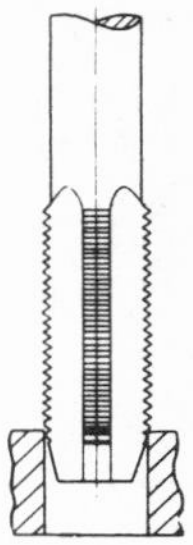
Temporary Storage	See *Buffer Storage*.	
Tool Function	A tape command identifying a tool and calling for its selection. The address is normally a "t" word and may be used in conjunction with a turret or automatic tool changer.	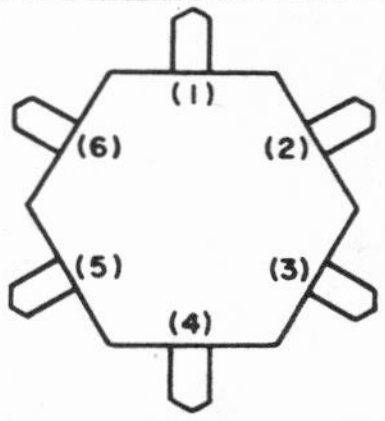 (1) (2) (3) (4) (5) (6) t01 would bring (1) position into operation.
Tool Offset	In order to avoid the time-consuming requirement for setting the depth of the tool to an exact dimension in a turret, the tool may be set to an approximate dimension (generally within 0.1 inch of the required setting) and manual switches can be used to make up the difference. Found extremely helpful with turrets on a lathe or adjusting the "z" or depth motion of a turret drill having three axes of control.	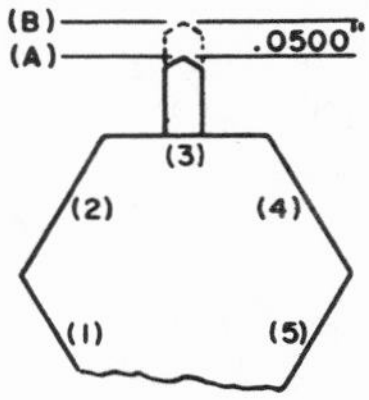 (B) (A) .0500" (3) (2) (4) (1) (5) When setting single-point tools, as in the turret of a lathe, it is often difficult, if not impossible, to determine cutter deflection or the net effect that machining has on the position of the cutting tool. An adjustment is most often necessary after the first cut. This may be readily accomplished with numerical control by setting the exact

TERM	DEFINITION	EXAMPLE
Tool Offset (Cont'd)		depth via manual switches. If, as an example, the tool is initially set at position (A) and after cutting it is realized that because of forces and deflection, the tool should be at position (B) the tool need not be touched but the difference dialed in. Whenever the t03 turret position is called for by the code on the tape, the turret will automatically adjust to position (B). Each tool face may be adjusted independently. Correct compensation is automatically adjusted as the tool face is indexed to the cutting position.
Track	Also known as level or channel. A path parallel to the edge of the tape along which information may be stored by the presence or absence of holes or magnetized areas. The EIA standard 1-inch-wide tape has eight tracks.	TRACKS
Transducer	A device which converts one form of energy into another form of energy. A thermocouple which converts temperatures into millivolts is a type of transducer. In numerical control design, it usually applies to a device for converting rotary motion into a varying sine wave voltage. A resolver is a type of transducer.	
Transistor	A solid state electronic device which functions in a manner similar to a triode vacuum tube. It consists of a small block of a semi-conductor material that has at least three "electrodes." It is light, almost unbreakable, long-lived, and highly efficient.	

TERM	DEFINITION	EXAMPLE
Undercut	A machine cut where the cutter fails to arrive at a programmed point following a command change in direction (in contrast to Overshoot).	
Variable Block Format	A tape format which allows the number of words in successive blocks to vary.	The word address system is a variable block format.
Windup	Lost motion in a mechanical system which is proportional to the force or torque applied.	
Word	An ordered set of characters which may be used to cause a specific action of a machine tool.	x, y, and z together with the numerical values would be described as dimension words. Other words with their appropriate numerical values would be the g and f words describing preparatory and feed-rate functions, respectively.
Word Address Format	Addressing each word in a block by one or more characters which identify the meaning of the word.	x would be the address for the word X 5.4693 which would call for a motion in the x direction of 5.4693 inches.
Zero Offset (Sometimes referred to as Zero Shift)	A means of shifting the coordinate zero point from a fixed known zero point. The zero offset "shift" is normally accomplished by a series of switches or dials.	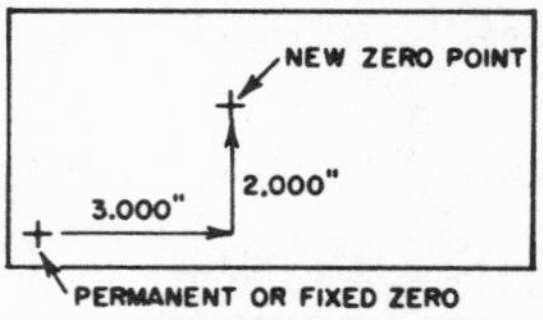 NEW ZERO POINT 2.000" 3.000" PERMANENT OR FIXED ZERO If it is required to shift the permanent zero position, say, 3.000 inches in the +X direction and 2.000 inches in the +Y direction, it would be necessary to dial in these dimensions. A new zero point will then have been established and all part dimensions, when programming, will be taken from the new zero point.

TERM	DEFINITION	EXAMPLE
Zero Reset (Also known as Zero Synchronization)	The ability to automatically realign each axis slide to the zero or target point by pressing a button when the slides are brought within close proximity of the target point.	TABLE Y X DIAMETER = .080" If only the X and Y axes are considered, it is not necessary to "manually" realign the cutting tool to the exact starting or target point. It is only necessary that the tool be brought within a circle whose radius is approximately .040 inch around the target point and the machine will automatically realign itself to the starting point after a button is pressed.
Zero-Suppression	The elimination of nonsignificant zeros *either* before or after the significant figures in a tape command. The purpose is to reduce the number of characters that are required to be read.	In the tape word x + 01000 A control system responding only to trailing zero-suppression would require: x + 01 A system responding only to leading zero-suppression would require: x + 1000 The preceding and trailing zeros may be shown if desired.

APPENDIX C

Geometry and Computation Statements

GEOMETRY STATEMENTS

The purpose of the geometry statement is to describe the configuration of the portion of the workpiece to be machined. A typical geometry statement is given below:

```
SYMBOL─┐                ┌─GENERAL DESCRIPTION
       P1 = PØINT/5.000,-3.000,4.000
                    └─SPECIFIC DESCRIPTION
```

The geometry statement has *three major parts:* First, a symbol (P1) must be given to the left of the equal-sign. Following this, a general description (PØINT) is given to the left of the slash (/). Finally, a specific description (5.000, −3.000,4.000) is written to the right of the slash.

The first part of the geometric description, or the symbol that appears ahead of the equal sign, is used to identify a specific geometry element of a part. This may be a point, a line, a circle, a plane, or other geometry configuration.

The geometry statements described here are the most common ones and will allow the part programmer to describe all but the more complex geometry configurations. A variety of statements are listed for the PØINT, LINE, CIRCLE, and PLANE. It is not unusual for a particular geometry to be expressed by a number of different statements. It is left to the part programmer to select the one that requires the least amount of calculation.

PØINT

1. By the x, y, and z coordinates of the point.
   Example:

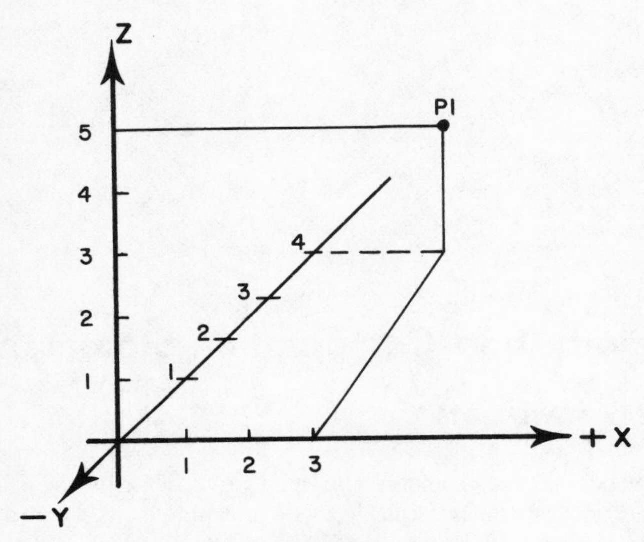

P1 = PØINT/3,4,5

2. By the intersection of two lines.
   Example:

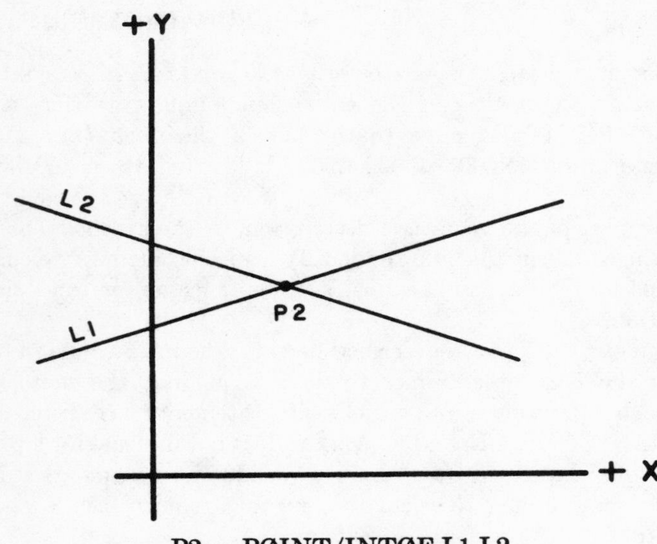

P2 = PØINT/INTØF,L1,L2

3. By the intersection of a line and a circle.
   Example:

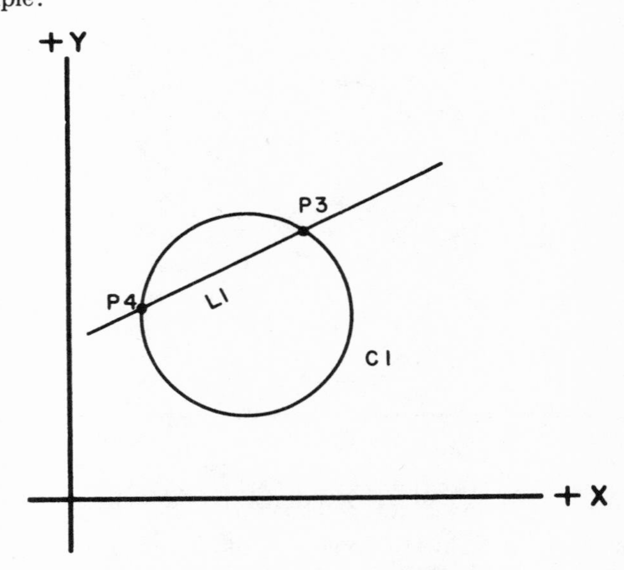

P3 = PØINT/XLARGE,INTØF,L1,C1

Note that there are two points where the line L1 intersects the circle C1. The word XLARGE, in the above statement, means that the x coordinate of the point P3 is larger than the x coordinate of the alternate point P4. The statement for P3 could also have been written as:

P3 = PØINT/YLARGE,INTØF,L1,C1

since the y coordinate of P3 is larger than the y coordinate of the alternate point P4. P4, on the other hand, could be written as:

P4 = PØINT/XSMALL,INTØF,L1,C1
or
P4 = PØINT/YSMALL,INTØF,L1,C1

The words LARGE and SMALL can also be used to apply to the z axis. For example, ZLARGE or ZSMALL.

4. By the intersection of two circles.

   Example:

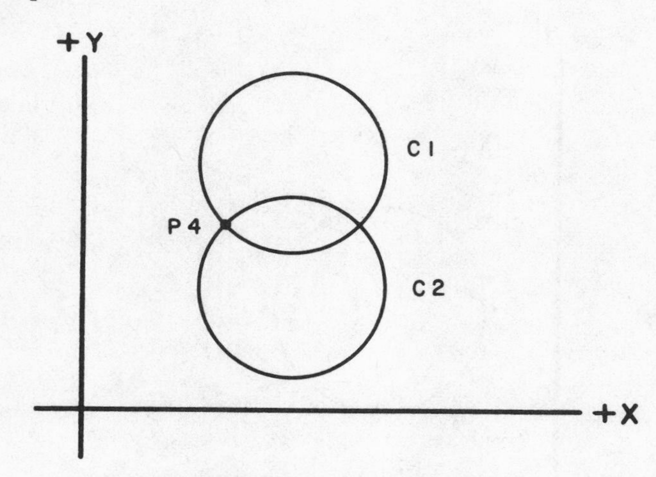

P4 = PØINT/XSMALL,INTØF,C1,C2

(See explanation of XSMALL in statement 3, above)

5. By the center of a circle.

   Example:

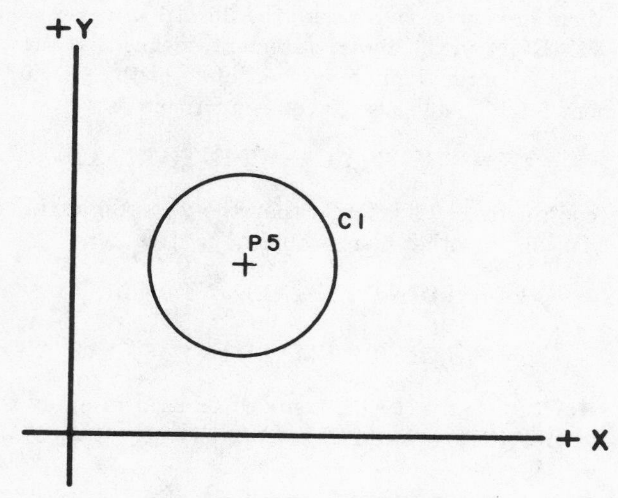

P5 = PØINT/CENTER,C1

LINE

1. By two points on the line.
   Example:

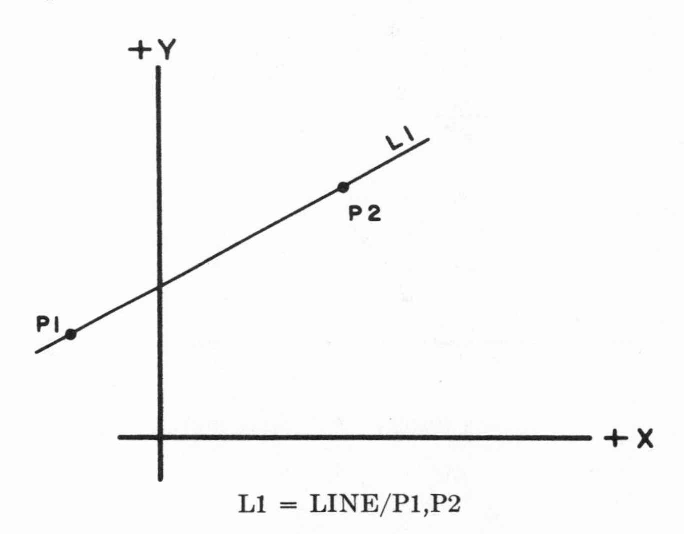

L1 = LINE/P1,P2

2. By a point on the line and the angle that the line makes with the *X* axis.
   Example:

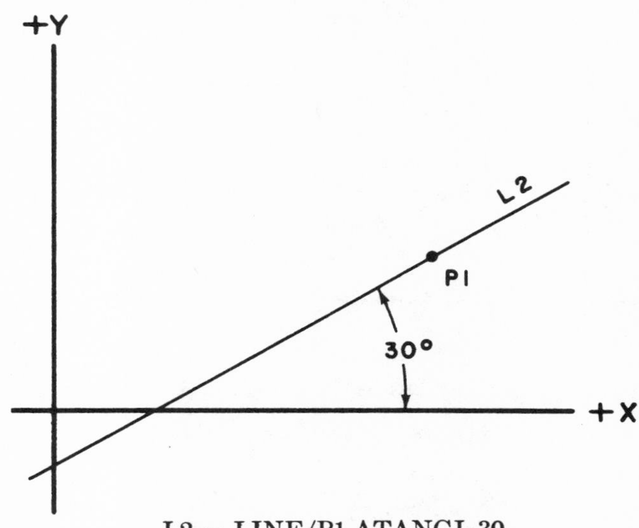

L2 = LINE/P1,ATANGL,30

3. By a point on the line and the angle that the line makes with another line.
   Example:

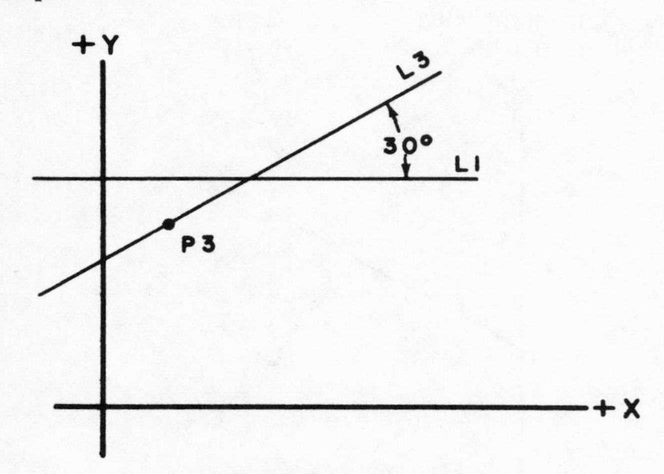

L3 = LINE/P3,ATANGL,30,L1

4. By a point on the line and being parallel to another line.
   Example:

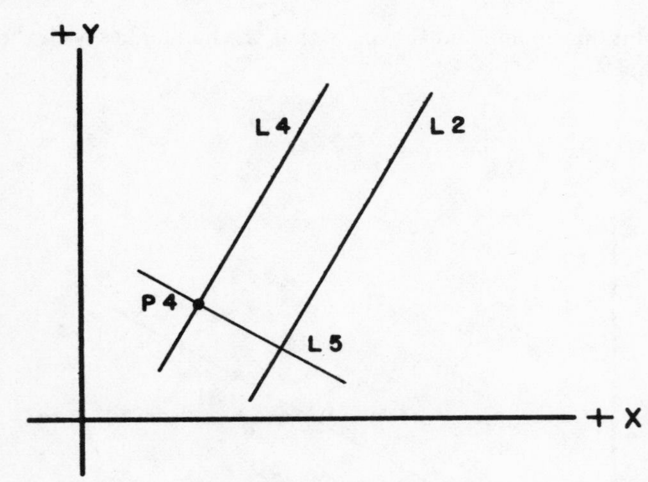

L4 = LINE/P4,PARLEL,L2

L5 = LINE/P4, PERPTO, L4
and
L5 = LINE/P4, PERPTO, L2

5. By a line parallel to another line and located at a specified distance from it.

   Example:

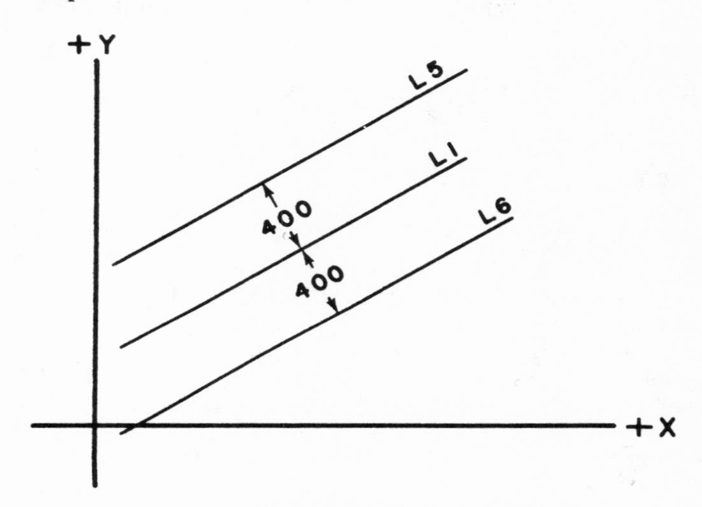

L5 = LINE/PARLEL,L1,YLARGE,.4

In this instance YLARGE means that L5 is larger in the *Y* direction than L1. The distance noted in the statement is the normal, or shortest distance, between the two lines.

The line L6 would be described as:

L6 = LINE/PARLEL,L1,YSMALL,.4

6. By a point on a line and the line being tangent to a circle.

   Example:

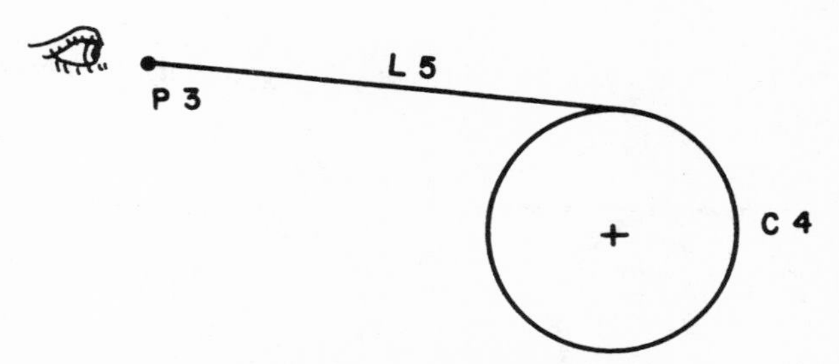

L5 = LINE/P3,LEFT,TANTØ,C4

In this case the part programmer imagines himself looking *from* the point *to* the circle. Here the line lies to the *left* of the center of the circle.

In the example shown below the line lies to the *right* of the center of the circle when looking from the point to the circle. There is no connection between these geometry statements and the direction in which the cutter moves.

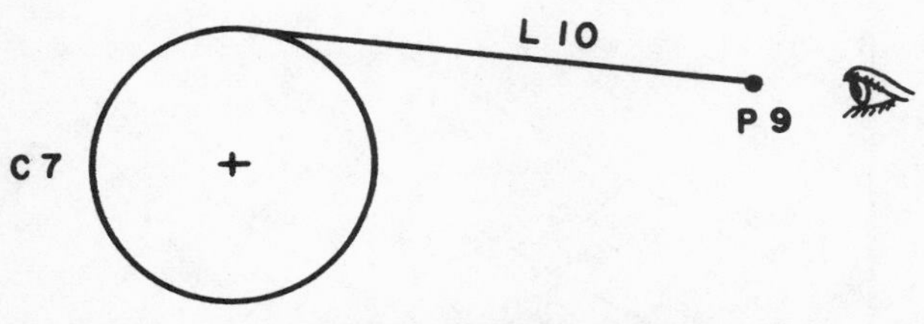

L10 = LINE/P9,RIGHT,TANTØ,C7

7. By a line tangent to two circles.
   Example:

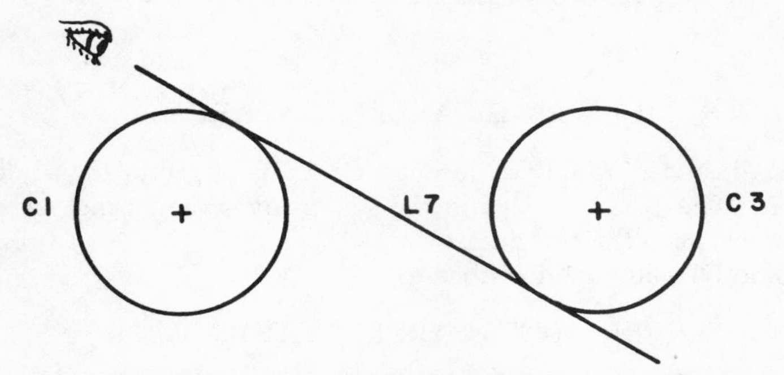

L7 = LINE/LEFT,TANTØ,C1,RIGHT,TANTØ,C3

The part programmer, in this case, is looking *from* the circle C1 *to* the circle C3.

In the example shown below the part programmer is looking *from* C3 to C1.

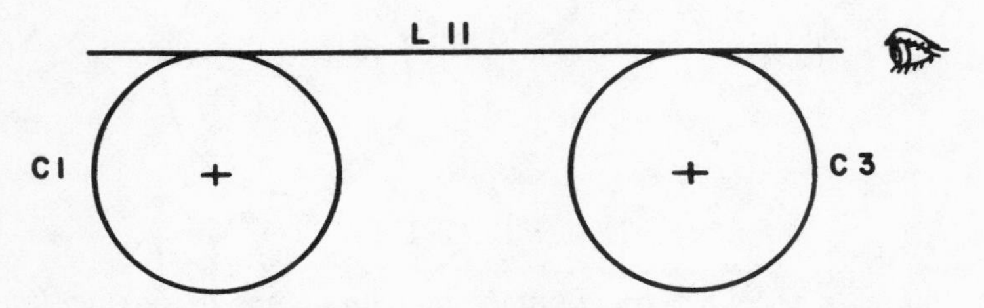

L11 = LINE/RIGHT,TANTØ,C3,RIGHT,TANTØ,C1

CIRCLE

1. By the x and y coordinates of the center and the radius.
   Example:

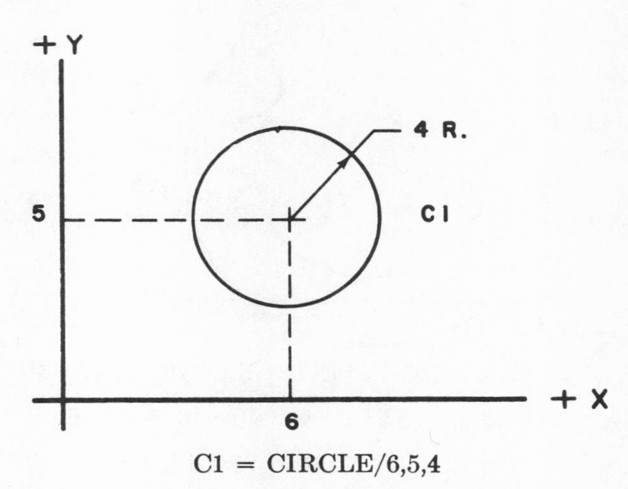

C1 = CIRCLE/6,5,4

(This could also be noted as C1 = CIRCLE/6,5,0,4 where 0 is the Z-axis coordinate.)

2. By the symbol for the center and the radius.
   Example:

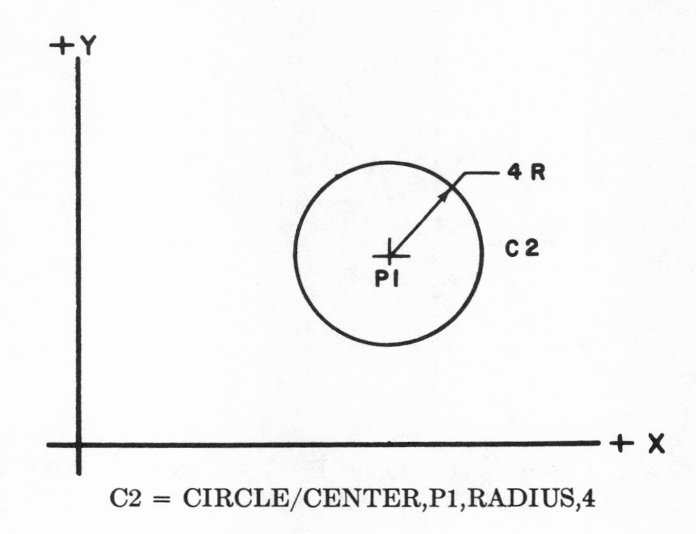

C2 = CIRCLE/CENTER,P1,RADIUS,4

3. By the symbol for the center and another tangent circle.

   Example:

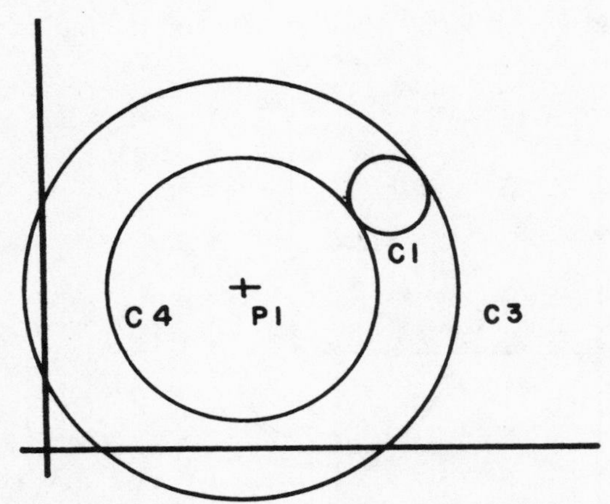

C3 = CIRCLE/CENTER,P1,LARGE,TANTØ,C1

In the same illustration shown above C4 could be described as:

C4 = CIRCLE/CENTER,P1,SMALL,TANTØ,C1

It will be noted that there are *two* circles which have a center P1 and which are tangent to C1. In order to distinguish which circle we are considering, the statement notes whether it is the larger or the smaller one.

4. By being tangent to two lines and having a given radius.

   Example:

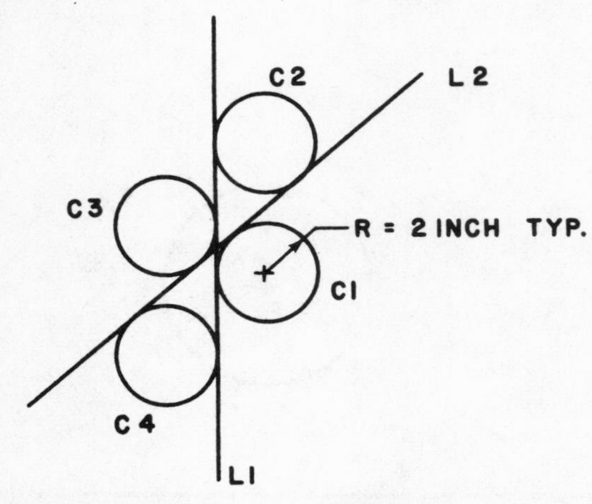

NOTE: Refer to Statement 3 under PØINT section for explanation of XLARGE, XSMALL, YLARGE, YSMALL.

Observe that there are four circles which can be tangent to L1 and L2 and have the same radius. In order to describe any particular circle we must eliminate three of the circles. First, by noting that C1 is YSMALL with respect to L2 we eliminate C2 and C3. Next, by noting that C1 is XLARGE with respect to L1, we eliminate C4 and again, C3. This leaves C1 as:

C1 = CIRCLE/YSMALL,L2,XLARGE,L1,RADIUS,2

C2, C3, C4 would be expressed as follows:

C2 = CIRCLE/YLARGE,L2,XLARGE,L1,RADIUS,2

C3 = CIRCLE/YLARGE,L2,XSMALL,L1,RADIUS,2

C4 = CIRCLE/YSMALL,L2,XSMALL,L1,RADIUS,2

5. By being tangent to a line, a circle, and having a given radius.
Example:

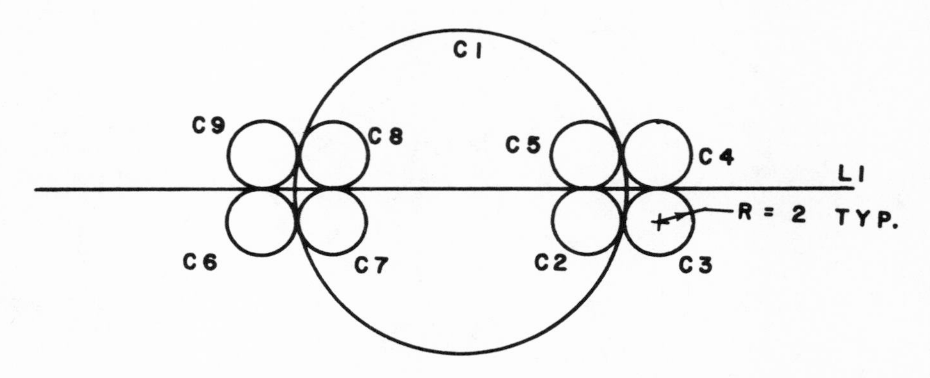

C4 = CIRCLE/YLARGE,L1,XLARGE,OUT,C1,RADIUS,2

NOTE: Refer to Statement 3 under PØINT section for explanation of XLARGE, XSMALL, YLARGE, and YSMALL.

There are *eight* circles which are tangent to L1 and C1 and have the same radius. The object is to eliminate all but the one desired. In the illustration above, if C4 is the circle desired, then by noting that it is YLARGE with respect to L1, we eliminate all of the circles lying below L1. Next, by noting that it lies outside C1 (by the word ØUT) we eliminate the two remaining circles lying *inside* the circle C1, namely C5 and C8. Now there are two circles left, namely C4 and C9; and by noting that C4 is XLARGE with respect to C9 we eliminate C9, thus leaving C4 as the only circle of the eight that conforms to the definition above. Statements for the other circles are as follows:

C2 = CIRCLE/YSMALL,L1,XLARGE,IN,C1,RADIUS,2

C3 = CIRCLE/YSMALL,L1,XLARGE,OUT,C1,RADIUS,2

C5 = CIRCLE/YLARGE,L1,XLARGE,IN,C1,RADIUS,2

C6 = CIRCLE/YSMALL,L1,XSMALL,OUT,C1,RADIUS,2

C7 = CIRCLE/YSMALL,L1,XSMALL,IN,C1,RADIUS,2

C8 = CIRCLE/YLARGE,L1,XSMALL,IN,C1,RADIUS,2

C9 = CIRCLE/YLARGE,L1,XSMALL,OUT,C1,RADIUS,2

6. By being tangent to two circles and having a given radius.
   Example:

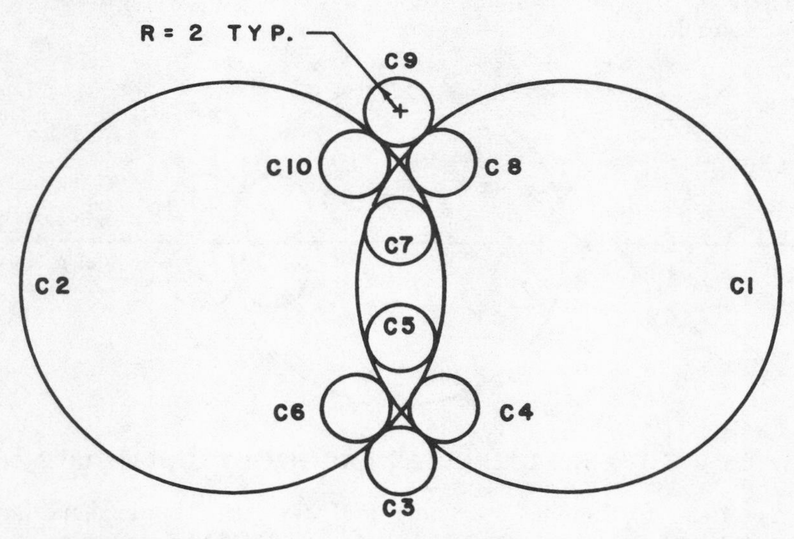

NOTE: Refer to Statement 3 under PØINT section for explanation of XLARGE,XSMALL,YLARGE,YSMALL.

There are *eight* circles that are tangent to both C1 and C2 which have the same radius. The object is to eliminate all but the one desired. In the illustration above, if C3 is the circle desired, then by noting that it lies outside *both* C1 and C2 we eliminate all but C9; and then by noting that C3 is YSMALL with respect to C9 we eliminate C9 and have C3 left. The statement for C3 would be:

C3 = CIRCLE/YSMALL,OUT,C1,OUT,C2,RADIUS,2

Statements for the other circles would be:

C4 = CIRCLE/YSMALL,IN,C1,OUT,C2,RADIUS,2

C5 = CIRCLE/YSMALL,IN,C1,IN,C2,RADIUS,2

C6 = CIRCLE/YSMALL,OUT,C1,IN,C2,RADIUS,2

C7 = CIRCLE/YLARGE,IN,C1,IN,C2,RADIUS,2

C8 = CIRCLE/YLARGE,IN,C1,OUT,C2,RADIUS,2

C9 = CIRCLE/YLARGE,OUT,C1,OUT,C2,RADIUS,2

C10 = CIRCLE/YLARGE,OUT,C1,IN,C2,RADIUS,2

PLANE

1. By three points, lying on the plane, which are not in a straight line.
   Example:

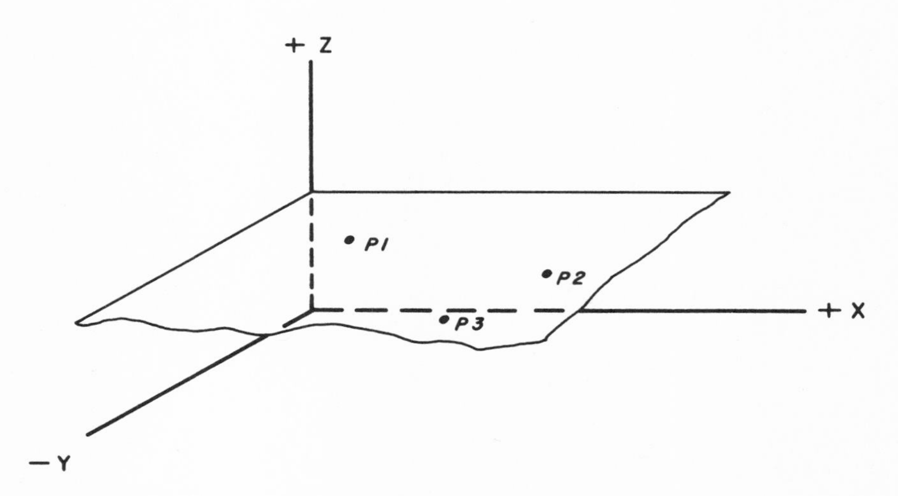

PL1 = PLANE/P1,P2,P3

2. By a point lying on the plane and the plane being parallel to another plane.
   Example:

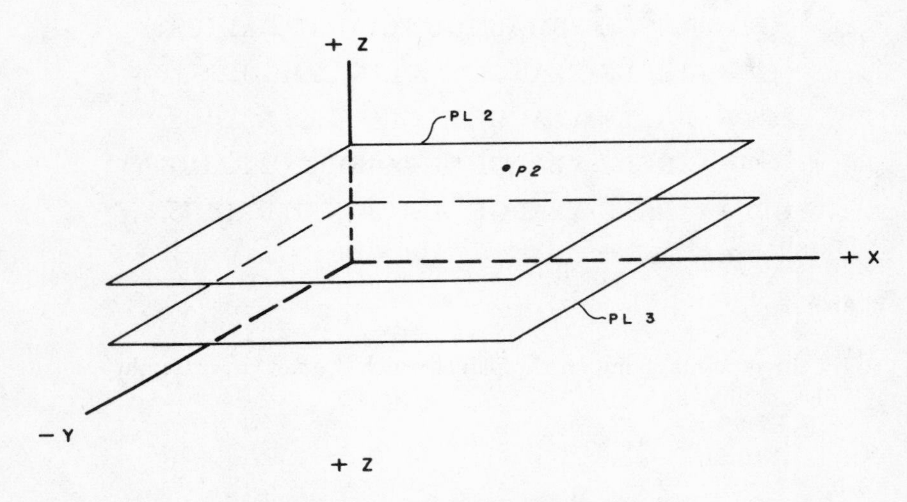

+ Z

PL2 = PLANE/P2,PARLEL,PL3

3. By being parallel to another plane and at a specified distance from it.
   Example:

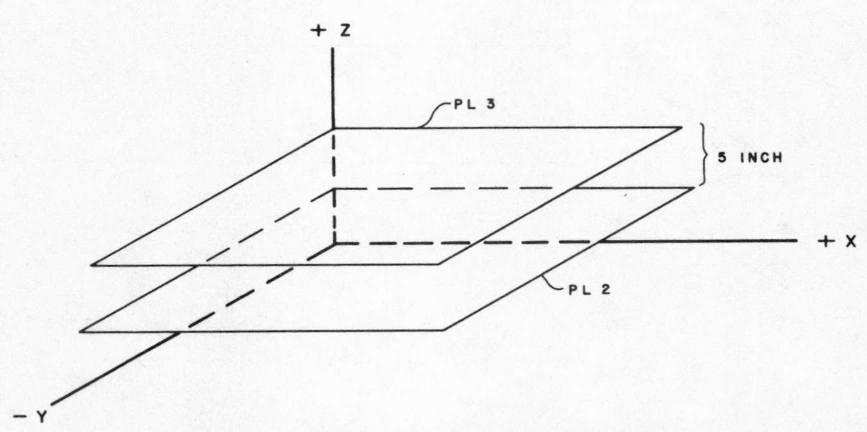

PL3 = PLANE/PARLEL,PL2,ZLARGE,5

NOTE: Refer to Statement 3 under the PØINT section for an explanation of ZLARGE. Also refer to Statement 5 under the LINE section.

4. By noting the distance from the XY plane, providing the plane being described is parallel to the XY plane.

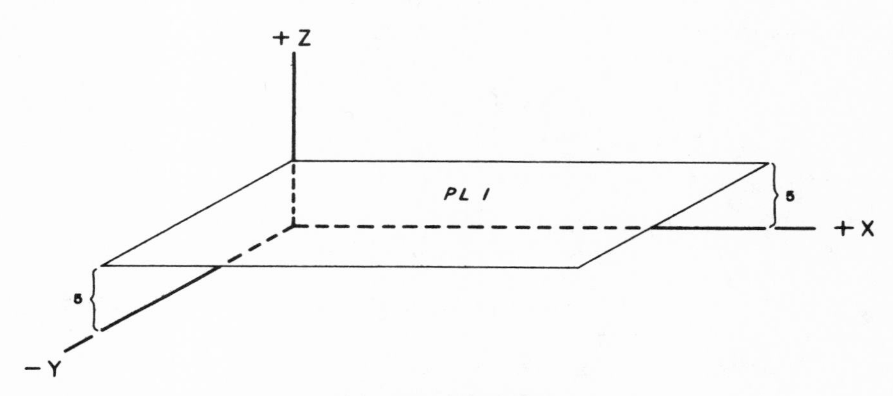

PL1 = PLANE/0,0,1,5

NOTE: The 0,0,1 are constants and stay as they are. The 5 denotes that the distance of the plane PL1 above the XY plane is 5 inches. If the plane were 5 inches *below* the XY plane the statement would be:

PL1 = PLANE/0,0,1,−5

COMPUTATION STATEMENTS

It is possible to perform algebraic and trigonometric calculations using the APT system.

Arithmetic form	APT form
25×25	25*25
$25 \div 25$	25/25
$25 + 25$	25 + 25
$25 - 25$	25 − 25
25^2	25**2
25^n	25**n
$\sqrt{25}$	SQRTF(25)
$\sin \theta$	SINF(θ)
$\cos \theta$	CØSF(θ)
$\tan \theta$	TANF(θ)
arctan .5000	ATANF(.5)

There are two ways in which computations may be used in the APT system. One is to let a factor equal the computation and then substitute the factor in a statement; the other is to put the computation directly into the statement. The following are a series of APT statements illustrating the first approach.

```
P1 = PØINT/0,0,1
 T = (25*2/3 + (3**2 - 1))
P2 = PØINT/T,0,0
```

The second way would be as follows:

```
P1 = PØINT/0,0,1
P2 = PØINT/(25*2/3 + (3**2 - 1)),0,0
```

NOTE: The parentheses have been used as they would be in an algebraic formula in order that the calculations be carried out in proper order. The operations within the inner parentheses would be carried out first. It is important for the total number of left-hand parentheses to equal the total number of right-hand parentheses; otherwise the program will fail.

APPENDIX D

APT Auxiliary and Post Processor Statements[1]

(Listed in the Approximate Order in Which They Might Appear in a Part Program)

APT AUXILIARY STATEMENTS

PARTNØ — This is the first statement in any part program. After PARTNØ the programmer may list any identification that he wishes. PARTNØ is one of the few exceptions where the P must begin in the first column of the programming sheet and there can be no spaces between the letters in the word.
Example: PARTNØ M 44 XYZ

CUTTER/ — Describe the *diameter* of the cutter. Must be used whenever a milling operation is to be performed and the center of the cutter is to be offset from the geometry being described.
Example: CUTTER/.250.
The diameter of the cutter is .250 inch.

INTØL/ — Notes the allowable *inside* tolerance of a straight line that approximates a curve.

[1] The statements shown represent only a partial list of the total available auxiliary and post processor statements. Complete lists would normally be furnished by the machine tool builder and/or the computer manufacturer or computer service company.

Example: In the sketch shown below the inside tolerance would be .0005 inch and the statement would be INTØL/ .0005. The computer calculates the length of the line.

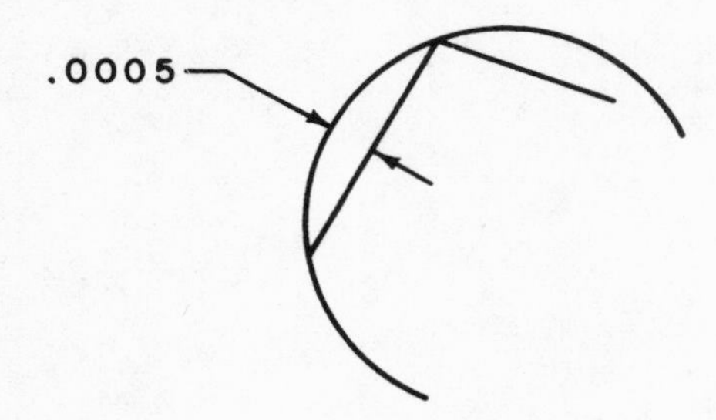

ØUTTØL/ Notes the allowable outside tolerance of a straight line that approximates a curve.

Example: In the sketch (A) shown below the outside tolerance would be .001 inch and the statement would be ØUTTØL/.001. The computer calculates the length of the line. The programmer may call for both an inside and an outside tolerance, and, in this case, the straight lines would appear as shown in (B).

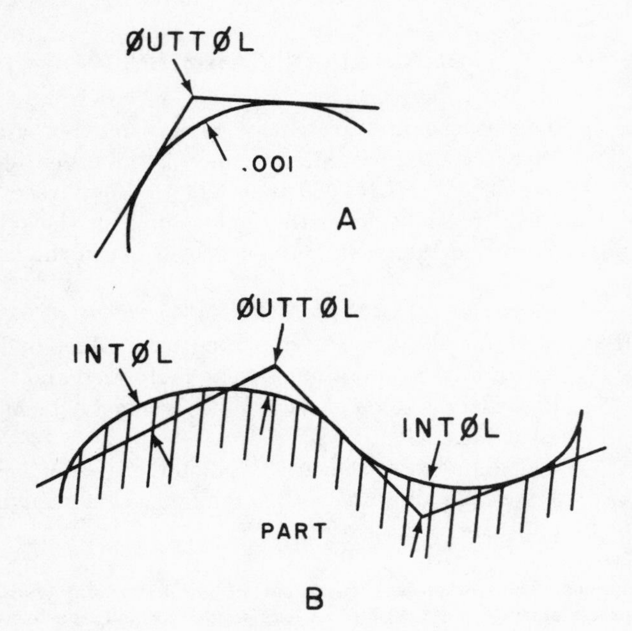

REMARK Means that the part programmer wants to make a note in the program, such as a particular tool to be changed. The REMARK statement does not affect the part program. As with PARTNØ, the REMARK word must begin in the first column and there can be no spaces between the letters.

$ This is a symbol used at the end of a line when it is necessary to run the statement on to another line.

CLPRNT Notes that the computer is to print out the coordinate listing of all the end points of the straight line moves.

FINI Must appear at the end of every program.

APT POST PROCESSOR STATEMENTS[2]

CØØLNT/ Commands the control system to turn the coolant either *on* or *off*. The statement for turning on the coolant would be CØØLNT/ØN. If the machine had both mist and flood coolant, the statement for turning on the mist would be CØØLNT/MIST; turning the flood coolant on would be CØØLNT/FLØØD.

CYCLE/ With most two-axis machines cycles such as drilling, tapping, or boring are performed as a fixed, or "canned," type operation; that is, the operator may set the speed, feed, rapid traverse distance, and feed distance. If this is the case, all the programmer need note is the function to be performed.

Examples: A drilling cycle would be initiated by the statement CYCLE/DRILL; a tapping cycle by the statement CYCLE/TAP.

For those two-axis machines where the depth can be controlled,[3] the post-processor statement may include distance movements and commands within the cycle statement.

Example: CYCLE/DRILL,RAPTØ,−1,FEDTØ,−1.3, IPM,3.

[2] The statements shown represent the more common types and are in the most usual form. However, this is not a complete listing. Also, post-processor statements may differ depending on the machine tool and control system builder. It is advisable to obtain a "post-processor listing" from the machine tool builder before attempting to utilize the APT system.

[3] Distinguished from simultaneous three-axis machines where all three axes can be controlled at the same time.

In this case the drill would move toward the workpiece at a rapid traverse until it reached the z coordinate of -1 inch. It would then move at a feed rate of 3 inches per minute (ipm) to a depth where the z coordinate is equal to -1.3. In this latter move it would presumably be drilling the hole. If we wanted to note the drilling feed rate in inches per revolution (ipr), which is the more normal practice, the IPM in the cycle statement would be changed to IPR, with the proper inches per revolution following it.

Example: CYCLE/DRILL,RAPTØ,−1,FEDTØ,−1.3, IPR,.008

The computer would convert the IPR to IPM which is the form of feed rate required by most NC machine tools. Again, it should be pointed out that the cycle statement, as with any post-processor statement, may differ, depending on the machine tool and control system manufacture.

END — Stops the machine but does not shut off the control system.
Example: Stopping the machine to change a cutting tool.

FEDRAT/ — Notes the feed rate. If only a number is shown after the slash (/) mark, then the computer assumes inches per minute, IPM. If we wish to express the feed rate in inches per revolution, then IPR would be noted. The computer will convert IPR to IPM, assuming that the speed has also been noted in the program.
Example: FEDRAT/.005,IPR

MACHIN/ — Calls in the proper post processor. Each post processor has an identity, such as UNIV,1 or PDG,5, or BESTL,6.
Example: MACHIN/LATH,8

PITCH/ — Required for threading on a lathe.
Example: If a ⅛-inch pitch thread is wanted, the statement would be PITCH/8 (meaning 8 threads/inch).
Of course, the machine has to have a thread cutting capability which is controlled by tape commands.

RAPID — Instructs the machine to move at a rapid traverse. Unlike the feed-rate statement FEDRAT, which holds until changed, the RAPID command must be noted before every motion statement in which a rapid traverse is desired. When the rapid traverse command is not shown before a motion statement the feed rate reverts to the last FEDRAT command.

SPINDL/ — Applies to those machines in which the spindle speed can be controlled by tape. This is normally the case with turret-type or tool-changing machines. The speed may be expressed directly in rpm and the statement would be SPINDL/800,RPM or SPINDL/800. If the speed range is handled in steps, as with most machines, the computer will select the closest step to that which is noted in the SPINDL/ statement.

The spindle speed may also be expressed in surface feet per minute (sfm)

Example: SPINDL/100,SFM.

The computer will calculate the corresponding rpm. In a drilling or milling operation this will depend on the diameter of the cutter. In a turning operation this will depend on the diameter of the part being machined, which is described in the geometry statements.

STØP — This shuts off both the machine tool and the control system.

THREAD/ — Used to start a threading cycle.

Example: THREAD/TURN

TURRET/ — Used to call a selected turret position on a turret drill or lathe or to select a tool from an automatic tool-changing machine.

APPENDIX E

Preparatory and Miscellaneous Functions and Other Address Characters

Preparatory (g) Functions

Word (Code)	*Explanation*
g00	Used with combination point-to-point and contouring systems for expressing a point-to-point operation.
g01 *g10* *g11*	Used to describe linear interpolation blocks and reserved for contouring. (Refer to Chapter 6 for explanation of linear and circular interpolation.) Not necessary with a system using linear interpolation exclusively and having the same dimension block. *g01* designates blocks of *normal* dimension, i.e., the maximum length of machine tool travel can be 9.9999 inches along any one axis. *g10* describes incremental distance machine movements up to 99.9990 inches and is designated as a *long* dimension block. In this instance, the fourth place decimal figure must be 0. *g11* describes a *short* dimension block with the maximum machine movement, in any one axis of .9999 inches.
g02	Used with *circular interpolation*, clockwise direction, normal dimension.
g20	Used with circular interpolation, clockwise direction, long dimension.
g21	Used with circular interpolation, clockwise direction, short dimension.
g03	Used with circular interpolation, counterclockwise direction, normal dimension.
g30	Used with circular interpolation, counterclockwise direction, long dimension.
g31	Used with circular interpolation, counterclockwise direction, short dimension.
g04	A calculated time delay during which there is no machine motion (dwell). It can be regulated by the feed rate number (*f* word). In this instance dimension words are set at zero. With some control systems the *g04* code is not required in order to accomplish a dwell.
g05	This is a *hold* code or command which restricts the machine's motion until terminated by an operator or interlock motion.

Preparatory (g) Functions (*Continued*)

Word (Code)	*Explanation*
g06 } *g07*	Unassigned. May be used at the discretion of the machine tool or system builder. Could also be standardized at a future date.
g08	Acceleration code.
g09	Deceleration code. Causes the machine to decelerate at a smooth exponential rate. (Assuming the control system and machine have this capability.)
g12	Unassigned.

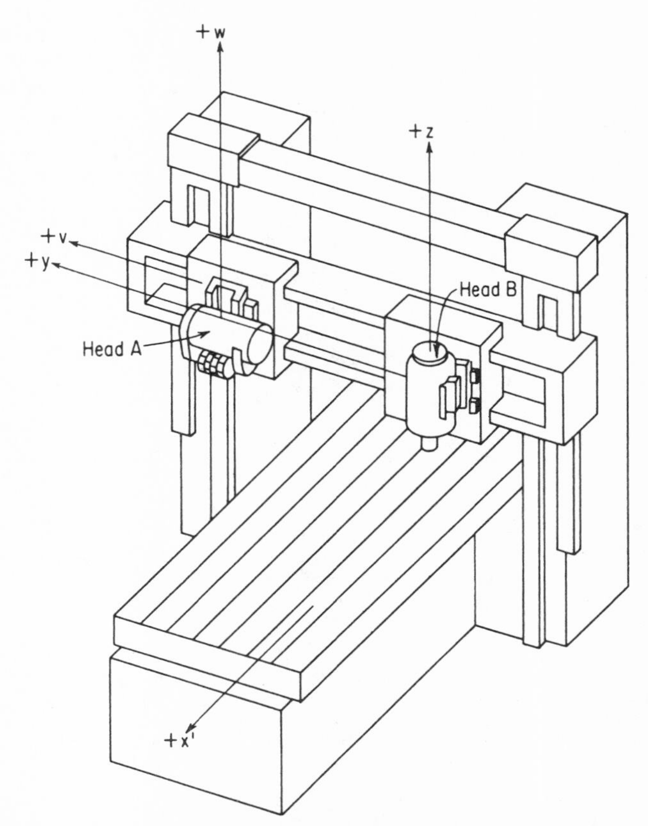

Fig. E–1. The *g13* through *g16* codes are generally used to direct the control system to operate a set of axes. In the example above, one 3-axis system can operate the two motions of either head y and z, or v and w, plus the third motion of the table x'. In this instance the heads are not operated simultaneously.

Preparatory (g) Functions (*Continued*)

Word (Code)	*Explanation*
g13 *g14* *g15* *g16*	Used to direct the control system to operate on a particular *set* of axes. An example would be a machine tool having two heads which are not to operate simultaneously. Refer to Fig. E–1. The set of axes for Head *A* consists of motions in the *v* and *w* directions in addition to the *x′* motion of the table. A three-axis control system would therefore be sufficient to machine a part in these three simultaneous dimensions. If it were required to operate head *A*, a *g13* code would precede the *x*, *v* and *w* words. The same 3-axis control system could also operate head *B* by preceding the *x*, *y* and *z* words with a *g14* word. Codes *g15* and *g16* may be utilized for other sets of axes, if required.
g17 *g18* *g19*	Used to identify, or select, a coordinate plane for such functions as *circular interpolation* or *cutter compensation*[1] which can only operate in two dimensions simultaneously. If the sample part shown in Fig. E–2. were to be cut in the *X–Y* plane, a *g17* preparatory code would be used. *g18* would denote the horizontal or *X–Z* plane and cutting in the *Y–Z* plane would require a *g19* preparatory code. The details of programming Circular Interpolation are discussed in Chapter 6.
g22 thru *g29*	Unassigned.
g32	Unassigned.

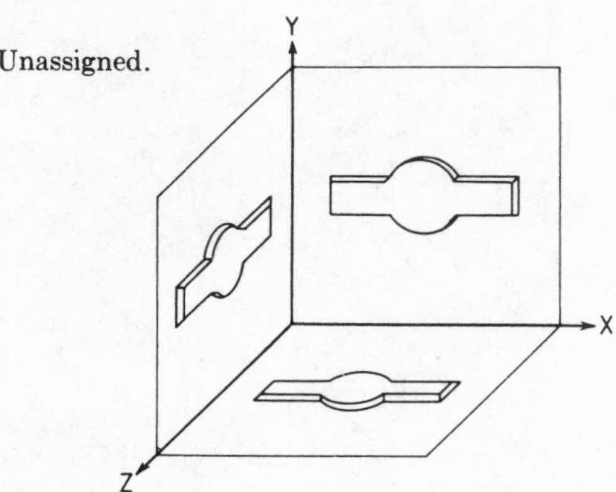

Fig. E–2. The *g17*, *g18* or *g19* preparatory codes describe the plane selected for functions such as circular interpolation or cutter compensation. If the sample part (here shown positioned in the three planes) shown were to be machined in the *X-Y* plane, the *g17* code would precede the other word instructions; *g18* would refer to the *X-Z* plane and *g19* to the *Y-Z* plane.

g33 *g34* *g35*	A mode selected for machines equiped with thread cutting facilities and generally referring to lathes. The *g33* code is used when a constant lead is sought. Code *g34* is noted when a constantly increasing lead is required and *g35* is employed to designate a constantly decreasing lead.

Preparatory (g) Functions (*Continued*)

Word (Code)	*Explanation*
g36 thru *g39*	Reserved for control use.
g40	A command which will *discontinue* any *cutter compensation.*
g41	A code associated with *cutter compensation* wherein the cutter is on the left side of the work surface, looking in the direction of cutter motion. (See Fig. E–3.)
g42	A code associated with *cutter compensation* wherein the cutter is on the right side of the work surface. (See Fig. E–3.)
g43 thru *g49*	Used with *cutter compensation* if this feature is incorporated in the system. Otherwise, unassigned.
g50 thru *g59*	Unassigned.
g60 thru *g79*	Unassigned and reserved for point-to-point systems.

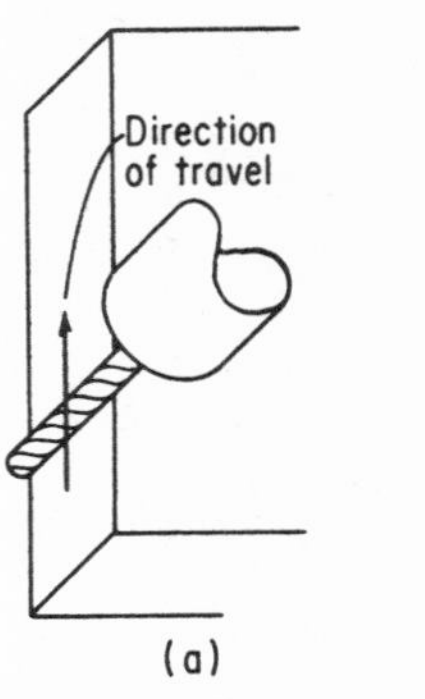

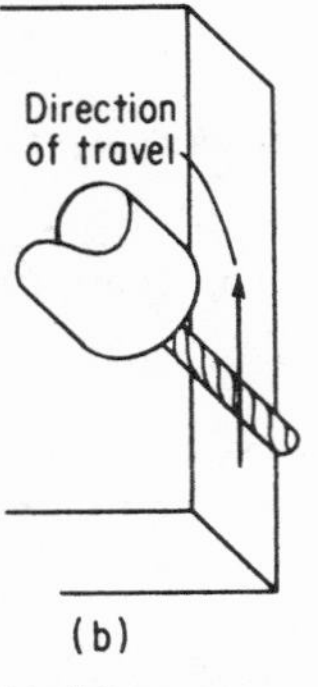

FIG. E–3. (a) The *g41* preparatory code used with cutter compensation denotes that the cutter is on the left side of the work. (b) The *g42* preparatory code used with cutter compensation denotes that the cutter is on the right side of the work.

g80	Cancel fixed cycle.
g81 thru *g89*	Reserved for fixed cycles. These commands normally initiate a series of events such as drilling or counterboring, where a slow feed *in* is required which may be followed by a dwell then by a rapid feed *out*.
g90 thru *g99*	Unassigned.

m or Miscellaneous Functions

m00 Automatically *stops* the machine. The operator must push a button in order to continue with the remainder of the program.

m01 Noted as an *optional stop* and acted upon only when the operator has previously signaled for this command by pushing a button. When the control system senses the *m01* code via the tape reader, the machine will automatically stop.

m02 At the completion of the workpiece, this *end of program* code stops the machine after completion of all commands in the block. May include rewinding of tape.

m30 Noted as an *end of tape* command and goes slightly further than the *m02* code in that this code *will* rewind the tape (assuming the control has this facility); and also automatically transfer to a second tape reader if incorporated in the control system.

m03 Start *spindle rotation* in a *clockwise* direction—looking out from the spindle face.

m04 Start *spindle rotation* in a *counter-clockwise* direction—looking out from the spindle face.

m05 *Stop* the spindle in a normal and efficient manner.

m06 Command to execute the *change of a tool* (or tools) manually or automatically, not to cover the selection of a tool as is capable with the *t* words.

m07 Code to turn a *coolant* on.

m08 Also a code for turning a coolant on. This may differ from *m07* in that *m07* may control *flood* coolant, and *m08* *mist* coolant.

m09 Automatically shuts the coolant off.

m10, *m11* *m10* applies to the automatic *clamping* of the machine slides, workpiece, fixture spindle, etc. *m11* is an *unclamping* code.

m12 Unassigned.

m13 Combines *clockwise spindle* motion and *coolant on* in the same command which will cause both to occur simultaneously.

m14 Combines *counter-clockwise* spindle motion and *coolant on* in the same command.

m15, *m16* Rapid traverse or feed motion in either the +(*m15*) or −(*m16*) direction.

m17, *m18* Unassigned.

m19 Oriented spindle stop. Causes the spindle to stop at a predetermined angular position.

m or Miscellaneous Functions (*Continued*)

m20 thru *m29* — Unassigned.

m31 — A command known as *interlock bypass* for temporarily circumventing a normally provided interlock.

m32 thru *m35* — Codes normally used with turning operations whereby it is desired to maintain a constant surface cutting speed by adjusting the speed of rotation of the workpiece inversely with respect to the distance of the cutting tool from the center of rotation. See Fig. E–4.

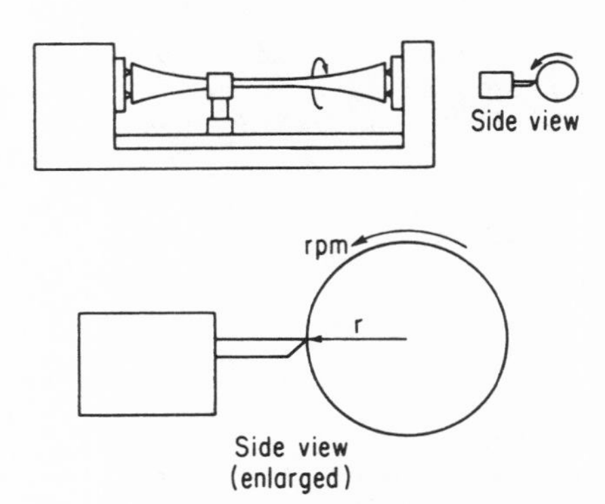

Fig. E–4. In this lathe the *m32* through *m35* codes would automatically initiate a control (interlock) that would adjust the rpm of the spindle inversely with respect to the distance, *r*, of the cutting tool from the center of rotation. The further the tool from the centerline, the slower the rpm. This feature insures a constant and optimum cutting feed, resulting in exceptionally smooth surface finishes. However, the machine as well as the control system, must incorporate this capability. The *m* codes only signal the machine's action.

m36 thru *m39* — Unassigned.

m40 thru *m45* — Used to signal gear changes if required at the machine; otherwise, unassigned.

m46 thru *m99* — Unassigned.

Other Address Characters

Address Character	*Explanation*
a	Angular dimension about the *X* axis. See Fig. E–5.
b	Angular dimension about the *Y* axis. See Fig. E–5.
c	Angular dimension about the *Z* axis. See Fig. E–6.
d	Can be used either to express an angular dimension around a special axis (See Fig. E–5) or a third feed function. Since only one type word may be expressed in any one block, it may be desirable, on occasion, to move combinations of different axes at different feed rates for each axis.
e	This address also may be used for expressing an angular dimension around a special axis or a second feed function.
h	Although technically unassigned according to the standards, this address has been employed for an extra set of auxiliary functions following *m* address assignments.

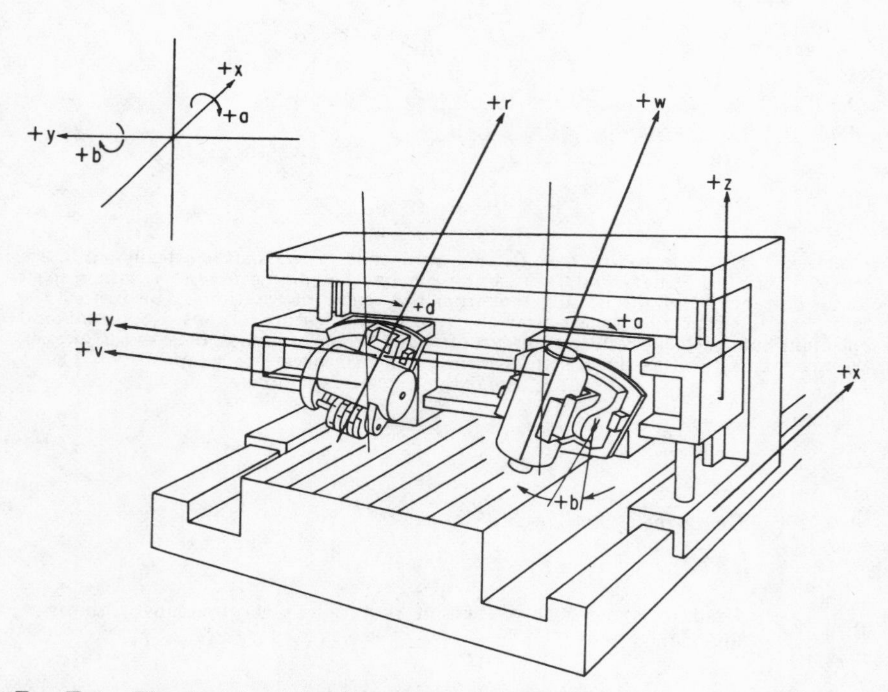

Fig. E–5. The dual headed milling machine shown above requires a number of axes designations over and above the normal *x*, *y* and *z* motions. Twist motions of the heads are designated as *a*, *b*, and *d* while the cross motions of each head require separate identities, namely, *y* and *v*.

Other Address Characters (*Continued*)

Address Character	*Explanation*
i, j, k	Used with circular interpolation. (See Chapter 6.)
l	Not to be used because of possible confusion with the numeral *1*.
o	Reference rewind stop. To stop manually initiated rewind and may be used in place of the customary "Sequence Number" word address "*n*".
p	A third rapid traverse code or tertiary motion dimension parallel to the X axis.
q	Second rapid traverse code or tertiary motion dimension parallel to the Y axis.
r	First rapid traverse code or tertiary motion dimension parallel (unless set at an angle, see Fig. **E–5**) to the Z axis.
u	Secondary motion dimension parallel to the X axis.

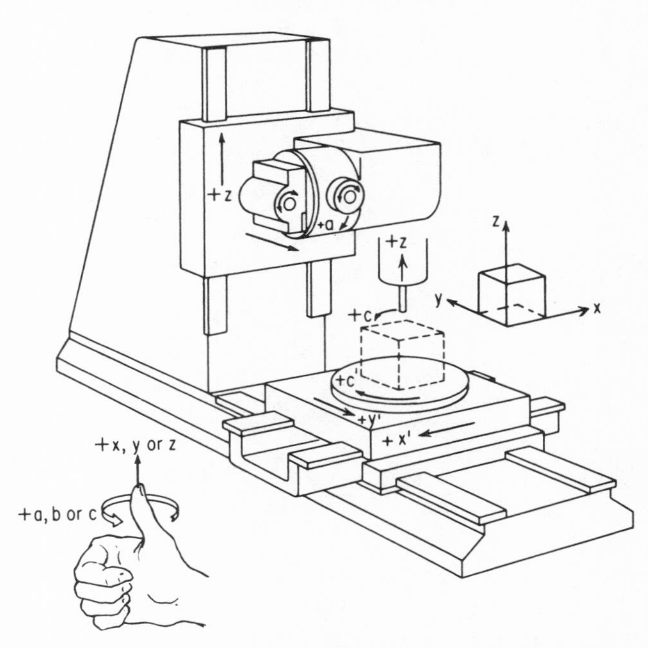

FIG. E–6. The c' motion is the rotary motion of the table about the Z axis while the c motion of the head is the circular traversing of the cutting head about the Z axis. Thus, a $+c'$ motion is equivalent to a $+c$ motion and vice versa. According to the right hand coordinate system rule (see insert) a c' motion is in a negative direction. This however is equivalent to a movement of the cutting head in a positive direction and is so used to indicate a positive cutting motion on the part.

Other Address Characters (*Continued*)

Address Character	*Explanation*
v	Secondary motion dimension parallel to the Y axis. Figure E–5 notes the cross motion of the left hand head as "v" which is a parallel motion to the cross motion of the right hand head noted as y motion.
w	Secondary motion dimension parallel (unless set at an angle, see Fig. E–5) to the Z axis.
end of record	This character is used to *stop* rewind motion.

APPENDIX F

Answers to Practice Exercises

CHAPTER 3

1. $x + 4.000$; $y + 2.000$.
2. Quadrant II or Quadrant III.
3. Counterclockwise.
4. $x = +2.500$; $y = -3.125$.
5. Quadrant III.
6. Above.
7. 1.375 inches.
8. $x = +20.000$; $y = +10.250$.
9. $x = +1.000$; $y = -3.000$; $z = +4.000$ (see below).

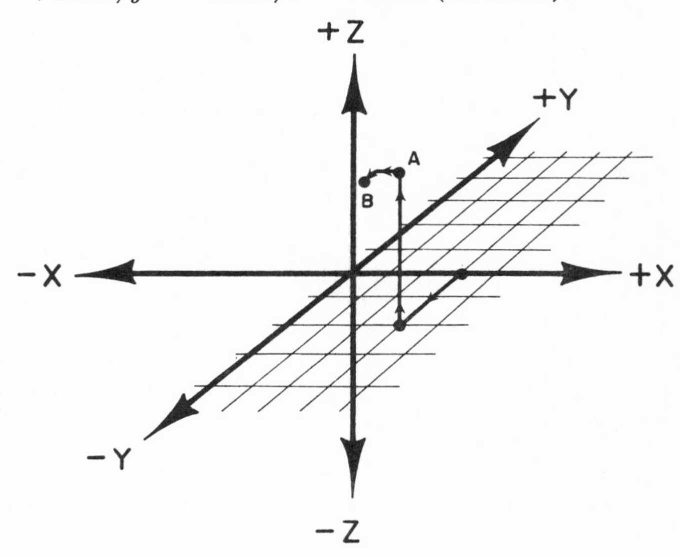

10. $x = +.700$; $y = -2.300$ (see below).

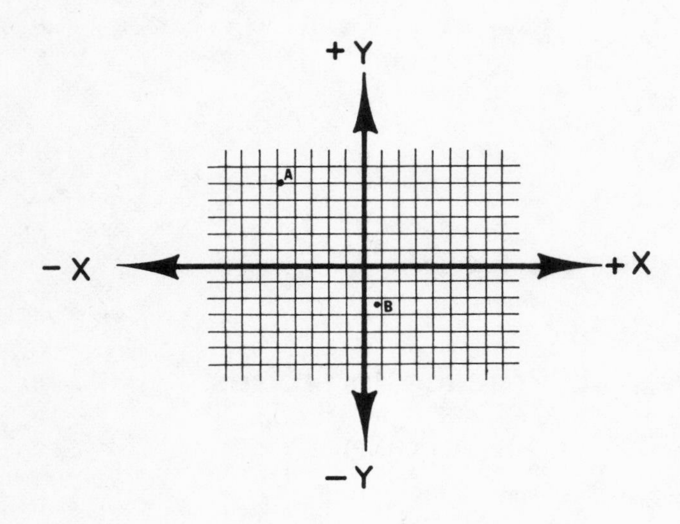

CHAPTER 4

1. 0.004 inches ±0.0003. (Refer to Fig. 4–1.)
2. Track and level.
3. Eight.
4. Four.
5. Eight.
6. 0 correct
   3 incorrect
   1 correct
   8 correct
   5 incorrect
   x correct
   + incorrect
   6 incorrect
7. The parity check with the ASCII code (RS-358) requires that an *even* number of holes be described across the tape whereas an odd number is required with the RS-244-A code.
8. The *sequence number* word.
9. In inches per minute (ipm).
10. a. $x19$ or $x+19$
    b. $y-00002$
    c. $y-0625$

d. $x-00065$
e. $y65666$ or $y+65666$

11. a. $x = +00.1650$
b. $y = -00.0001$
c. $x = -00.0025$
d. $z = +00.0625$
e. $y = +05.1650$
12. Three.
13. Revolutions per minute (rpm).
14. The *sequence number* for the particular block.
15. *f*0020 (5/2.5 is satisfactory since the .0001 distance is insignificant and would not affect the *f* word.)
16. a. *f*610
b. *f*525
c. *f*517
d. *f*225
e. *f*432
f. *s*736
17. The length of the block, or number of characters, does not change.
18. A manual tape preparation unit resembling a typewriter.
19. Word address, tab sequential, and fixed block.

PRACTICE EXERCISE NO. 1 CHAPTER 5

1. He must determine the sequence of operations.
2. A part requiring contouring operations is usually more difficult to program.
3. The operator must be sure that the part is properly aligned with respect to the X and Y axes, as called for by the part programmer.
4. Point A: $x = 3.000$, $y = 3.000$; Point B: $x = -5.000$, $y = 6.000$
Point C: $x = -3.000$, $y = -6.000$; Point D: $x = 2.000$, $y = -3.000$
5. The target or set-up point may be at any convenient location over the full range of travel of the machine, and all four quadrants may be used.
6. No. Any convenient location within the operating range of the machine may be selected. Very often it is more practical to select a park location away from the set-up point so that the operator can remove the finished piece and replace it more readily with a new piece of material.

7. Part program for drilling hole:

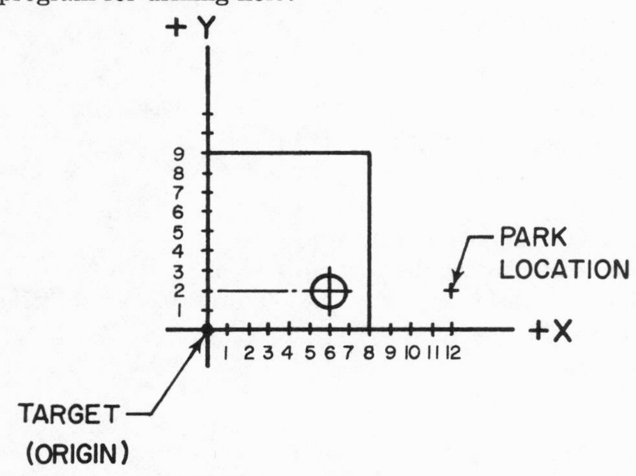

PART NO. 12350 PART NAME FLAT PLATE	MANUSCRIPT XXXXX MACHINE	DATE: x/x/xx PREPARED BY: [illegible] CHECKED BY: [illegible]

SEQUENCE NO.	TAB or EOB	X COORDINATE	TAB or EOB	Y CORRDINATE	TAB or EOB	M WORD	TAB or EOB	COMMENTS
RWS								
001	TAB	+6.000	TAB	2.000	EOB			TARGET AT LEFT HAND BOTTOM CORNER.
								USE ½" DRILL. SET DEPTH TO DRILL
								THROUGH.
002	TAB	+12.000	TAB		TAB	30	EOB	

NOTE: The y coordinate for sequence number 002 need not be repeated since it does not change. Decimal points are shown for clarity although they are not punched on the tape.

8. The words are lined up in orderly columns.
9. It means that the coordinates of all points are measured from the X, Y, and Z axes.

PRACTICE EXERCISE NO. 2 CHAPTER 5

1. In a full floating zero system the part may be positioned anywhere on the table. In a fixed zero system the part must be located with respect

to a fixed origin, or point, where $x = 0$ and $y = 0$. The origin may be shifted with most fixed zero systems.

2. Since there is no letter address for the words nor tab character to separate the words, the control system determines the proper word from its *location* in the block, and all characters must therefore be described.
3. The cutter will move directly to the point above the first hole.
4. Sketch as the part would appear with respect to the X and Y axes.

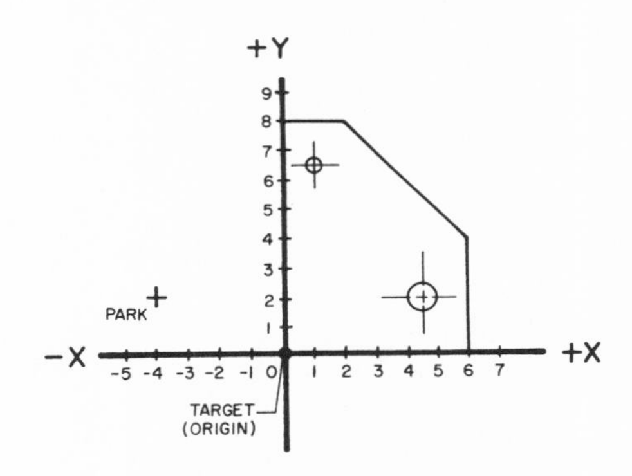

PART NO. 12351
PART NAME M FLAT PLATE

MANUSCRIPT
XXXXX MACHINE

DATE :
PREPARED BY :
CHECKED BY :

SEQUENCE NO.	TAB or EOB	x COORDINATE	TAB or EOB	y COORDINATE	TAB or EOB	m WORD	TAB or EOB	COMMENTS
RMS								
001	TAB	+4.500	TAB	+2.000	EOB			SPOT DRILL TWO HOLES. SET DEPTH MANUALLY.
002	TAB	+1.000	TAB	+6.500	TAB	02	EOB	
003	TAB	-4.000	TAB	+2.000	TAB	02	EOB	TO PARK POSITION & STOP.
004	TAB	+1.000	TAB	+6.500	EOB			INSERT ½" DRILL & DRILL ½" HOLE.
005	TAB	-4.000	TAB	+2.000	TAB	02	EOB	TO PARK POSITION & STOP.
006	TAB	+4.500	EOB					INSERT 59/64 DRILL & DRILL HOLE.
007	TAB	-4.000	TAB		TAB	30	EOB	TO PARK POSITION & STOP. TAPE REWINDS. REMOVE PART AND INSERT SPOT DRILL FOR
								SEQUENCE NO. 001.

PRACTICE EXERCISE NO. 3 CHAPTER 5

1. All of the coordinate values will be plus (+), which will reduce a common source of error due to noting the incorrect sign.
2. There is no Z-axis control from the tape instructions.

PRACTICE EXERCISE NO. 3 CHAPTER 5 (*Cont'd*)

3. The standard word for stopping is *m*02; the word for stopping the machine and rewinding the tape is *m*30.
4. The speed at which the cutter travels, which is called the feed rate.
5. For the two-axis machine, the feed rate is generally set by the operator. For three-axis machines the feed rate is often controlled directly by a command on the tape, such as an *f* word.
6. A sketch as the part would appear in the first quadrant:

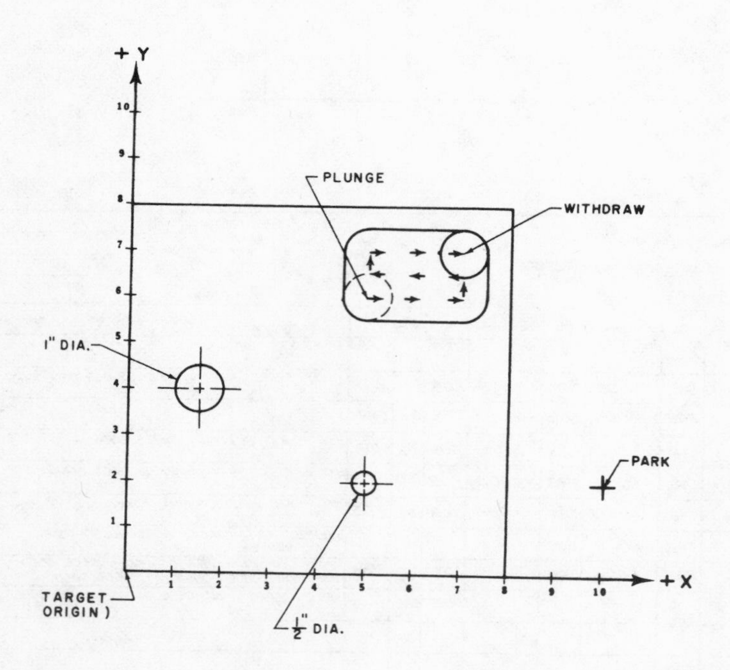

The program as prepared on a manuscript form is given at the top of page 337.

PRACTICE EXERCISE NO. 3 CHAPTER 5 (*Cont'd*)

6 (*cont.*)

PART NO. 12352

PART NAME M POCKET PLATE

MANUSCRIPT
XXXXX MACHINE

DATE : ____________

PREPARED BY : ________

CHECKED BY : ________

SEQUENCE NO.	TAB or EOB	X COORDINATE	TAB or EOB	Y COORDINATE	TAB or EOB	Z COORDINATE	TAB or EOB	f WORD	TAB or EOB	m WORD	TAB or EOB	COMMENTS
RWS												TARGET AT LEFT-HAND BOTTOM CORNER. USE .500 INCH
												FEELER BLOCKS TO SET Z AXIS.
001	TAB	1.500	TAB	4.000	TAB	0.000	TAB	1500	EOB			CUTTER MOVES TO POINT ABOVE 1" DIA HOLE.
002	TAB		TAB		TAB	-2.250	TAB	20	EOB			DRILL 1" DIA HOLE.
003	TAB		TAB		TAB	0.000	TAB	1500	EOB			WITHDRAW FROM 1" DIA HOLE.
004	TAB	10.000	TAB	2.000	TAB		TAB		TAB	02	EOB	MOVE TO PARK LOCATION & STOP. CHANGE TO 1/2" DRILL
005	TAB	5.000	EOB									PUSH "START" BUTTON AND CUTTER MOVES TO POINT
												ABOVE 1/2" HOLE.
006	TAB		TAB		TAB	-1.125	TAB	20	EOB			DRILL 1/2" DIA HOLE.
007	TAB		TAB		TAB	0.000	TAB	1500	EOB			WITHDRAW FROM 1/2" DIA. HOLE.
008	TAB	10.000	TAB		TAB		TAB		TAB	02	EOB	MOVE TO PARK LOCATION. & STOP. CHANGE TO 1" DIA
												FLAT END MILL CUTTER.
009	TAB	5.000	TAB	6.000	EOB							CUTTER MOVES TO POINT ABOVE LEFT-HAND BOTTOM
												CORNER OF POCKET OUTLINE.
010	TAB		TAB		TAB	-1.500	TAB	20	EOB			PLUNGE FOR START OF POCKET CUT.
011	TAB	7.000	TAB		TAB		TAB	50	EOB			
012	TAB		TAB	6.500	EOB							
013	TAB	5.000	EOB									
014	TAB		TAB	7.000	EOB							
015	TAB	7.000	EOB									
016	TAB		TAB		TAB	0.000	TAB	1500	EOB			WITHDRAW FROM POCKET.
017	TAB	10.000	TAB	2.000	TAB		TAB		TAB	30	EOB	CUTTER MOVES TO PARK LOCATION & STOPS. TAPE
												REWINDS. CHANGE TO 1" DRILL.

7. $x = -3.000$, $y = +2.700$.
8. The most logical location for the origin:

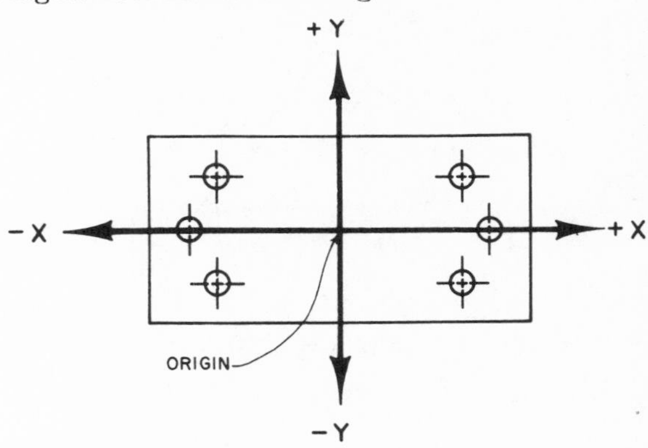

PRACTICE EXERCISE NO. 3 CHAPTER 5 (*Cont'd*)

9. The x and y coordinates of points A, B, C, and D:

Point A

$x = +2.000$
$y = \quad 0.000$

Point C

$$x = 2.000 + 3.500 + 2.500 + 2.000 \cos 60°$$
$$= 9.000$$
$$y = -2.000 \sin 60°$$
$$= -1.732$$

Point B

$x = +5.500$
$y = \quad 0.000$

Point D

$$x = 9.000 + 4.000 \cos 60°$$
$$= 11.000$$
$$y = -1.732 + 4.000 \sin 60°$$
$$= -5.196$$

PRACTICE EXERCISE NO. 1 CHAPTER 6

1. The *path* of the cutter must be calculated in a contouring system, whereas only the *points* where an operation is to be performed must be calculated in a point-to-point system.
2. Linear interpolation means approximating a curve with a series of straight-line segments.
3. The allowable calculation tolerance and the radius of the arc are the two factors governing the length of linear interpolation lines.
4. Calculations for the linear interpolation line are as follows:

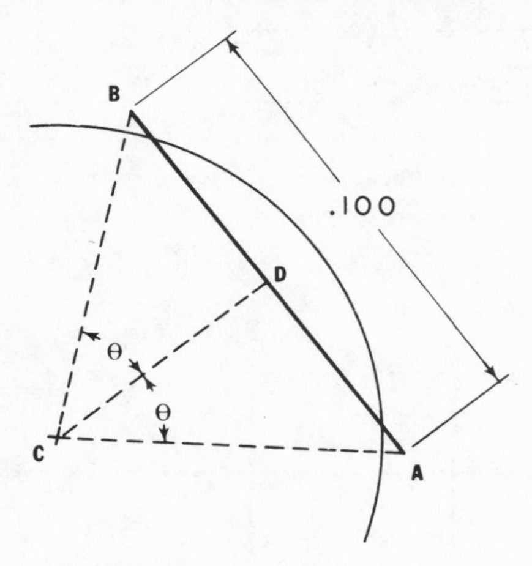

$$\text{Line } CD = R - .003 = .250 - .003 = .247$$
$$\text{Line } CB = R + .002 = .250 + .002 = .252$$

$$\cos\theta = \frac{CD}{CB} = \frac{.247}{.252} = .98016$$

$$\text{Angle } \theta = 11°26'$$

$$\text{Line } BD = CB \sin\theta = .252 \sin 11°26'$$

$$= .252 \times .19823 = .050$$

The length of the line AB is then equal to $2 \times .050 = .100$

5. To calculate the Δx and Δy distances it is first necessary to know either the angle ϵ or α. (Refer to Fig. 6–9.) In this case α has been chosen and is equal to the sum of $\beta + \phi$. Also, ϕ is equal to ω, which is noted as 10° in the problem.

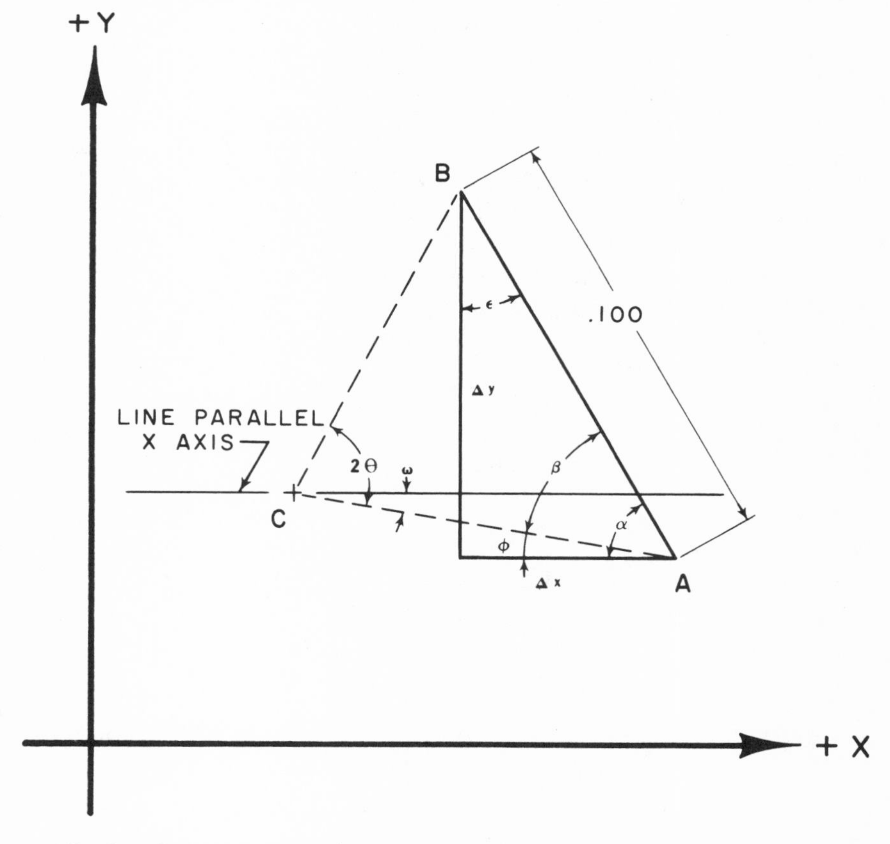

If, therefore, $\phi = \omega$ and $\omega = 10°$, then $\phi = 10°$. $\beta = 90° - \theta$, and θ was calculated in the answer to Problem No. 4, above. Then $\Delta x = .100 \cos\alpha$ and $\Delta y = .100 \sin\alpha$.

Using numbers: $\beta = 90° - 11°26' = 78°34'$

$$\alpha = 78°34' + 10° = 88°34'$$

Therefore, $\Delta x = .100 \cos 88°34' = .100 \times .02443 = .002$

$$\Delta y = .100 \sin 88°34' = .100 \times .99970 = .100$$

It should again be pointed out that the dimensions shown are not proportional to the actual measurements—i.e., certain angles and lines have been exaggerated in the diagram for greater clarity.

6. *g02*
7. Four.
8. In the row beginning with sequence *001*, *g01* is noted since the block is to represent a straight line. If the preceding block had also been for a straight-line movement, it would not have been necessary to note the *g01* word in this block. The word shown in the x word column describes the Δx movement from point A to point B, which is 3.000 inches. Since the Δy movement is zero, nothing is shown in the y word column. Also, because this is a linear interpolation block, i and j words are not required. The f word would be shown if there is to be a change from the previous f word listed or if no f word had been listed previously. The third block, which begins with *n003*, changes the mode from linear interpolation to circular interpolation and instructs the cutter to move from point C to the top of the arc at point D. The *x1000* word represents the Δx distance and the *y300* word represents the Δy distance, which is .300 inch. The i distance is 1.000 inch and the j distance is 1.000 inch; and both these distances are shown in the i and j column. The line starting with *n004* instructs the cutter to

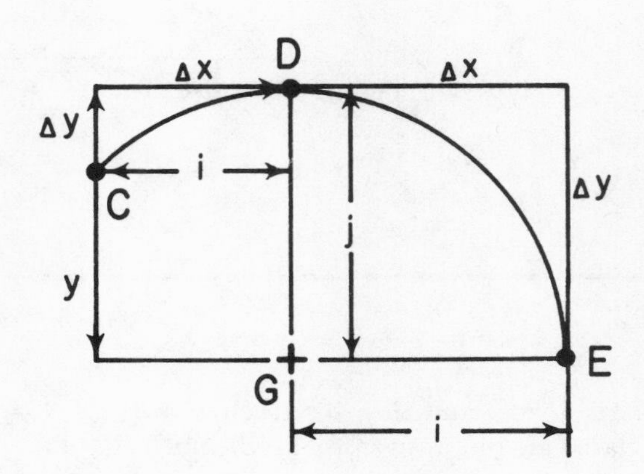

MANUSCRIPT
CXXXXX MACHINE

PART NO :
PART NAME :

DATE :
PREPARED BY :
CHECKED BY :

SEQUENCE NO.	g WORD	x WORD	y WORD	i WORD	j WORD	f WORD	m WORD	EOB	COMMENTS
RWS									
n001	g 01	X3.000				f50		EOB	MOVE FROM POINT A TO B
n002		X4.000	y3.200					EOB	MOVE FROM POINT B TO C
n003	g 02	X1.000	y.300	i 1.000	j1.000			EOB	MOVE FROM POINT C TO D
n004		X1.300	y1.300		j1.300			EOB	MOVE FROM D TO E
n005	g 01		y8.200					EOB	MOVE FROM E TO F
↓	↓	↓	↓	↓	↓	↓	↓	↓	
↓	↓	↓	↓	↓	↓	↓	↓	↓	

move from point D to point E, also in a circular interpolation mode. It will be noted, in this case, that the i dimension is *0*, and the j is equal to the distance from point G to point D, or the radius. The next line beginning with *n005* moves the cutter from point E to point F, in a minus direction. There is no incremental x move, and the incremental y movement is -8.200 inches.

PRACTICE EXERCISE NO. 2 CHAPTER 6

1. The radius of the cutter.

2. Point A: $x = 9.000$
   $y = 3.500$
   Point B: $x = 9.000$
   $y = 8.000$
   Point C: $x = 3.000$
   $y = 8.000$
   Point D: $x = 3.000$
   $y = 3.500$

3. Movement from A to B: $\Delta x = 6.000$
   $\Delta y = 0.000$
   from B to C: $\Delta x = 2.000$
   $\Delta y = 2.000$
   from C to D: $\Delta x = 0.000$
   $\Delta y = -4.500$

4 and 5. These answers are given with problems 4 and 5 on pages 157–160.

PRACTICE EXERCISE NO. 1 CHAPTER 8

1. APT has the capability to calculate the motion of a cutter so that it can move in three dimensions at the same time.

2. No.

3. (b) GØLFT
   (c) GØRGT
   (d) GØRGT
   (e) GØLFT
   (f) GØRGT
   (g) GØLFT
   (h) GØBACK

4. (a) Position a: Drive surface is SURF A.
   Check surface is SURF B.
   (b) Position b: Drive surface is SURF B.
   Check surface is SURF C.
   (c) Position c: Drive surface is SURF C.
   Check surface is SURF D.
   (d) Position d: Drive surface is SURF D.
   Check surface is SURF E.
   (e) Position e: Drive surface is SURF F.
   Check surface is SURF G.

5. FRØM/SP
   GØTØ/P1
   GØTØ/P2
   GØTØ/P3
   GØTØ/SP

6. Drive surface, part surface, check surface.

7. Yes; the program will not work.

8. CUTTER/.5
   CUTTER/2.25

   (Note that the diameters must be expressed in decimal form.)

9. FRØM/SP
   GØ/TØ,ASURF,TØ,PSURF,PAST,CSURF
   GØLFT/ASURF,PAST,BSURF
   GØRGT/BSURF,PAST,CSURF
   GØRGT/CSURF,PAST,ASURF
   GØTØ/SP

10. Drilling Operation:
   FRØM/SP
   GØTØ/P1
   GØDLTA/0,0,−3
   GØDLTA/0,0,3
   GØTØ/P2
   GØDLTA/0,0,−3
   GØDLTA/0,0,3
   GØTØ/P3
   GØDLTA/0,0,−3
   GØDLTA/0,0,3
   GØTØ/SP
   STØP

   Milling Operation:
   GØ/TØ,ASURF,TØ,PSURF,TØ,DSURF
   GØRGT/ASURF,PAST,BSURF
   GØLFT/BSURF,PAST,CSURF
   GØLFT/CSURF,PAST,DSURF
   GØLFT/DSURF,PAST,ASURF
   GØTØ/SP

PRACTICE EXERCISE NO. 2 CHAPTER 8

1. (a)
2. (a) FRØM/SP
   GØ/TØ,DSURF
   (b) FRØM/SP
   GØ/ØN,DSURF
   (c) FRØM/SP
   GØ/PAST,DSURF
3. FRØM/SP
   GØ/TØ,BSURF
   GØLFT/BSURF,TØ,CSURF
   GØLFT/CSURF,TØ,DSURF
   GØTØ/SP

PRACTICE EXERCISE NO. 1 CHAPTER 9

1. a. Incorrect. Does not contain a letter.
   b. Correct.
   c. Correct.
   d. Incorrect. Exceeds six characters.
   e. Correct.
2. Line = L9
   Vertical line = VL7
   Circle = CIR9
   Plane = PL5
3. P1 = PØINT/5,5,0
   P2 = PØINT/9,−5,0
   P3 = PØINT/11,12,0
   P4 = PØINT/−10,5,0
   P5 = PØINT/−6,−7,0
4. a. L1 = LINE/P1,P3
   b. L2 = LINE/P2,ATANGL,15
   c. L3 = LINE/P5,ATANGL,90
   d. To describe a statement for L4, it is necessary to write a statement for a point that lies on the line. The most convenient point occurs where the line intersects the *X* axis. The coordinates of this point are $x = 22$, $y = 0$, $z = 0$. If we assign a symbol to the point, such as P4, then the statement for the point would be:

      P4 = PØINT/22,0,0

   The statement for the line would then be:

      L4 = LINE/P4,ATANGL,90

   e. In order to define L5, two points are needed. P5 is one, and the other can be readily obtained since it lies on the *Y* axis and is a noted distance in the minus-Y direction. If we assign the symbol P6 to the second point, then the statement for the point would be:

      P6 = PØINT/0,−10,0

   The statement for the line would be:

      L5 = LINE/P5,P6

PRACTICE EXERCISE NO. 2 CHAPTER 9

1. C2 = CIRCLE/5,5.5,0,3
   C3 = CIRCLE/0,3,0,1.5

PRACTICE EXERCISE NO. 2 CHAPTER 9 (*Cont'd*)

C4 = CIRCLE/−5,−2,0,3.5
C5 = CIRCLE/5,−4,0,2

2. C1 = CIRCLE/4.130,1,0,.5
C2 = CIRCLE/4.130,3,0,.5
C3 = CIRCLE/1,3,0,.5
C4 = CIRCLE/1,1,0,.5

NOTE: The arcs are described as full circles in the geometry statements. The motion statements that would follow will determine the lengths of the arcs. This is a calculation performed by the computer. Also note that the *z* coordinate, which is zero, is shown. And following the general rule in APT of dropping zeros, we may also drop the zero in this case. C1, for example, might therefore be

C1 = CIRCLE/4.130,1,.5

3. PLN7 = PLANE/P1,P2,P3
4. A statement describing PLN8 would be:

PLN8 = PLANE/P6,P7,P8

The symbols for the points P6, P7, and P8 were chosen at random and could have been any symbols conforming to the rules noted earlier in this chapter. Before being used, P6, P7, or P8 would have to be defined in three statements, as follows:

P6 = PØINT/0,0,−4.5
P7 = PØINT/1,1,−4.5
P8 = PØINT/6,6,−4.5

Again, it should be noted that *any* three points, lying on the plane and not in a straight line, could have been used and their *x*, *y*, and *z* coordinates noted in the point statements. The above statements, as they would appear in a part program, are as follows:

P6 = PØINT/0,0,−4.5
P7 = PØINT/1,1,−4.5
P8 = PØINT/6,6,−4.5
PLN8 = PLANE/P6,P7,P8

5. The fourth, fifth, and sixth statements are incorrect.
The fourth statement should be:

P3 = PØINT/6,5,7

The fifth statement should be:

L1 = LINE/P1,P3

PRACTICE EXERCISE NO. 2 CHAPTER 9 (*Cont'd*)

The sixth statement is incorrect since the symbol P1 is repeated, but for a different point.

6. PARTNØ
7. FINI
8. This means that the tolerance allowable for approximating a curve on the inside is .0005 inch. See sketch below:

   NOTE: This applies only to linear interpolation, wherein the curve is approximated with straight-line segments. It would not apply to circular interpolation.

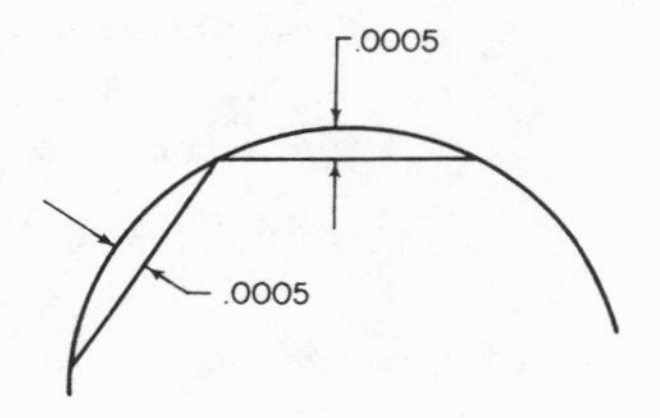

9. a. L1 = LINE/P1,P2
   b. L2 = LINE/P1,ATANGL,20
   c. L1 = LINE/P6,PERPTØ,L2
   d. L2 = LINE/P4,PARLEL,L3
   e. L1 = LINE/P5,RIGHT,TANTØ,C1
   f. L2 = LINE/P1,LEFT,TANTØ,C2

PRACTICE EXERCISE NO. 1 CHAPTER 10

1. a.

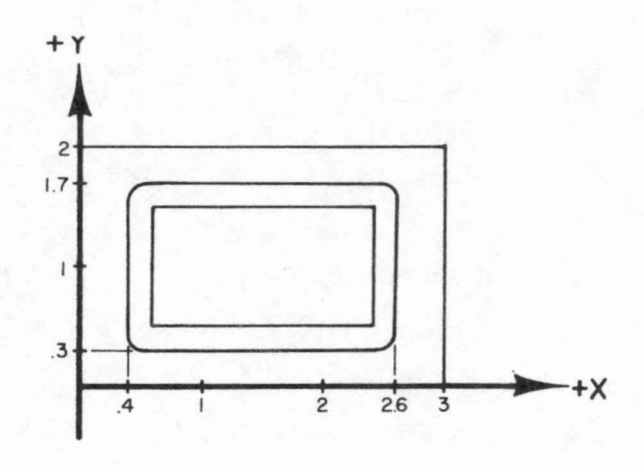

PRACTICE EXERCISE NO. 1 CHAPTER 10 (*Cont'd*)

1. b.

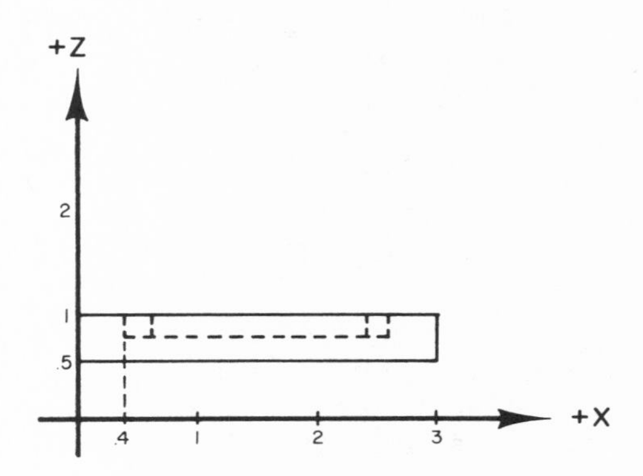

1. c. Programming the part: The symbols assigned to the points and lines shown in the following sketch were selected arbitrarily. The student may assign any legal symbols; also it would be permissible to use other statements when defining the lines providing they conform to the APT requirements:

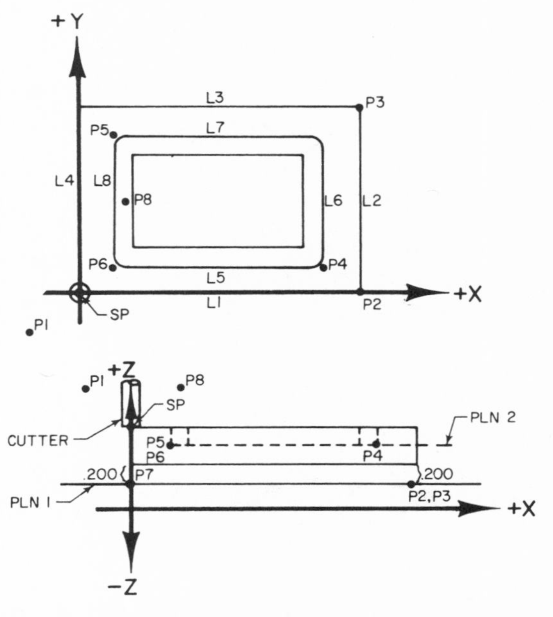

Lines L5, L6, L7, and L8 identify the outside portions of the groove and, since the diameter of the cutter is to be the same dimension as

PRACTICE EXERCISE NO. 1 CHAPTER 10 (*Cont'd*)

the width of the groove, it is only necessary that *either* the outside *or* inside lines be described. P2, P3, and P7 will be the points used in the statement to define PLN1; and P4, P5, and P6 will be used in the statement to define PLN2. These points may also define the lines referring to the sides of the part and to the sides of the groove, because such lines are actually surfaces as far as the computer is concerned, and the use of any point on the surface would be satisfactory.

PLN1 will be the part surface when the perimeter is being machined, and will guide the bottom of the cutting tool as it moves around the outside surface of the part. PLN2 will be the part surface when the tool is cutting the groove.

Referring to the sketch below, in cutting the outside surface, the cutter is to move from the set-point (SP) to the point P1. Next it is to move to the intersection of the drive surface, part surface, and check surface, which are noted by the symbols L1, PLN1, and L4. After cutting the permiter of the part, the cutter is to move back to P1.

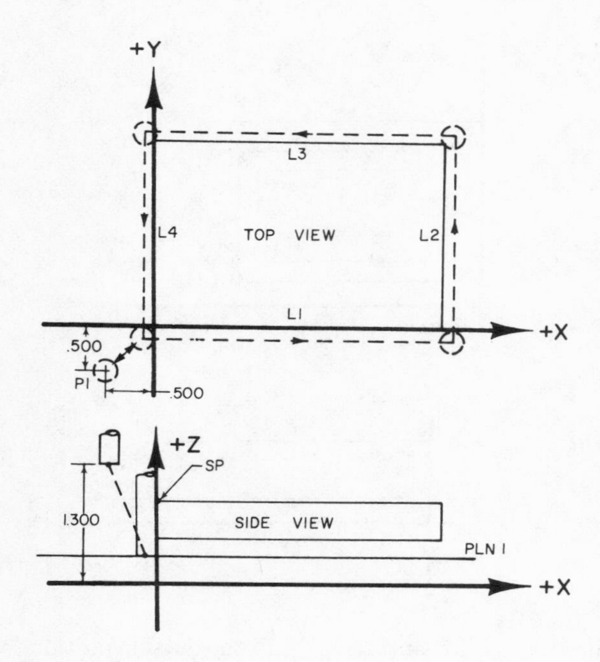

The auxiliary, geometry, and motion statements for this portion of the

PRACTICE EXERCISE NO. 1 CHAPTER 10 (*Cont'd*)

program are as follows:

```
      PARTNØ  FLAT PLATE WITH GRØØVE
      MACHIN/ABC
      CUTTER/.20
      CØØLNT/ØN
      CLPRNT
      FEDRAT/5,IPM

      SP = PØINT/0,0,1
      P1 = PØINT/-.5,-.5,1.3
      P2 = PØINT/3,0,.3
      P3 = PØINT/3,2,.3
      P4 = PØINT/2.6,.3,.8
      P5 = PØINT/.4,1.7,.8
      P6 = PØINT/.4,.3,.8
      P7 = PØINT/0,0,.3
      P8 = PØINT/.5,1,1.3

      L1 = LINE/SP,ATANGL,0
      L2 = LINE/P2,PERPTØ,L1
      L3 = LINE/P3,PARLEL,L1
      L4 = LINE/SP,PERPTØ,L3
      L5 = LINE/P4,ATANGL,0
      L6 = LINE/P4,PERPTØ,L5
      L7 = LINE/P5,PERPTØ,L6
      L8 = LINE/P5,PARLEL,L6
    PLN1 = PLANE/P2,P3,P7
    PLN2 = PLANE/P4,P5,P6

      FRØM/SP
      RAPID
      GØTØ/P1
      GØ/TØ,L1,TØ,PLN1,TØ,L4
      GØRGT/L1,PAST,L2
      GØLFT/L2,PAST,L3
      GØLFT/L3,PAST,L4
      GØLFT/L4,PAST,L1
      RAPID
      GØTØ/P1
```

Next, the cutter is directed to move to P8, which is directly above the

PRACTICE EXERCISE NO. 1 CHAPTER 10 (*Cont'd*)

center line of the groove. After moving to P8, the cutter is then directed by a GØ/TØ statement to move to the intersection of L5 and PLN2, which will become the drive and part surfaces respectively. Also, since the cutter is to move at a right angle to L5, a two-surface start-up statement will be satisfactory. After moving to the lower left-hand corner of the groove, the cutter then goes along L5 to L6, left along L6, and left twice more until it arrives back where it started from, or the lower left-hand corner of the groove. Next the cutter is withdrawn from the groove, and moves back up to point P8. Then it returns to P1, where it stops. The statements would be as follows:

```
RAPID
GØTØ/P8
GØ/TØ,L5,TØ,PLN2
GØLFT/L5,TØ,L6
GØLFT/L6,TØ,L7
GØLFT/L7,TØ,L8
GØLFT/L8,TØ,L5
RAPID
GØTØ/P8
RAPID
GØTØ/P1
CØØLNT/ØFF
FINI
```

Index